Analysis of Residual Stress by Diffraction using Neutron and Synchrotron Radiation

Analysis of Residual Stress by Diffraction using Neutron and Synchrotron Radiation

Edited by

M. E. Fitzpatrick and A. Lodini

Taylor & Francis
Taylor & Francis Group

LONDON AND NEW YORK

First published 2003
by Taylor & Francis
11 New Fetter Lane, London EC4P 4EE

Simultaneously published in the USA and Canada
by Taylor & Francis Inc,
29 West 35th Street, New York, NY 10001

Taylor & Francis is an imprint of the Taylor & Francis Group

© 2003 Taylor & Francis

Typeset in Times New Roman by
Newgen Imaging Systems (P) Ltd, Chennai, India
Printed and bound in Great Britain by
TJ International Ltd, Padstow, Cornwall

Every effort has been made to ensure that the advice and information
in this book is true and accurate at the time of going to press. However,
neither the publisher nor the authors can accept any legal responsibility
or liability for any errors or omissions that may be made. In the case of
drug administration, any medical procedure or the use of technical
equipment mentioned within this book, you are strongly advised to consult
the manufacturer's guidelines.

British Library Cataloguing in Publication Data
A catalogue record for this book is available from the British Library

Library of Congress Cataloging in Publication Data
A catalog record for this book has been requested

ISBN 0–415–30397–4

Contents

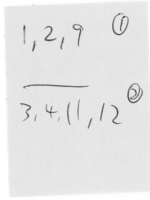

Contributors

R. Coppola
ENEA-Casaccia
INN-FIS, CP 2400
I-00100 Roma
Italy

M. R. Daymond
ISIS
Rutherford Appleton Laboratory
Chilton, Didcot
Oxfordshire, OX11 0QX
UK

C. H. de Novion
Laboratoire Léon Brillouin (CEA-CNRS)
CEA/Saclay
F-91191 Gif-sur-Yvette
France

L. Edwards
Department of Materials Engineering
The Open University
Walton Hall Milton Keynes MK7 6AA
UK

A. Ezeilo
TWI Ltd
Granta Park
Great Abington
Cambridge CB1 6AL
UK

M. E. Fitzpatrick
Department of Materials Engineering
The Open University, Walton Hall
Milton Keynes, MK7 6AA
UK

T. M. Holden
Los Alamos Neutron Science Center
Los Alamos National Laboratory
Los Alamos
New Mexico 87545
USA

M. W. Johnson
ISIS
Rutherford Appleton Laboratory
Chilton, Didcot
Oxfordshire, OX11 0QX
UK

A. Lodini
University of Reims (LACM-UFR
 Sciences), France, and
Laboratoire Léon Brillouin
Saclay
91191 Gif-sur-Yvette
France

T. Lorentzen
DanStir ApS
Park Allé 345
PO Box 124
DK-2605 Brøndby
Denmark

D. Möller
Hahn-Meitner-Institut
Bereich Strukturforschung
Glienicker Straße 100
D-14109 Berlin
Germany

C. Nardi
ENEA-Frascati, ERG-FUS
CP 2400, I-00100 Roma
Italy

L. Pintschovius
Forschungszentrum Karlsruhe
Institut für Festkörperphysik
P.O.B. 3640
D-76021 Karlsruhe
Germany

H.-G. Priesmeyer
GKSS
Max-Planck Straße 1
D-21502 Geesthacht
Germany

A. Pyzalla
TU Berlin, Metallphysik BH 18
Ernst-Reuter-Platz 1
D-10587 Berlin
Germany

C. Riekel
European Synchrotron Radiation Facility
B.P. 220
F-38043 Grenoble Cedex
France

W. Reimers
TU Berlin, Metallphysik BH 18
Ernst-Reuter-Platz 1
D-10587 Berlin
Germany

S. Spooner
Oak Ridge National Laboratory
Oak Ridge
TN 37831
USA

J. M. Sprauel
Laboratoire MécaSurf
ENSAM
F-13617 Aix-en-Provence Cedex 1
France

R. I. Todd
University of Oxford
Department of Materials
Parks Road
Oxford OX1 3PH
UK

P. J. Webster
Centre for Materials Research
Division of Civil and Environmental
Engineering
University of Salford
Salford M5 4WT
UK

R. A. Winholtz
University of Missouri
Department of Mechanical & Aerospace
  Engineering
Columbia
MO65211
USA

P. J. Withers
Unit for Stress & Damage Characterisation
Manchester Materials Science Centre
Grosvenor St.
Manchester, M1 7HS
UK

Preface

The presence of residual stresses in materials and components can have an important influence on their behaviour. Residual stresses can be introduced during manufacture and/or use by such processes as mechanical forming operations, welding and heat treatments. Generally those processes that result in near surface residual compression are beneficial and enhance resistance to failure whereas those that produce surface tension usually aid the onset of cracking which can lead to premature fracture.

Certain manufacturing operations, such as machining and welding, can produce undesirable surface residual tension. Often it is necessary to apply heat treatments to relax these stresses prior to use. For high-performance applications in, for example, the aerospace, automotive and chemical process industries where cyclic loading is involved, manufacturing operations such as shot peening, autofrettage, chemical surface treatment and laser surface hardening are often employed deliberately to introduce beneficial surface compression to enhance resistance to fatigue failure. In order to predict the influence of residual stress on material or component behaviour, it is essential to have an accurate knowledge of the magnitude of the stresses generated.

Residual stresses can be determined by experimental methods or by calculation from models of the processes responsible for their production. The experimental methods can be destructive or non-destructive. The destructive methods usually involve cutting or drilling operations to relax the residual stresses. The residual stresses are then calculated from the resulting dimension changes. Mostly the non-destructive methods use diffraction techniques. With this approach elastic strains are measured and residual stresses calculated from the elastic properties of the materials concerned. Diffraction techniques can employ conventional X-rays, synchrotron radiation or neutrons. In the former case only surface determinations of stress, to less than $100\,\mu m$ deep, can be made unless progressive polishing away of the surface is used. With synchrotron radiation and neutrons, penetration depths of a few millimetres or centimetres, respectively, can be achieved in specific materials. The determination of residual stresses from models requires an accurate understanding of the processes responsible for the stresses and of the elastic and plastic properties of the materials being investigated. This book concentrates on the experimental determination of residual stress by the synchrotron radiation and neutron diffraction methods.

The book consists of individual chapters which are authored by an international collection of the foremost specialists in the field of residual stress measurement. The book is effectively split up into four parts dealing with, respectively, the principles of the techniques, how stresses are extracted from the strain measurements, the measurement procedures employed and applications to particular fields of study. The main emphasis is on the neutron diffraction method since this is the most established technique.

Both methods can only be applied to crystalline materials. Initially each method is discussed in turn. The theory of the interaction of neutrons with specific crystallographic planes to produce lattice strains is reviewed and procedures for generating neutrons covered. There is then a similar presentation of the synchrotron radiation method. It is described how the process provides a very intense beam of high-energy X-rays which can be used to determine elastic strains by diffraction in the same way as with neutrons.

Part 2 deals mainly with the factors that have to be taken into account in calculating stresses from the experimental measurements. It is described how three types of residual stress can be identified. These are separated into what are commonly called macro residual stresses (type I) and micro residual stresses (types II and III). The macro residual stresses are those which exist over the scale of the dimensions of a sample or component and are of most interest for engineering stress analysis and continuum mechanics applications. The type II stresses are those which exist across grains as a result of the anisotropy of grains and the constraints to deformation that occur between them. They can also exist between different phases in materials or composites. Type III microstresses are those residual stresses that are produced at the atomic level due to the presence of defects and precipitates in materials. The choice of crystallographic planes on which to make measurements for particular applications is discussed. Methods of extracting macro- and microstresses from the data are presented. The most appropriate elastic constants for use in calculating stresses in mono- and polycrystals are considered, as is the role of texture. The information needed to determine the full residual stress sensor is also discussed.

Specific experimental measurement techniques for obtaining residual stresses are examined in Part 3. Several types of instrument have been developed. Designs which employ a monochromatic beam of neutrons and those which use a pulsed polychromatic beam are both described. Detector systems and data analysis procedures are discussed. The principles of obtaining residual stresses from neutron transmission measurements of Bragg edges are also presented. Recent advances in the development of the synchrotron radiation technique are then outlined.

The final part considers a range of applications of the techniques. It is shown that neutron diffraction can be employed successfully for measuring residual stresses in single crystal and large grained materials, metals, ceramics and composites. It is demonstrated that it can also be used to measure stresses close to and through surfaces although precautions are required to avoid surface aberration effects. Manufacturing processes that are examined include shot peening and welding. It is noted that with welding, and other situations where metallurgical changes and/or gradients can exist, considerable care is needed in establishing the appropriate stress-free crystallographic lattice spacing to adopt when determining strains. The role of synchrotron radiation for performing rapid area strain scans and for identifying steep stress gradients is also examined.

Synchrotron radiation and neutron diffraction are complementary non-destructive methods for measuring residual stresses. The book is particularly timely. It is appearing at a time when the neutron diffraction technique is becoming established and when an international standard is to be issued recommending the experimental and data analysis procedures to be adopted to ensure reliable determinations of residual stress. It is also appearing at a time when dedicated synchrotron X-ray diffraction facilities are becoming available and more accessible for making residual stress measurements. The book is timely, particularly, because of the capability of modern computer analysis procedures. Sufficient computing power now exists to allow advanced stress analysis and modelling calculations to be carried out. They can be applied to components to account for the influence of residual stress on performance.

In addition they can be used in material science studies for predicting the residual stresses that are generated by different processes and within individual grains in polycrystalline materials. These stresses can then be compared to experimental measurements to validate the models.

The book should prove to be a valuable reference source for a wide audience. It is relevant to the fields of engineering and materials science. It is particularly appropriate to all students, researchers and industrialists with an interest in residual stress.

Prof. George A. Webster
Department of Mechanical Engineering
Imperial College, London SW7 2BX

Acknowledgements

In this book it has been our intention to provide a state-of-the art review of the application of diffraction techniques to the determination of residual stress in materials and components. Neutron diffraction has become well established over the last decade, and we are now beginning to see the rapid emergence of synchrotron X-rays as a useful and attractive technique.

For a book of this nature, thanks are due primarily to the authors who have supplied the chapters; the book would not exist without their efforts.

We wish to acknowledge the support of the Department of Materials Engineering at The Open University and the Laboratoire Léon Brillouin, Saclay, for secretarial support in aid of the production of the book. Particular thanks are due to Chantal Pomeau for styling and collating the final draft.

The cover images were provided by Dr J. Santisteban of The Open University.

Dr Michael Fitzpatrick
Department of Materials Engineering
The Open University
Walton Hall
Milton Keynes MK7 6AA
UK

Prof. Alain Lodini
Laboratoire Léon Brillouin
CEA-Saclay
91191 Gif-sur-Yvette
France

Part 1

General applications of neutron and synchrotron radiation to materials research

1 The use of neutrons for materials characterization

C. H. de Novion

1.1 Introduction and historical background

The neutron was discovered by Chadwick in 1932. The fission of the uranium nucleus, that is, its splitting in two fragments after having absorbed a neutron, was demonstrated in 1938 by Joliot-Curie and Hahn. The fission process leads to the emission of a few neutrons, making a chain reaction possible: nuclear reactors work on this principle.

As any moving particles, neutrons have a wave character (the "matter waves" first put forward by de Broglie around 1923). Because their mass is close to that of individual atoms, slow neutrons in thermodynamical equilibrium with a medium at ambient temperature (such as the moderator in a fission reactor), called "thermal" neutrons, have altogether a wavelength of the order of interatomic distances in condensed matter ($\sim$0.1 nm), and kinetic energies close to atomic vibration energies ($\sim$10^{-2} eV). They then give rise, like X-rays, to diffraction by crystalline solids; moreover, within the solid, they also can absorb or emit quanta of collective vibration energy ("phonons"), raising the possibility of spectroscopic studies (e.g. measurement of elastic wave dispersion curves) by inelastic neutron scattering. Both phenomena, diffraction and inelastic scattering, are observed by analysing the change of direction and speed of thermal neutrons "scattered" by a sample of solid or liquid matter.

Thermal neutron diffraction and inelastic scattering by crystals, theoretically predicted, were experimentally discovered in the late 1940s around the reactors built in the frame of the Manhattan project in the US. The 1994 Nobel prize in physics was awarded to C. Shull and B. Brockhouse for the development of neutron scattering techniques. Neutron scattering was found at that time to be a unique method to study lattice dynamics and (because the neutron bears a "spin" and an individual magnetic moment) to reveal the ordered magnetic structures of crystalline solids [1]. Dedicated single crystal and powder spectrometers for elastic (diffractometers) and inelastic ("triple-axis" instruments) scattering were installed on multipurpose reactors built in the 1950s and the 1960s, the largest of which being those in Oak Ridge, Brookhaven and Chalk River in North America.

A major progress was reached with the construction of the High Flux Reactor (HFR) of the Laue-Langevin Institute (ILL) in Grenoble (resulting from a French–German initiative, rapidly joined by the British), which opened in 1972, entirely dedicated to fundamental research with thermal neutrons, and specially designed for this purpose. ILL initiated an external user program, which progressively allowed it to receive a large number (>1000 per year) of scientists from many laboratories. The principle of an external user facility was then applied in several other centers (e.g. the NIST Center for Neutron Research, Gaithersburg, in the US).

Very important technical progress was made in the 1960s and 1970s (e.g. "cold" and "hot" neutron moderators, neutron guides, neutron polarization, etc.), and new neutron scattering techniques developed: small-angle neutron scattering (SANS), quasi-elastic scattering by time-of-flight (TOF), backscattering or spin-echo techniques, neutron reflectivity, etc. Progressively, neutron scattering broadened its applications to larger scientific domains: solid state chemistry, liquids, soft matter, materials science, geosciences, biology, etc. (for a general overview, see Ref. [2]).

Neutrons can also be produced by another type of nuclear reaction, called "spallation," obtained by hitting a target with protons (see below). Unlike the reactors, which produce continuous neutron flux, spallation sources generate pulses of neutrons. Spallation sources for neutron scattering were developed in the 1980s (first in Tsukuba, Japan and in Argonne, US); presently, the most intense is ISIS, at Abingdon in the UK. Generally, it is estimated that the potential of progress of pulsed spallation sources is much larger than that of reactors. Several regional new pulsed spallation sources are either in construction (SNS at Oak Ridge in the US), decided (in Japan) or planned (in Europe): they should allow major progress in the structural and spectroscopic study of materials in the coming decades.

1.2   Characteristics of the neutron particle

The neutron is a particle of zero electric charge, of mass $m = 1.67 \times 10^{-24}$ g, of radius $r_0 = 6 \times 10^{-16}$ m, with a spin of 1/2 and a magnetic moment $\mu = -1.9\mu_N$ (nuclear magnetons). One may note that the mass of the neutron is practically equal to that of the hydrogen atom (the proton and the neutron are two different charge states of the "nucleon"), and its radius is 5–6 orders of magnitude smaller than the average size of an atom ($\sim 10^{-10}$ m).

The following relationships can be written between the properties of the neutron, considered either as a particle or as a wave: velocity $\mathbf{v}$, energy E, momentum $\mathbf{p}$, wavelength λ, wavevector $\mathbf{k}$:

$$\hbar\mathbf{k} = m\mathbf{v} = \mathbf{p}, \qquad E = \frac{1}{2}m\mathbf{v}^2 = \frac{h^2}{2m\lambda^2} = \frac{\hbar^2\mathbf{k}^2}{2m}$$

($\hbar = h/2\pi$: reduced Planck constant).

For example, thermal neutrons with a wavelength of 0.4 nm (=4 Å) have a speed of 10^3 m/s and a (kinetic) energy of 5 meV.

It is interesting to note that the wave nature of the neutron has been directly demonstrated by neutron interferometry experiments, quite analogous to light interferometry [3].

1.3   Production of neutrons

Neutrons are particles difficult and expensive to produce: one reason for this is the necessity of biological protection and of safety procedures to prevent accidents.

Presently, two types of sources produce beams of thermal neutrons: nuclear fission (steady state) reactors and neutron spallation (pulsed) sources. They are schematically shown in Figures 1.1 and 1.2, respectively. The major neutron sources for fundamental research, and some of their characteristics, are listed in Table 1.1.

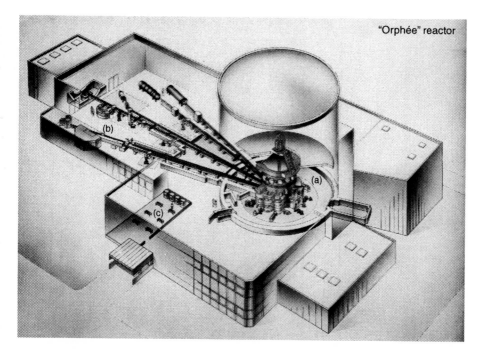

Figure 1.1 Perspective view of a fission reactor for thermal neutron beam production (the Orphée reactor, located in Saclay, France; note that not all instruments have been drawn, for clarity): (a), reactor hall: (b), guide hall: (c), computing facilities. (See Colour Plate I.)

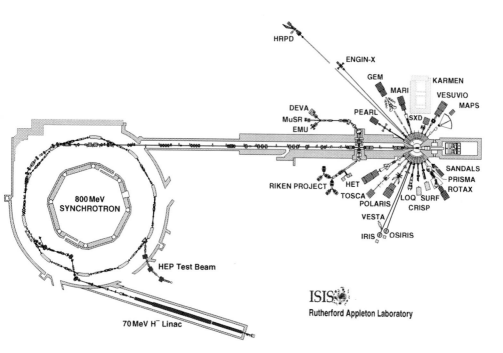

Figure 1.2 Plan of the ISIS facility (spallation source and instruments), located in Abingdon, UK. (See Colour Plate II.)

Table 1.1 Major thermal neutron beam sources

Continuous sources

Country	Source	Location	First operation	Power (MW)	Thermal flux (10^{14} n/cm^2/s)	Special moderator[a]	Scattering instruments	Stress diffractometer
Australia	HIFAR	Lucas Heights	1958	10	1.4	No	7	0
Canada	NRU	Chalk River	1957	120	3.0	No	6	1
Denmark[b]	DR3[b]	Risø	1960	10	1.5	1 C	8	1
France[c]	HFR-ILL[c]	Grenoble	1972–1994	58	12.0	2 C, 1 H	32	[d]
France	Orphée	Saclay	1980	14	3.0	2 C, 1 H	25	1
Germany	BER-2	Berlin	1973–1986	10	2.0	1 C	16	1
Germany	FRJ-2	Jülich	1962	23	2.0	1 C	16	0
Hungary	BNC	Budapest	1959	10	1.6	1 C	7	1
Japan	JRR-3	Tokai	1962	20	2.0	1 C	23	1
Korea	Hanaro	Taejon	1996	30	2.8	No	6	1
Sweden	R-2	Studsvik	1960	60	1.0	No	5	1
Switzerland	SINQ	Villigen	1996	1000 kW (spallation source)	2.0	1 C	13	1
USA[b]	HFBR[b]	Brookhaven	1965	30	4.0	1 C	14	1
USA	HFIR	Oak Ridge	1960	85	12.0	1 C	9	1
USA	NBSR-NIST	Gaithersburg	1969	20	2.0	1 C	17	1

Under construction

Country	Source	Location	First operation	Power (MW)	Thermal flux (10^{14} n/cm^2/s)	Special moderator[a]	Scattering instruments	Stress diffractometer
Germany	FRM-II	München	2003[e]	20	7.0	1 C, 1 H	17	1

Pulsed sources

Country	Source	Location	First operation	Beam power (kW)	Pulse length (proton) and repetition rate	Thermal peak flux (10^{14} n/cm^2/s)	Moderators[a]	Scattering instruments	Stress diffractometer
Japan	KENS-KEK	Tsukuba	1980	3	0.1 μs; 20 Hz	3	1 C, 1 T	16	0
Russia	IBR2	Dubna	1984	2000 (fission)	305 μs; 5 Hz (thermal neutrons)	100	1 C, 3 T	11	1
UK	ISIS	Abingdon	1985	160	0.4 μs; 50 Hz	20–100	2 C, 2 T	19	1
USA	LANSCE	Los Alamos	1985	56	0.27 μs; 20 Hz	34	1 C, 3 T	7	[d]
USA	IPSN	Argonne	1980	7	0.1 μs; 30 Hz	5	3 C	13	[d]
Under construction									
USA	SNS	Oak Ridge	2006[e]	2000	1 μs; 60 Hz	200	2 C, 2 T	10	1

Source: Richter and Springer [7].

Notes

Two other new reactors have been approved and should start in the next decade: the PIK reactor in Gatchina, St. Petersburg (Russia), with performances close to those of ILL, and a new medium flux multipurpose reactor HIFAR II (which will replace the present HIFAR) in Lucas Heights, Australia.

Apart from the Spallation Neutron Source (SNS) in the US presently in construction, two other large regional spallation sources are planned, with a power of 5 MW, a neutron scattering facility within the JHP/NSRP multipurpose Joint Project in Tokai (Japan), which has recently been decided and should start to operate in 2007, and the European Spallation Source (ESS) which is not decided yet, but has an anticipated start date around 2013. A medium-flux spallation source in Central Europe is also in project (AUSTRON, 0.5 MW).

It is worth mentioning that high-quality stress diffractometers are also found in four continuous sources not mentioned in the table: at the High Flux Reactor in Petten (The Netherlands), at the FRG reactor in Geesthacht (Germany), at the Rez reactor near Prague (Czech Republic) and at the reactor MURR of the University of Missouri (Columbia, USA).

a Moderator – C, cold; T, thermal; H, hot.

b The Brookhaven and Risø reactors have been shut down since 1998 and 2000, respectively.

c The Institut Laue-Langevin (ILL), located in Grenoble (France), is a multinational institute, managed by a consortium of three major associates (France, Germany, UK) and several minor Scientific Member countries (Switzerland, Spain, Italy, Russia, Austria and Czech Republic).

d Stress experiments performed on a general purpose high-resolution powder diffractometer.

e For sources in construction (FRM-II reactor and SNS spallation source), these are anticipated starting dates.

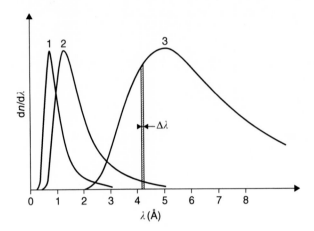

Figure 1.3 Wavelength distribution of thermal neutrons for several thermalization temperatures T: (1) $T = 1000\,\mathrm{K}$ ("hot" neutrons), (2) $T = 300\,\mathrm{K}$ ("thermal" neutrons),
(3) $T = 20\,\mathrm{K}$ ("cold" neutrons). The wavelength range of width $\Delta\lambda$ selected by a
monochromator is indicated.

A neutron source contains several components:

- It should produce the highest possible neutron flux (usually expressed in number of neutrons/cm^2/s). The neutrons produced by the elementary nuclear reactions are high energy (MeV range) particles: the "fast" neutrons.
- They must then be "moderated" to obtain neutrons with energy and wavelength in the required order of magnitude. In the moderator, which is constituted of light atoms (e.g. water, heavy water, graphite, beryllium), the fast neutrons are slowed down by successive collisions, to form a Maxwellian distribution of particles in equilibrium at the temperature of the moderator: the so-called "thermal neutrons" (see Figure 1.3).
- The various types of neutron spectrometers may require neutrons with different energies and/or wavelengths; by varying the thermalization temperature, one can displace the center of the wavelength distribution (see Figure 1.3). High energy (i.e. short wavelength) neutrons can be obtained with a "hot source" (typically a solid graphite moderator heated to 1000–2000°C). Low energy (i.e. long wavelength) neutrons are obtained with a "cold source" (usually liquid hydrogen or deuterium, but can be solid or liquid methane or liquid helium). The available wavelengths range from 0.05 nm ("hot" neutrons) to 2 nm ("cold" neutrons). Spallation sources allow production of epithermal neutrons as well, which are interesting for spectroscopy in the eV range.
- Then, thermalized neutron beams in a given direction are extracted from the source. An optimized flux-resolution compromise results from the choice of the collimation.

"Cold" neutrons can be transported to large distances (10–100 m), using *neutron guides* which are tubes where the neutrons propagate by total reflection along the internal surfaces. These are generally nickel coated, which have a critical angle of 0.5° for a wavelength of 0.5 nm (the critical angle θ_c is proportional to the neutron wavelength). Non-periodic multilayers (e.g. Ni–Ti), called "supermirrors," have recently been developed to increase the critical angle and therefore, the efficiency of the neutron guides.

The various neutron scattering instruments are installed either as close as possible to the source (to benefit from the highest possible flux), or at the end or along the guides to reduce the background due to fast neutrons and γ-rays (see Figures 1.1 and 1.2).

1.3.1 Fission reactor (steady state) sources

Most of the present neutron sources are fission reactors (see Table 1.1).

In a modern reactor, optimized to produce the highest neutron flux, the core is a very compact (e.g. $25 \times 25 \times 90\,cm^3$ for Orphée, Saclay) assembly of uranium alloy plates, highly enriched in ^{235}U isotope. Belonging to this design are the HFR of ILL in Grenoble (the world-leading neutron source, built in 1972 and refurbished in 1994); the Orphée reactor in Saclay (built in 1980, sketched in Figure 1.1); and the FRM-II reactor in Munich, currently under construction.

Neutron beam tubes point towards the required moderator, tangentially to the core in order to minimize fast neutrons and γ-rays which increase the background. The neutron beams emitted by the reactor have a broad distribution of wavelengths, according to the temperature of thermalization (see above and Figure 1.3).

Many of the spectrometers on steady sources require a monochromatic incident beam: this is obtained by Bragg scattering from a crystal installed in the beam tube or in the neutron guide, or (in the case of cold neutrons) by a periodic multilayer or by a mechanical velocity selector. The selected wavelength bandwidth $\Delta\lambda$ depends directly on the monochromator characteristics (e.g. mosaicity of the crystal, etc.).

Spin-polarized neutron beams, useful for studying magnetic materials, can be obtained from ferromagnetic monochromators (crystals or multilayers), but also, quite recently, by transmission through spin-polarized helium-3 filters.

1.3.2 Spallation (pulsed) sources

In a neutron spallation source (see Figure 1.2), sharp pulses of high energy protons accelerated in a synchrotron hit a target of high atomic mass (Pb, W, Ta or U). As a consequence of the subsequent nuclear reactions, the target nuclei are put in a highly excited state, and then decay by "evaporating" neutrons, some of which trigger further reactions on other nuclei. Because of the pulsed character of the incident proton beam, the spallation (γ, n) reaction produces sharp pulses of neutrons of very high energy (in the MeV range). The spallation process has the advantage of a much lower heat production in the target, and of a high neutron brightness in the pulse. The target is surrounded by various thermal or cold moderators in order to thermalize the neutrons produced.

Typical values for a spallation neutron source are a pulse rate between 10 and 50 Hz and a proton pulse length in the microsecond range (see Table 1.1). The neutron brightness in the pulse of an intense spallation source such as ISIS is comparable to that of the highest flux steady state reactors (e.g. ILL), but of course the average flux is much lower (by approximately two orders of magnitude).

Presently, steady state and pulsed sources are complementary, in the sense that each is optimized for different types of experiment. Instruments on pulsed sources generally use a polychromatic ("white") neutron beam and are based on TOF techniques; pulsed sources are therefore very efficient when all the neutrons in the pulse are utilized, that is, for powder diffraction and TOF inelastic scattering (see below). Steady state reactors are still preferred for many techniques, such as the use of polarized neutrons, spin-echo, single-crystal diffraction and spectroscopy.

It is worth mentioning that the next-generation high intensity neutron sources are all proposed to be spallation sources (SNS in the US, NSRP/JHF joint project in Japan, ESS in Europe), aiming for the very highest fluxes, one to two orders of magnitude above ISIS.

1.3.3 Other types of sources

There are currently two non-conventional neutron sources in operation: the pulsed fission reactor IBR-2 in Dubna (Russia), and the new quasi-continuous spallation source SINQ in Switzerland (which began operation in 1996).

1.4 Interaction between neutrons and matter

The neutron–matter interaction is generally weak, and neutron scattering is very well described by the lowest-order ("kinematic") perturbation theory (i.e. the Born approximation). Higher-order perturbation theory produces *multiple scattering* corrections. The theory of neutron scattering by matter can be considered as well established. A complete presentation is found in the book by Lovesey [4].

The only exceptions to the kinematic theory are scattering by surfaces in grazing incidence (reflectivity near the "plateau" for total reflection), Bragg scattering by very perfect crystals (neutron topography and interferometry), and very small-angle scattering close to the transmitted beam. In these cases, important for neutron optics, where the scattered beam is comparable in amplitude with the incident beam, one should take into account a "dynamical" theory [5].

1.4.1 Interaction with a single atom

As it carries no electric charge, the neutron has no electrostatic interaction with the electron cloud of the atom. The atom–neutron interaction potential contains two terms:

(i) A very short-range *nuclear interaction* between the neutron and the nucleus, via nuclear forces; this is characterized by a form factor (scattering amplitude) b independent of the scattering angle θ (in the case of X-rays, which interact with the electron cloud of the atom, the atomic form factor f decreases with $\sin\theta/\lambda$, see Figure 1.4b). The nuclear scattering amplitude b has the dimension of length, and a value of the order of 10^{-14} m. b can vary strongly between neighboring elements of the periodical table (e.g. Mn versus Fe) or between different isotopes of a given element (e.g. deuterium versus hydrogen, or ^{62}Ni versus ^{58}Ni); compared to X-rays, for which f is proportional to the atomic number Z, neutrons are much more sensitive to light elements (H, D, Li, C, N, O) (see Figures 1.4a and 1.4c).

For a given element, b depends on the mass number (i.e. on the isotope) and on the orientation of the nuclear spin relative to the neutron spin. In order to obtain information on the element from the scattering data, one has to consider the average scattering amplitude $\langle b \rangle$: this is called the *coherent* scattering amplitude (or coherent scattering length). The fluctuations of b around $\langle b \rangle$ allow the definition of a quantity $4\pi(\langle b^2 \rangle - \langle b \rangle^2)$ called the *incoherent* scattering cross-section. It has a very high value in the case of hydrogen (protons).

(ii) A *magnetic interaction* between the spins of the neutron and of the electrons, which is important only in strongly magnetic systems; its atomic form factor b_m decreases with $\sin\theta/\lambda$.

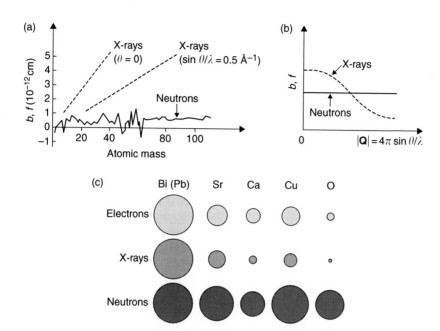

Figure 1.4 (a) Dependence of the neutron scattering length b with atomic mass (full line). Comparison with the atomic form factor f for X-rays (dotted lines). (b) Dependence of b and f with scattering vector $\mathbf{Q} = 4\pi \sin\theta/\lambda$. (c) Compared relative scattering powers of electrons, X-rays and neutrons for a few elements (for $Q = 1\,\text{Å}^{-1}$).

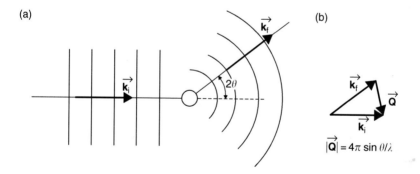

Figure 1.5 Elastic interaction between thermal neutrons and matter. (a) Scattering by an individual atom. (b) Definition of the scattering vector $\mathbf{Q}$.

If an incident neutron beam (e.g. described by a plane wave of amplitude e^{ikz}) is directed on a sample in the z direction, each atom re-radiates ("scatters") the incident radiation in all directions, that is, as a spherical wave of amplitude $(b + b_m)e^{ikr}/r$ (see Figure 1.5a).

Neutrons also interact with atoms via nuclear reactions (such as $^{10}\text{B} + \text{n} \rightarrow {}^{11}\text{B} + \gamma\text{-ray}$). These can be expressed as imaginary terms in the nuclear scattering length b, and lead to a true *absorption* cross-section, which is generally weak. The consequences for bulk materials are described below.

A compilation of the bound scattering lengths and cross-sections (coherent and incoherent scattering, and absorption) for low-energy neutron scattering by the different isotopes of each element, is given by Sears [6]. The variation of b between different isotopes of a given element gives the advantage in neutron scattering studies to vary the contrast by the use of isotopically substituted samples.

1.4.2   The scattering function for an array of atoms

As stressed in Section 1.1, because the wavelength and kinetic energy of thermal neutrons are comparable, respectively, to interatomic distances and internal excitation energies in condensed matter, the scattering phenomena are described within the wave vision of the neutron (and in the Born approximation, because the neutron–matter interaction is very weak).

Let us consider a parallel and monochromatic neutron beam, incident on a sample. In this incident beam, the neutrons are defined by their momentum or wavevector $\vec{k}_i$ and their energy E_i. A scattered neutron is then defined by $\vec{k}_f$ and E_f. Within the scattering process, the neutron has exchanged with the sample the momentum $\mathbf{Q} = \vec{k}_i - \vec{k}_f$ and the energy $\Delta E = \hbar\omega = E_i - E_f$.

The amplitude of the neutron beam scattered by the array of atoms is the coherent superposition of the amplitudes of the elementary spherical waves scattered by each individual atom. This leads to interference and diffraction phenomena. The quantity of scattered neutrons in the direction $\vec{k}_f$ with energy E_f is directly related to the differential cross-section $d\sigma(Q, \omega)/d\Omega = S(Q, \omega)$, called the *scattering function*, which is a characteristic function of the sample.

The scattering process can be *elastic* (no change of the energy of the neutron, $\Delta E = \hbar\omega = 0$), or *inelastic* ($\Delta E = \hbar\omega \neq 0$).

- In the case of elastic scattering, one has:
  $|\mathbf{k}_i| = |\mathbf{k}_f| = 2\pi/\lambda$, with $|\mathbf{Q}| = |\vec{k}_i - \vec{k}_f| = 4\pi \sin\theta/\lambda$ (see Figure 1.5b). $\vec{Q}$ is called the scattering vector. In the case of a crystalline material, the coherent elastic scattering cross-section contribution to $S(Q, \omega)$ corresponds to Bragg diffraction (see below); scattered intensity is observed only in specific directions θ given by the Bragg law:

$$2d^{hkl} \sin\theta = \lambda$$

  where d^{hkl} is the interplanar distance between planes of Miller indices $(h\,k\,l)$.
- On the other hand, during their transmission through a solid sample, neutrons can exchange, for example, a quantum of atomic vibration energy ("phonon"). This is the case of inelastic scattering, where :

$$|\mathbf{k}_i| \neq |\mathbf{k}_f| \quad \text{and} \quad \Delta E = \hbar\omega = \hbar^2(k_i^2 - k_f^2)/2m$$

1.4.3   Absorption

The true absorption of thermal neutrons by matter is due to nuclear reactions: for most elements, it is weak (in particular for C, O, Al, Fe, Zr, Nb, Pb, etc.). For a bulk material, the linear absorption coefficient μ is usually of the order of 10–$100\,\mathrm{m}^{-1}$ and varies proportionally to λ. In some cases, the absorption by nuclear reactions is very large (e.g. ^{10}B, ^{113}Cd, Gd).

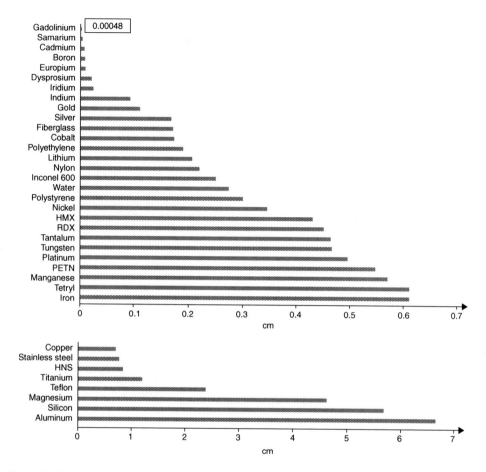

Figure 1.6 Material thicknesses required to attenuate a thermal neutron beam by 50% ($\lambda = 0.18$ nm).

The total apparent absorption of neutrons by a material includes the true absorption and all the scattering processes.

Because the total neutron–matter interaction (scattering + absorption) is generally weak, neutron penetration in condensed matter specimens is large. Contrary to conventional X-rays, which sample only the region close to the surface (10^{-6}–10^{-4} m, depending on the average atomic number), neutrons allow study of bulk specimens. The typical penetration depth of thermal neutrons in materials is of the order of several mm to several cm (see Figure 1.6), for example 10 cm in Al and 1 cm in Fe for $\lambda = 0.18$ nm. This property is particularly useful in materials and engineering science.

1.5   Measurement and detection equipment

1.5.1   General

In all neutron scattering instruments, an incident (monochromatic or white) beam impinges on the sample, and the scattered neutrons are counted with detectors (via nuclear reactions such as $^3\mathrm{He} + {}^1\mathrm{n} \to {}^3\mathrm{H}\,(0.20\,\mathrm{MeV}) + \mathrm{p}\,(0.55\,\mathrm{MeV})$ in helium-3 counters), as a function of momentum transfer Q and/or transferred energy ΔE.

Neutron scattering instruments can be classified into two main categories: *diffractometers* for structural determinations by elastic scattering, and *spectrometers* which give information on atomic motions by inelastic scattering. Typical schematic spectra obtained by the various scattering techniques are shown in Figure 1.7.

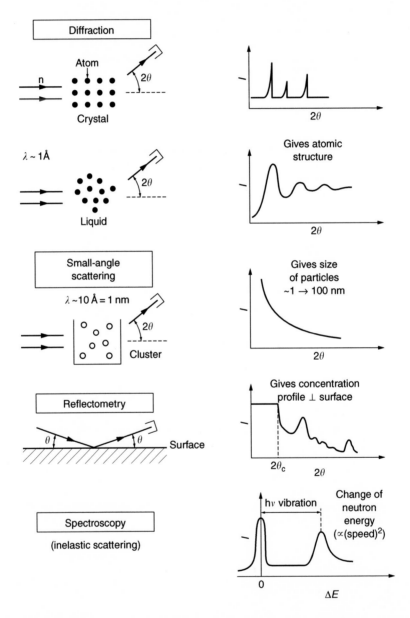

Figure 1.7 Typical schematic spectra (scattered intensity versus scattering angle θ or energy transfer ΔE) obtained by the various neutron scattering techniques.

1.5.1.1   Diffractometers

The scattered intensity is plotted versus the momentum transfer $Q = 4\pi \sin\theta/\lambda$ and gives information on the time-averaged structure. Neutron diffractometers include:

- powder or single crystal instruments for determination of crystalline and magnetic atomic structures, residual stresses and crystallographic textures, by elastic Bragg scattering;
- diffuse scattering instruments for determination of correlation functions and local atomic arrangements in atomically disordered systems (liquids, amorphous materials and glasses, etc.);
- small-angle diffractometers for the study of nanometer-scale heterogeneities;
- reflectometers for the study of surfaces, thin films or multilayers.

1.5.1.2   Spectrometers

These include triple-axis instruments (mainly for single crystals), TOF, backscattering or spin-echo instruments. The scattered intensity is plotted versus the energy transfer ΔE; it allows determination of electronic, vibrational or magnetic excitations (in particular dispersion curves of collective excitations $\omega(q)$), and diffusional processes in solids and liquids.

1.5.2   Powder diffractometers and strain scanners

The precise interplanar distances in a crystalline material are obtained from Bragg's law: $2d^{hkl} \sin\theta = \lambda$, and the positions of atoms in the unit cell are calculated from the intensities of the Bragg peaks. Two cases are possible:

(a) Either select a monochromatic incident beam (λ fixed) and measure the scattered intensity versus the scattering angle θ. This is the usual method at a continuous source (reactor). A typical diffractometer for elastic scattering on a steady state source includes a crystal monochromator, collimators, a goniometer for the sample and single or position-sensitive detectors (see Figure 1.8).

(b) Or, send polychromatic ("white") pulses to the sample, select a given scattering angle (θ fixed), and measure the scattered intensity versus the wavelength λ. This is the technique used on pulsed spallation sources. Practically, one measures the TOF of neutrons: $t = L/v = (\lambda m/h)L$, where t is the time for a neutron of mass m and speed v to travel along the distance L between the moderator and the detector (h = Planck's constant).

Strain scanners are particular powder diffractometers devoted to the three-dimensional (3D) mapping of internal strains and stresses in materials, by measuring very precisely the local lattice spacing deviations $\Delta d^{hkl}/d^{hkl}$ induced by these strains and stresses (see below). For this, use is made of the differentiation of Bragg's law:

$$\Delta d^{hkl}/d^{hkl} = (\Delta\lambda/\lambda) - \cot\theta\,\Delta\theta.$$

Constant wavelength strain scanners correspond to $\Delta\lambda = 0$ and $\Delta d^{hkl}/d^{hkl} = -\cot\theta\,\Delta\theta$, and pulsed instruments to $\Delta\theta = 0$ and $\Delta d^{hkl}/d^{hkl} = \Delta\lambda/\lambda = \Delta t/t$.

1.5.3   Comparison between instruments on steady state reactors and on pulsed spallation sources

This discussion is delicate and controversial. The situation is very different for the different types of instruments. A classification has been made by Richter and Springer [7] in terms of the different ability of the various instrument classes to utilize efficiently all the neutrons in

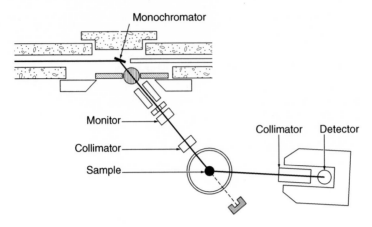

Figure 1.8 Schematic diagram of a powder diffractometer on a steady state neutron source (monochromatic beam).

the incident polychromatic beam propagating in a guide or a beam tube. In summary:

(i)   Powder diffractometers and inelastic TOF spectrometers use efficiently all the neutrons in the polychromatic pulse of a spallation source, whereas the corresponding instruments on reactors lose the majority of the neutrons by monochromatizing or pulsing the incident white beam. This is why the instruments of this sort on a pulsed source like ISIS are competitive with those on the best continuous reactors, although the average flux of the sources are in the ratio 1 : 100.

(ii) On the contrary, the spectrometers for single crystal spectroscopy are only sensitive to the average flux: therefore, the triple-axis spectrometers on reactors offer much larger effective flux on the sample. On spallation sources, dynamics of single crystals are measured by the TOF method and position sensitive detectors; the fixed geometry of these instruments allows the use of large area detectors to partially compensate for the lack of incident intensity. In the present state of technology, TOF instruments on spallation sources are very competitive for mapping single crystal excitations simultaneously in a wide q-range at high energy, but are much less efficient than triple-axis instruments on reactors for high q-resolution or low-energy (cold neutrons) measurements.

(iii) In between, one finds SANS, reflectometers and spin-echo instruments, which in a reactor use a broad wavelength band: here, the benefit on pulsing is not very large.

The quality of measurements does not depend only on intensity arguments. In several aspects, instruments on pulsed spallation sources suffer from drawbacks, which are expected to be overcome in the future:

•   instabilities of the source;
•   more complicated data corrections, because of the polychromatic nature of the incident beam: extinction for single crystal diffractometry, inelastic effects for small-angle scattering and diffuse scattering;
•   inability to polarize polychromatic neutron beams (this should be resolved by the success of ^{3}He polarizing filters).

On the other hand, instruments on spallation sources can obtain a whole spectrum (very large q range) in a single shot, which is very interesting for kinetic studies, but the rescaling of the data obtained on the different detector banks is a delicate procedure. Also, because of the fixed geometry, instruments based on the TOF technique are more favorable for experiments with complex sample environment (such as high pressures).

Concerning powder diffraction (which is the most favorable case for pulsed sources), a detailed comparison between instruments of ISIS and ILL was made recently (ILL Workshop, Grenoble, 22–23 March 1999). The present leading high-resolution instruments, HRPD (at ISIS) and D2B (at ILL), were found to have, on average, similar resolution and intensity (D2B gains flux because of more efficient focussing, but this is compensated because of the mechanical constraints due to moving detectors). HRPD gives access to a larger q range, and has better resolution at high q; D2B remains better for magnetic studies (low q), has a simpler peak shape and a lower background.

In pulsed instruments, all Bragg peaks are measured simultaneously; therefore, concerning strain scanning, the technique is favorable when information from several Bragg peaks is required (textured or multiphase samples).

1.6   Principal applications in materials science

The main applications of neutron scattering and imaging in materials science are summarized below. An overview of the recent research performed by neutron scattering, in particular in materials science, and also of instrumentation development, can be found in the Proceedings of recent International and European Conferences on Neutron Scattering, held respectively in 1997 in Toronto [8] and in 1999 in Budapest [9]. A detailed review of the industrial and technological applications of neutrons has been published by Fontana *et al.* [10].

1.6.1   Bragg diffraction

Conventional powder and single crystal Bragg neutron scattering has many various applications in materials science [1]. One of the great advantages of neutron diffraction is that it allows the study of large ($\sim$cm^3) volumes. Such studies include:

- precise determination of the position of light atoms (H, D, C, N, O) in crystalline structures, in particular in the presence of heavy atoms;
- crystallographic studies in alloys made of neighbouring elements of the periodic table, which cannot (or barely) be distinguished by X-rays (e.g. Fe–Cr, Fe–Mn, etc.);
- precise determination of crystalline phase contents or of amorphous/crystalline ratio in multiphase samples;
- determination of *residual stresses*, which is the object of this book (see below);
- quantitative characterization of the bulk crystallographic *textures* of large ($\sim$1 cm^3) samples with a large grain size (10^2–10^3 μm);
- (antiferro-)magnetic ordered structures.

High-temperature *in situ* studies, owing to the transparency of furnace material to neutrons, are of particular interest: materials synthesis and processing, phase transformations, texture modification during recrystallization or annealing, crystallization of glasses, kinetic reactions (down to time ranges of the order of seconds on the best powder diffractometers), thermal expansion of lattice parameters, etc.

Let us mention a few significant recent results in the field of materials science, obtained by powder neutron diffraction:

- the determination of the precise positions of oxygen atoms in the non-stoichiometric high-temperature superconducting cuprate $YBa_2Cu_3O_x$, which were shown to form a layered structure [11] (X-rays simply showed a disordered perovskite); this allowed the development of the concept of charge reservoirs, and stimulated a successful search for new layered high-T_c superconductors;
- the determination of the different allotropic crystal structures of solid C_{60}, showing that the individual C_{60} molecules are locked into orientationally-optimal configurations, resulting from a compromise between electrostatic, van der Waals and close-packing requirements [12];
- the observation by *in situ* powder neutron diffraction, using a large position sensitive detector, of new transitory metastable phases (such as stoichiometric "YFe") with novel structures and magnetic properties, formed during the crystallization of amorphous metallic alloys [13];
- the observation at high temperature of the structural evolutions of zircaloy tubes utilized in pressurized water nuclear power plants, that is, precipitation/dissolution of hydride phases, and variation of the hydrogen content in solid solution in the matrix (see Figure 1.9).

1.6.2   *Residual stresses*

Macroscopic internal stresses in polycrystalline materials, which are homogeneous over a large number of crystalline grains (micrometer to millimeter size), lead to local strains and

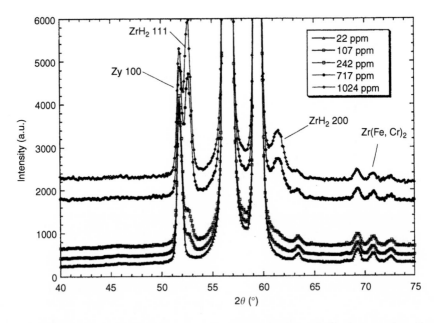

Figure 1.9 Powder neutron diffraction diagrams measured for several Zircaloy samples. One can observe the increase of the background with hydrogen content, and the weak diffraction lines of hydride δ-ZrH_2 and Laves phase $Zr(Fe, Cr)_2$ (F. Couvreur and G. André, unpublished).

therefore to displacements of the Bragg peaks. Indeed, the lattice strain $\varepsilon_{\phi\psi}$ along a general direction (ϕ, ψ) is directly related to the change of interplanar distance d^{hkl} between crystal planes (h, k, l) oriented along (ϕ, ψ). One has:

$$\varepsilon_{\phi\psi} = [d^{hkl}(\phi, \psi) - d_0^{hkl}]/d_0^{hkl}$$

where d_0^{hkl} is the value of d^{hkl} in the absence of internal stresses. To obtain the complete strain tensor, it is necessary to measure the strain $\varepsilon_{\phi\psi}$ in several (six) orientations, which can be made by neutron diffraction. Residual stresses σ_{ij} are then calculated from the measured internal strain tensor by the laws of elasticity (generalized Hooke's law).

Conventional X-ray diffraction can only measure strains close to the surface (a few micrometers); because of absorption, one cannot have access to the complete (ϕ, ψ) scanning; the stress is then calculated by assuming that the hydrostatic component normal to the surface, σ_{33}, is equal to 0, to overcome the difficulty of precise knowledge of d_0 (see Chapter 4).

Neutron diffraction is therefore necessary for in-depth determination of residual stresses [14]. The method, shown in Figure 1.10, relies on a precise evaluation of Bragg peak positions. In the present state of the art, neutron diffraction allows mapping of internal strains (stresses) in bulk samples (several centimeters in size), with the use of slits, translation tables and goniometers, with a linear resolution of the order of 1 mm (0.3 mm in 1D problems) (see Chapters 8 and 9). The precision on measured strains is usually 5×10^{-4}; that on the residual stresses depends on the studied material and is typically ± 20 MPa.

Residual stress measurements by neutron diffraction on industrial components is a growing field. Studies have been made on turbine blades for the aeronautics industry, on brazed or welded assemblies, on pieces of composite materials, thick coatings, steel and aluminum sections, etc. (see Part 5 of this book). Most of the neutron scattering centers have now a diffractometer dedicated to residual stress measurements (see Table 1.1).

For example, Figure 1.11 shows the stress profile measured in an automotive gear, which had undergone a surface treatment [15].

"Microstresses" (varying spatially on distances of the order or smaller than the grain size) induce a broadening of the diffraction peaks (see Chapter 5). Little neutron work has been made in this field, although important information can, in principle, be obtained from the

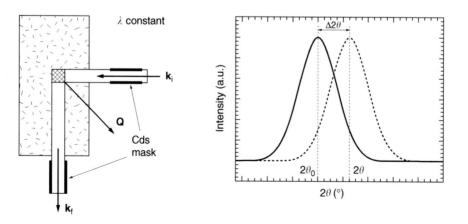

Figure 1.10 Principle of strain measurement by neutron diffraction.

(a)

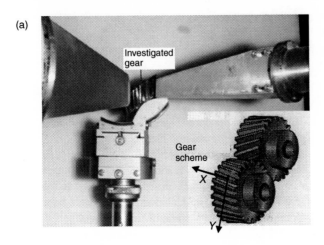

(b)

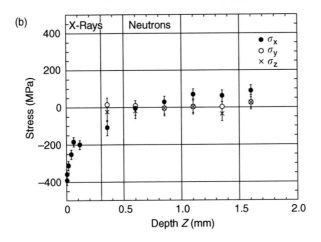

Figure 1.11   Neutron diffraction determination of the stress field in an automotive gear. (a) Photograph showing the gear geometry and the reference principal axes. (b) Evolution of the three stress components measured by X-ray and neutron diffraction (from Ceretti *et al.* [15]).

analysis of the peak profile, such as the shape of the plastic zone and the value of the plastic strain at the tip of a crack [16].

1.6.3   Elastic diffuse scattering

Liquids or solids with atomic disorder show a continuous distribution of (elastic) intensity in their diffraction spectra $I(Q, \Delta E = 0)$. The analysis of this "diffuse scattering" gives important, statistically averaged, structural information. The advantage of performing elastic diffuse scattering studies of structural disorder by neutrons rather than by X-rays, are: (i) the independence of the scattering length b with $\sin \theta / \lambda$; (ii) the possibility of experimentally separating the thermal diffuse scattering due to atomic motion by energy analysis of the scattered beam; (iii) the possibility of separating different pair correlation functions in polyatomic materials, by studying samples with different isotopic concentrations, displaying different nuclear scattering contrasts; and (iv) the greater ease to perform high-temperature measurements in thermodynamic equilibrium conditions.

Such studies with neutrons include:

- The determination of pair correlation functions and therefore of the local atomic arrangements in liquids, glasses and metallic amorphous systems; a good example is the study of the amorphous alloy $Ni_{63.7}Zr_{36.3}$, where the use of isotopic substitution allowed determination of the three partial structure factors S_{Ni-Ni}, S_{Ni-Zr} and S_{Zr-Zr}, and showed that a hard sphere model was inadequate [17].

- The local environments (chemical disorder) in disordered solid solutions and non-stoichiometric compounds, for which one determines effective pair interaction potentials by inverse Monte-Carlo simulation [18]. These pair potentials can be used to calculate theoretically "coherent" equilibrium phase diagrams and various metallurgical properties, such as antiphase boundary energies.

- The strain field of impurities in metals (down to atomic concentrations of 10^{-2}–10^{-3} in the most favorable cases, for example, oxygen in niobium [19]). This information is necessary to calculate impurity–impurity and impurity–dislocation interactions.

- The analysis of local atomic arrangements in complex systems presenting altogether strong chemical and topological disorders, such as Ca-stabilized cubic zirconia, interesting for its ceramic and ion conductive properties; this requires a combined experimental and numerical simulation approach (see, e.g. Ref. [20]).

Neutron diffraction allows also quantitative determination of the total hydrogen content in bulk samples, thanks to the high value of the incoherent scattering cross-section of the proton: the hydrogen content is proportional to the measured background of the spectrum (see Figure 1.9).

1.6.4 Small-angle scattering

If a parallel monochromatic incident thermal neutron beam is directed onto a solid sample containing nanometer-sized particles, the transmitted beam shows a broadening which is roughly inversely proportional to the average size of the particles (see Figure 1.7).

Small-angle scattering of thermal neutrons (SANS) by solid materials has proven to be a very powerful technique in the detection and study of fine inhomogeneities or particles in the nanometer range [21]. For metallic materials, the SANS technique is complementary to transmission electron microscopy (TEM), in the sense that it gives quantitative statistical information (gyration radius, size distribution, volume fraction, etc.) on particles of nanometer to micrometer size in a large sample (millimeter size), and can in some cases provide information at a somewhat finer scale (1 nm) than TEM, in particular in ferromagnetic materials. Types of SANS studies in the field of materials science are precipitation, cavities induced by fatigue or irradiation, nanometer size particles in composite materials, fine powders and sintering, etc. [22].

For example, SANS has given important information on the microstructural evolution induced by neutron irradiation and subsequent annealing of ferritic steels constituting the vessel of pressurized water-cooled nuclear reactors (PWR). These materials, which receive low-level radiation flux over a period of many years, at a temperature of around 300°C, harden and embrittle, but nothing can be seen using TEM. On the pressure vessel steel of the

Franco-Belgian reactor CHOOZ-A, SANS showed clearly the appearance of small clusters of radius $\sim$1.5 nm. With increasing flux, their size remains approximately constant, but their number increases nearly linearly with flux [23]. These clusters were subsequently shown to redissolve by annealing around 450°C (see Figure 1.12): this result gives the hope that in the future suitable thermal treatments will enable reactors to be used safely for longer times.

Other important applications of SANS are in the field of soft matter and nanocomposites, such as reinforced elastomers, in relation with their modes of deformation (see, e.g. Rharbi *et al.* [24]).

1.6.5   Reflectometry

Most materials have a refraction index for thermal neutrons slightly smaller than 1. There is total reflection for an incident angle $\theta < \theta_c$, where the critical angle θ_c has a very small value (0.4° for Ni at $\lambda = 0.4$ nm). For $\theta > \theta_c$, the reflectivity decreases rapidly and may show oscillations due to reflection at internal interfaces in the case of thin films or multilayers (see Figure 1.7).

Neutron reflectivity is a rapidly growing field, mainly applied to liquid or soft matter surfaces (e.g. grafted polymers, adhesion between polymers), or magnetic materials. In the latter case, the use of spin-polarized neutrons and polarization analysis of the scattered beam has created new investigative possibilities: it allows measurement of vectorial magnetic profiles in magnetic thin films and multilayers [25].

Structural changes in thin films or multilayers can be monitored by real-time neutron reflectometry measurements of their reflectivity profile. The self-diffusion coefficient of nickel in special isotope-enriched $^{nat}Ni/^{62}Ni$ multilayers of pure nickel or amorphous $Ni_{55}Zr_{45}$ has been measured, which is two orders of magnitude smaller and at temperatures 100°C lower than the minimum values obtained by competing techniques [26].

1.6.6   Inelastic and quasi-elastic scattering

The numerical relationship between the energy E of the neutron and its associated wavelength λ and their order of magnitude makes it an ideal probe to determine completely the dispersion curves of atomic vibrations (phonons) in crystalline materials. This is usually made on triple-axis type spectrometers.

For example, low-frequency and strongly damped phonon bands ("soft" modes) found in the bcc high temperature phases of Ti, Zr and Hf, have been interpreted as dynamical precursors of displacive (martensitic) transformations, and are at the origin of the anomalous atomic self-diffusion observed in these metals [27].

If diffusional motions occur in a sample, this leads to a broadening in energy of an incident monochromatic neutron beam. This slightly inelastic scattering, centered at $\hbar\omega = 0$, is called quasi-elastic neutron scattering (QENS) (for a review of principles and applications of QENS, see Bée [28]). The analysis of this QENS (e.g. with a backscattering spectrometer which attains energy resolutions of $\sim$1 μeV) gives access to diffusion coefficients down to 10^{-8} cm^2/s, and (on single crystals) to microscopic information on the atomic jumps, such as their frequency, their direction and their length. Important information has been obtained in this way on hydrogen diffusion in metals and intermetallic compounds, and on the role of trapping by impurities [29].

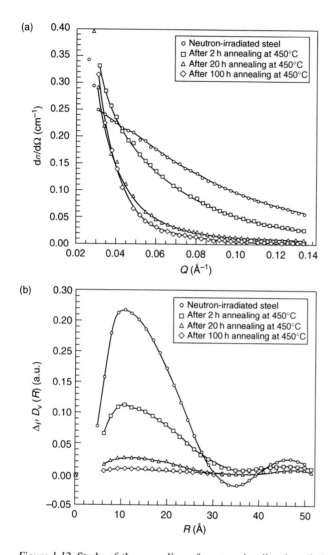

Figure 1.12 Study of the annealing of neutron-irradiated steels by SANS (performed on D11, ILL). (a) Difference intensities of irradiated steels before and after annealing for different times as a function of momentum transfer. The curves show the SANS intensities measured in the direction perpendicular to a magnetic field (1.4 T) after subtraction of that from a non-irradiated reference sample. The scattering contrast is mainly between the magnetic matrix and less magnetic clusters. (b) Contrast-weighted volume distribution functions (Fourier transforms of the curves in (a)) of irradiated steels before and after annealing for different times. The oscillations at larger dimensions are artifacts caused by the mathematical procedure (S. Miloudi and R. May, unpublished).

1.6.7 Neutron topography

This technique allows imaging of magnetic and crystalline inhomogeneities (defects, domains, phases) in single crystals: the defect contrast (due to structure factor, misorientation or lattice distortion) results in an image on the Bragg diffracted beam visualized on

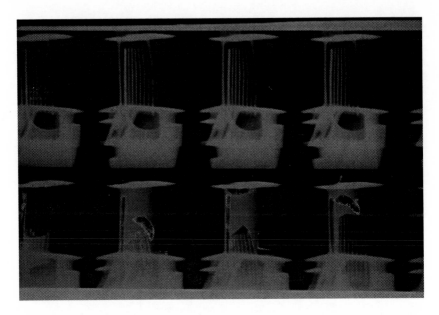

Figure 1.13 Neutron radiography check of a series of eight paddle turbines. The ribs of the four lower elements exhibit manufacturing defects (CEA and European Gas Turbine Ltd, unpublished).

a 2D position sensitive detector. Neutron topography is a unique technique to image directly antiferromagnetic or helimagnetic (chiral) domains [30].

1.6.8   Neutron radiography

Neutron radiography is a transmission imaging technique for heterogeneous materials, taking advantage of the scattering and/or absorption contrast between different elements. Because of the deep penetration of thermal neutrons in metallic structures, and of their specific sensitivities (strong scattering cross-section of hydrogen, strong absorption cross-section of some elements such as Ag or B), this technique is preferable to X-ray radiography, or even unique, for some applications such as non-destructive detection and visualisation of hydrogenated elements embedded in metallic structures (glue, coatings, gaskets, rubber seals, filling with resin, etc.) or various metallurgical controls (corrosion of Al, silver braze, etc.). High-quality images (with a resolution attaining $\pm 30\,\mu m$) can be obtained with a few minutes exposure time in a polychromatic cold neutron beam [31].

A typical image of industrial interest, presented in Figure 1.13, shows manufacturing defects in several paddle turbine ribs (G. Bayon, unpublished).

3D tomography with thermal neutrons is presently under development at FRM/ Garching [32].

1.7   Conclusions and future prospects

The detailed knowledge of atomic positions and motions in materials is necessary to understand their properties, and even more to "tailor" them in view of specific applications. This requires the use of complementary techniques, for example, high-resolution observation in direct space, diffraction techniques and various spectroscopies, if possible in operating conditions.

Neutron scattering has been shown to be very powerful for this, because of the unique properties of the neutron–matter interaction which have been described above, of its easy interpretation ("kinematic" theory), and of the reliability of the quantitative informations it provides. A look at the possible future of neutron scattering is found in the Proceedings of an ESF Exploratory Workshop held in Autrans, 11–13 January 1996, edited by Lander [33].

Indeed, the use of neutron scattering techniques by the scientific community, coming from many fields of fundamental and applied research, has considerably developed in the last few decades. The number of neutron beam users worldwide is estimated to about 6000. This number is still expected to increase, in particular in North America and in the Pacific Region [7]. It is important to emphasize that for most scientific domains, nearly half of these users are occasional.

Presently, about 25% of the total number of neutron scattering users come from the fields of materials science (i.e. understanding of applied materials) and engineering sciences. Among these, the number of scientists in the field of residual stress measurements regularly or occasionally using neutrons is probably of the order of 150: permanent staff of the neutron centers, scientists and students coming from Universities or technical centers, engineers coming from industry (presently mainly from large companies: space and aeronautics, transport, energy, etc.).

Compared to X-rays, the flux of neutron beams is low (this disadvantage is partially compensated by the weak absorption and the better signal-to-noise ratio); therefore, and this is particularly the case of strain scanning experiments, the time of measurement is long (up to several days), especially if one aims to measure the complete strain tensor. This is not attractive for industrialists, who will use neutrons only when they are by far the "best" technique or even the unique one. In residual stress determinations, neutron diffraction is often utilized as a tool to test the validity of (and to refine) finite element models, which cannot take into account all the metallurgical complexity (i.e. welds, cracks, coatings, etc.) and the thermomechanical history of the studied device.

This is why any improvement of the efficiency of existing instruments (neutron optics, detectors, etc.) is very important. Strain scanning is one of the techniques which should largely benefit of future high flux pulsed spallation sources. For example, with the European Spallation Source (ESS), it is planned to have a diffractometer which will be able to perform simultaneously, 3D mapping of residual strains/stresses, of phase contents (down to 0.1%) and of textures in materials or engineering components with a gauge volume of $0.1 \times 0.1 \times 0.1$ mm^3 [34]; this improvement by a factor 10 of the present linear resolution for residual stress measurements will have an important technological impact.

Also, increased flux will allow higher angular resolution in neutron diffraction spectra, and therefore to enter strongly in the field of the study of lineshape and line broadening due to strains and structural fluctuations at the microscopic scale, Huang scattering by defects, subjects which are presently in their infancy, compared to X-rays; this is of considerable potential scientific and technological interest. *In situ* and real-time studies (e.g. stress relief mechanisms and procedures) on model samples as well as on large industrial components will be an important future development.

Acknowledgments

The author wishes to thank Dr Monica Ceretti for useful comments and discussions.

References

[1] Bacon G. E., *Neutron Diffraction*, 1975, 3rd edition, Oxford University Press, Oxford.

[2] Baruchel J., Hodeau J. L., Lehmann M. S., Regnard J. R. and Schlenker C. (eds), *Neutron and Synchrotron Radiation for Condensed Matter Studies*, 1993, Springer, Berlin.

[3] Rauch H., in *Neutron Interferometry*, 1979, U. Bonse and H. Rauch (eds), p. 162, Oxford Science Publishers, Oxford.

[4] Lovesey S. W., *Theory of Neutron Scattering from Condensed Matter*, 1984, Clarendon Press, Oxford.

[5] Sears V. F., *Neutron Optics*, 1989, Oxford University Press, New York.

[6] Sears V. F., Thermal-neutron scattering lengths and cross-sections for condensed matter research, 1984, Report AECL-8490, Chalk River Nuclear Laboratories, Chalk River, Ontario, Canada.

[7] Richter D. and Springer T., A twenty years forward look at neutron scattering facilities in the OECD countries and Russia, 1998, Technical Report, European Science Foundation, Strasbourg (ISBN 2-912049-03-2).

[8] Jorgensen J. D., Shapiro S. M. and Majkrzak C. F. (eds), Proceedings of the International Conference on Neutron Scattering (ICNS '97), *Physica B*, **241–243**, (1998)

[9] Cser L. and Rosta L. (eds) Proceedings of the Second European Conference on Neutron Scattering (ECNS '99), *Physica B*, **276–278**, (2000).

[10] Fontana M., Rustichelli F. and Coppola R. (eds), *Industrial and Technological Applications of Neutrons*, 1992, Enrico Fermi Course CXIV, North-Holland, Amsterdam.

[11] Cava R. J., Hewat A. W., Hewat E. A., Batlogg B., Marezio M., Rabe K. M., Krajewski J. J., Peck W. F. and Rupp L. W., Structural anomalies, oxygen ordering and superconductivity in oxygen deficient $Ba_2YCu_3O_x$, *Physica C* **165**, 419, (1990).

[12] David W. I. F., Ibberson R. M., Dennis T. J. S., Hare J. P. and Prassides K., Structural phase transitions in the fullerene C_{60}, *Europhys. Lett.* **18**, 219, (1992).

[13] Kilcoyne S. H., Manuel P. and Ritter C., Synthesis and characterization of a novel Y–Fe phase via kinetic neutron diffraction, *J. Phys. Condensed Matter* **13**, 5241, (2001).

[14] Hutchings M. T. and Krawitz A. (eds), *Measurement of Residual and Applied Stress using Neutron Diffraction*. NATO ASI Series E, **216**, (1992).

[15] Ceretti M., Magli R. and Vangi D., Neutron diffraction study of stress field distribution in automotive gears. *Mater. Sci. Forum* **321–324**, 847, (1999).

[16] Hirschi, K. (1999) *Analyse des contraintes résiduelles et des paramètres microstructuraux par diffraction de neutrons dans un acier inoxydable austénitique* (in French), PhD Thesis, University of Reims Champagne-Ardenne, 11 October 1999.

[17] Lefebvre S., Quivy A., Bigot J., Calvayrac Y. and Bellissent R., A neutron diffraction determination of short-range order in a $Ni_{63.7}Zr_{36.3}$ glass. *J. Phys. F: Metal Phys.* **15**, L99, (1985).

[18] Le Bolloch D., Caudron R. and Finel A., Experimental and theoretical study of the temperature and concentration dependence of the short-range order in Pt–V alloys. *Phys. Rev. B* **57**, 2801, (1998).

[19] Barbéris P., Beuneu B. and de Novion C. H., Local lattice distortions around interstitial oxygen in niobium, *J. Phys. I (France)* **2**, 1051, (1992).

[20] Proffen Th., Analysis of the diffuse neutron and X-ray scattering of stabilized zirconia using the reverse-Monte-Carlo method, *Physica B*, **241–243**, 281 (2000).

[21] Kostorz G., X-ray and neutron scattering, in *Physical Metallurgy*, 1996, R. W. Cahn and P. Haasen (eds), 4th edition, p. 1115, Elsevier, Amsterdam.

[22] Cotton J. P. and Nallet F. (eds), *Diffusion de Neutrons aux Petits Angles*, 1999, EDP Sciences, Paris.

[23] Van Duysen J. C., Bourgoin J., Moser P. and Janot C., ASTM STP 1170, 1993, L. E. Steele (ed.), p. 132. ASTM, Philadelphia.

[24] Rharbi Y., Cabane B., Vacher A., Joanicot M. and Boué F., Modes of deformation in a soft/hard nanocomposite: A SANS study, *Europhys. Lett.* **40**, 472, (1999).

[25] Fermon C., Miramond C., Ott F. and Saux C., Polarisation analysis of neutron reflectivity from magnetic thin films, *J. Neutron Res.* **4**, 251, (1996).

[26] Speakman J., Rose P., Hunt J. A., Cowlam N., Somekh R. E. and Greer A. L., The study of self-diffusion in crystalline and amorphous multilayer samples by neutron reflectometry. *J. Magn. Magn. Mater.* **156**, 411, (1996).

[27] Petry W., Heiming A., Trampenau J., Alba M., Herzig C., Schober H. R. and Vogl G., Phonon dispersion of the bcc phase of group-IV metals. I. bcc titanium. *Phys. Rev. B* **43**, 10933, (1991).

[28] Bée M., *Quasielastic Neutron Scattering*, 1988, Adam Hilger, Bristol and Philadelphia.

[29] Hempelmann R., Jump diffusion of H in metals: quasi-elastic neutron scattering, in *Neutron Scattering from Hydrogen in Metals*, 1988, A. Furrer (ed.), p. 201, World Scientific, Singapore.

[30] Baruchel J., Neutron and synchrotron radiation topography, in *Neutron and Synchrotron Radiation for Condensed Matter Studies*, 1993, Vol. 1, p. 399, Les Editions de Physique/Springer-Verlag, Paris/Berlin.

[31] Barton J. P. (ed.), *Proceedings of the 4th World Conference on Neutron Radiography* (4th WCNR), 1992, Gordon and Breach Science Publishers, Yverdon.

[32] Schillinger B., 3D computer tomography with thermal neutrons at FRM Garching. *J. Neutron Res.* **4**, 57, (1996).

[33] Lander G. H. (ed.), Scientific prospects for neutron scattering with present and future sources, 1996, European Science Foundation, Strasbourg (ISBN 2-903148-90-2).

[34] Finney J. L., Steigenberger U., Taylor A. D., Carlile C. J. and Kjems J. (eds), *ESS, A Next Generation Neutron Source for Europe*, 1997, Vol. II 'The Scientific Case' (ISBN 0902376608/500).

2 The use of synchrotron radiation for materials research

C. Riekel

2.1 Introduction

X-ray photons interact with matter in various ways (Figure 2.1). An analysis of these inter-actions allows the development of microscopic models, which serve to provide a better understanding of the physical properties of materials. In the following text, emphasis will be put on synchrotron radiation (SR) experiments with "hard" X-rays. "Hard" X-rays have, for the purpose of this text, wavelength of $\lambda \leq 0.25$ nm, which corresponds roughly to the spectral cut-off of a beryllium vacuum-window, which separates the SR source (see below) from the beamline optics. These wavelengths are, furthermore, on the same length scale as interatomic distances which allows high resolution structural studies.

The specific properties of SR can be summarized as:

- high brilliance;[1]
- pulsed radiation with pulse length down to about 0.1 ns range;
- extended spectral range;
- polarization.

These properties allow experiments which are complementary to experiments at laboratory X-ray sources.

2.2 Production of synchrotron radiation [1–3]

Synchrotron radiation is generated by bunches of electrons (or positrons) which circulate at relativistic energies close to the speed of light. In the simplest case, the electrons follow a circular orbit, which is defined by dipole magnets. As shown in Figure 2.2, the emitted radiation covers a range from the infrared to γ-rays. It is concentrated in a cone with opening angle Ψ, which is centered on the tangent to the orbit (Figure 2.3). Ψ can be expressed at the critical wavelength – λ_c (equation 2) – approximately by the so-called γ-ratio:

$$\Psi \approx \gamma^{-1} = (1957E)^{-1} \tag{1}$$

where E is the energy of the electrons (GeV). The critical wavelength, λ_c (nm), is defined at the wavelength where half of the total power is emitted according to:

$$\lambda_c = 0.559(R/E^3) \tag{2}$$

---

[1] Defined as: photons/s/mrad^2/mm^2/0.1% bandwidth.

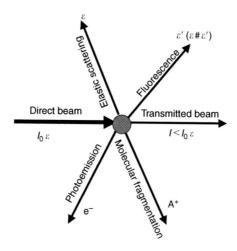

Figure 2.1 Schematic description of interaction modes of an X-ray photon beam with matter (I_0, incident beam intensity; I, transmitted beam intensity; ε, energy; A^+, ionized molecular fragment; e^- electron).

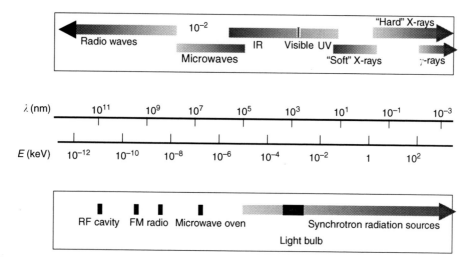

Figure 2.2 Spectral range of SR as compared to other radiation sources. The spectral range of soft X-rays is considered to start at wavelengths larger than about 0.25 nm, which corresponds roughly to the spectral cutoff of a Be-window separating the machine vacuum from the beamline. Conversion: $\lambda(\text{nm}) \approx 1.24/E(\text{keV})$.

where R (m) is the radius of curvature of the electrons in the bending magnet. The variation of λ_c with energy implies that several GeV of electron energy are required in order to obtain a sufficiently powerful beam in the hard X-ray regime.

Synchrotron radiation is usually generated in an electron storage ring, which is shown schematically in Figure 2.4. The polygon shape is characteristic for third generation SR sources (see Figure 2.8). The electron beam is guided from one straight section into the next by dipole magnets. Further, correcting magnets (quadrupole/hexapole magnets), are

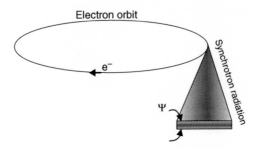

Figure 2.3 Schematic design of SR emission by a beam of electrons circulating at relativistic energies on a circular orbit. Ψ is the natural opening angle.

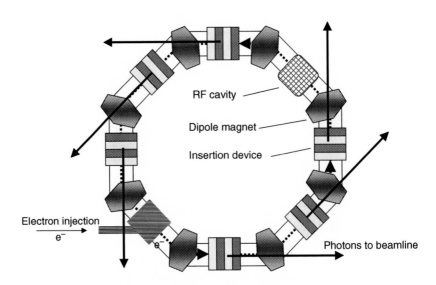

Figure 2.4 Schematic design of a SR storage ring. The main radiation sources are insertion devices, which are installed in the straight sections. Dipole magnets guide the electron beam from one straight section into the next.

not shown. The straight sections contain so-called *insertion devices* as principal radiation sources. Energy lost by the electron beam through SR is compensated by radio frequency (RF) cavities.

Insertion devices consist of a periodic array of magnets, which force the electrons to follow an (approximately) sinusoidal path in the orbital plane of the storage ring (Figure 2.5). They can be characterized by the so-called deflection (or magnetic field) parameter) K:

$$K = \delta\gamma = 93.4 B_0 \lambda_0 \tag{3}$$

where B_0 is the peak magnetic field (Tesla), λ_0 the spatial period of the magnet (m) and δ the maximum deflection angle of the electron trajectory (Figure 2.5). Depending on the strength

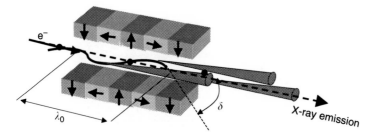

Figure 2.5 Periodic magnetic structure (period: λ_0) of an insertion device. Each arrow depicts the orientation of the magnetic field. The electron bunches follow a sinusoidal trace with a maximum deflection angle δ. Photon emission parallel to the electron path is indicated schematically at the inflection points. The opening angle of the emission corresponds to γ^{-1}.

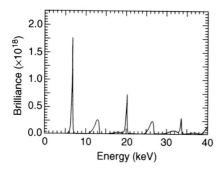

Figure 2.6 An undulator spectrum corresponds to a series of harmonics (example: undulator A at *Advanced Photon Source*). Brilliance: photons/s/mm^2/mrad^2/0.1% bandwidth.

of the magnetic field, and therefore, on the curvature of the sinusoidal track one can distinguish two operation regimes of insertion devices which are called *wigglers* or *undulators*:

Wiggler: High magnetic field with $\delta > \gamma^{-1}$. Only the points at the peak of the curve contribute to an emission of radiation into the observation direction. The distant observer will therefore "see" a multiplication of dipole sources.

Undulator: Weaker magnetic field with $\delta < \gamma^{-1}$. For this case, all points on the curve are emission points. This results in an interference phenomenon, which implies an emitted spectrum composed out of a series of harmonics (Figure 2.6).

Synchrotron radiation will also be observed from dipole magnets (not shown). For a dipole magnet or wiggler, the half-opening angle of the radiation in the vertical direction is γ^{-1} (equation 1) which corresponds to 85 μrad for a 6 GeV source such as the ESRF. In the horizontal plane, the photon emission has a spread of several milliradians. In contrast, the emission angle of an undulator is several tenth of microradians in both planes for the ESRF. Figure 2.7 shows the image of an undulator beam on a fluorescent screen at a distance of 31 m from the source [4]. The small emission angle results in an about 1 mm^2 spot size.

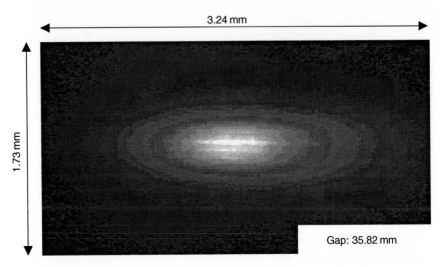

Figure 2.7 Image of an undulator photon beam on a fluorescence screen at 30.5 m from the undulator source point [4] (fifth harmonics; $\lambda_0 = 46$ mm; 35.82 mm gap between magnetic poles). (See Colour Plate III.)

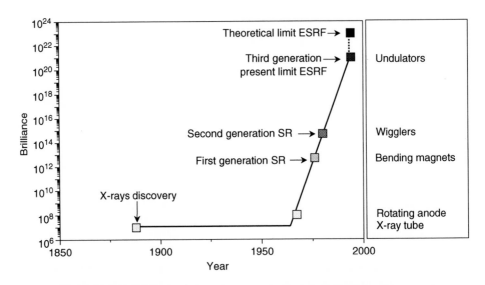

Figure 2.8 Increase of brilliance of X-ray sources since the discovery of X-rays. The principal radiation sources, which define the brilliance of the source, are indicated.

Figure 2.8 shows the nearly linear increase of brilliance of SR sources with time. Thus, the radiation from a dipole magnet at LURE is already a factor 10^4 times more brilliant than a rotating anode generator. Wigglers and, in particular, undulators have allowed further increases in the brilliance by a factor of 10^6, and the development is not finished as free electron laser sources are under development which are predicted to be up to a factor 10^6 times more brilliant and completely coherent [5].

2.3 Optics and instrumentation [6, 7]

Most SR experiments are performed with monochromatic, narrow bandwidth radiation. Wavelength (λ) and bandwidth ($\Delta\lambda = \lambda_{max} - \lambda_{min}$) selection relies on a monochromator, which is usually made out of one or more perfect crystals of silicon or germanium [8]. Bragg's law determines the wavelength at the exit of the monochromator:

$$n\lambda = 2d \sin\theta \qquad (4)$$

where n is the order of a reflection, d the distance between the lattice planes and θ is the Bragg angle. Thus, the reflection condition is fulfilled at $\lambda = 0.15$ nm for the (111) lattice plane of a Si-monochromator with $d = 0.313$ nm at $\theta = 13.9°$. The bandwidth is $\Delta\lambda \approx 10^{-4}$ at this wavelength. One should note, however, that some experiments require polychromatic radiation with a large bandwidth. Both monochromatic and polychromatic radiations are for example used in diffraction experiments (Figure 2.9). For experiments with monochromatic radiation, the detector has to cover a large 2θ-range. This implies either a zero-dimensional (0D) detector, which is rotated through the 2θ-range or a 1D, or 2D detector, which covers the 2θ range for a single setting. For polychromatic radiation, all harmonics allowed by Bragg's law will be superimposed. A separation is, however, possible by a 0D detector with energy resolution. Polychromatic radiation is for example of interest for experiments with restricted access to the sample, such as high-pressure or chemical reaction cells. This compensates the generally lower order resolution and the limited 0D detector counting rate.

In order to increase the flux at the sample position one can focus the beam by a curved monochromator [8]. Another possibility is to use a mirror, which is curved in 1D (e.g. cylindrical shape) or 2D (e.g. toroid).

Mirrors reflect X-rays by total reflection below a critical angle Θ_c:

$$\Theta_c = (20/E)\sqrt{\rho}\,(\text{mrad}) \qquad (5)$$

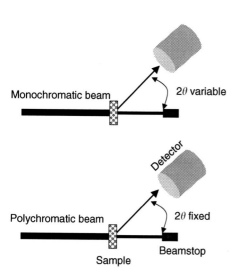

Figure 2.9 Schematic display of diffraction setup for monochromatic and polychromatic radiation. The Bragg angle θ is fixed for the latter case.

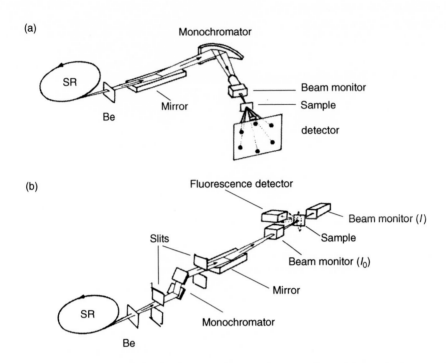

Figure 2.10 (a) SR beamline optimized for flux density. Be corresponds to the beryllium window separating the machine vacuum from the beamline. (b) SR beamline optimized for beam stability over an extended spectral range as required for the recording of EXAFS spectra with monochromatic radiation. (Adapted from [7].)

where E is the radiation energy (keV) and ρ the surface density (g/cm^3). For a glass surface, Θ_c corresponds to about 2.5 mrad at 0.1 nm wavelength [9]. An advantage of a mirror as compared to a focusing monochromator is the greater flexibility with respect to different wavelength and bandwidth.

A large number of combinations of mirrors and monochromators exist for specific applications. Only two typical examples will be mentioned. Figure 2.10(a) shows a double-focusing beamline, which has been optimized for flux density at the sample position [7, 10]. It uses a cylindrical mirror for focusing in the vertical plane and a cylindrically curved monochromator for focusing in the horizontal plane. Such optics are typically installed at a dipole magnet in order to capture a large angular range of the photon beam in the horizontal plane. The scattering pattern is recorded by a 2D detector. Depending on the application this could be a gas-filled detector, an image plate, or a charge-coupled device (CCD). An ionization chamber is used to monitor the flux on the sample continuously. Such a beamline can be used, for example, for protein crystallography [11] or real-time diffraction experiments during polymer crystallization [12].

Figure 2.10(b) shows a beamline optimized for absorption spectroscopy (EXAFS: extended X-ray fine structure analysis) [7]. The aim is to determine X-ray absorption for a more or less extended spectral range. In order to change the wavelength, while keeping the beam position at the sample constant, a double monochromator is used. A toroidal mirror is used for focusing in the horizontal and vertical planes. The sample absorption is determined by two ionization chambers, one upstream and another downstream from the sample. A third ionization chamber allows measurement of the fluorescence signal.

2.4   Applications of synchrotron radiation in materials research

A number of applications making use of the brilliance and extended spectral range of SR will be mentioned below. Special emphasis has been put on imaging and real-time experiments using the hard X-ray spectrum, where the high brilliance of an SR source is of particular importance. A more complete list of SR applications in materials research can be found in the Appendix.

2.4.1   Computed microtomography (μ-CT) [13, 14]

Computed tomography (CT) with monochromatic radiation permits distinguishing of the composition and density of an object by the attenuation of an X-ray beam. The "image" of the object can be spatially reconstructed from measurements performed at different rotation angles. This method is used in medical applications but also to verify the integrity of structural materials.

Microtomography (μ-CT) extends tomography to high spatial resolution. This becomes possible by a high brilliance source, which allows use of an intense microbeam. Spatial resolution at SR sources using a parallel, monochromatic beam is limited by the detector resolution to about 1 μm. One can, in addition visualize selectively certain elements by making use of their absorption edges. Figure 2.11 shows schematically a μ-CT setup, based on a parallel beam and a 2D detector [13].

An interesting use of μ-CT for the petrol industry is visualization of the pore structure of materials. This gives important information for the modelling of macroscopic properties, such as the transport of liquids. Figure 2.12 shows the reconstructed image of a Fontainebleau stone [15]. A 5×5 μm^2 beam was used. The resolution is about 10 μm. The fluid permeability calculated from these results agrees quite well with experimental data.

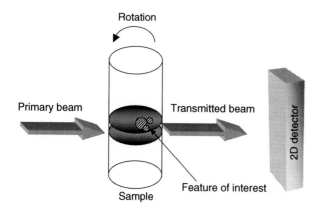

Figure 2.11   Schematic design of a microtomography setup. The beam transmission is measured for different rotation angles and at different heights. Image reconstruction allows localizing the feature of interest. (Adapted from [13].)

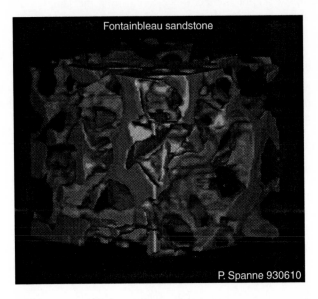

Figure 2.12 Example of image reconstruction based on a microtomography experiment: pore structure of a Fontainebleau stone [15]. (See Colour Plate VIII.)

2.4.2   *Time-resolved diffraction [16]*

Time-resolved (or real-time) diffraction provides a method of visualizing a phase transformation or chemical reaction and, thus, allows development of a microscopic model for the structural evolution being studied. The temporal length of the photon pulses, which are down to 60 ps for the ESRF storage ring, determines the ultimate temporal resolution. This resolution is, for example, required to study fast structural processes in proteins, which are initiated by a laser flash [17].

The formation of Portland cement, a technologically important material, takes place on a much slower time-scale. The overall reaction continues for several hours, but a fast onset reaction takes place within several seconds. This fast reaction could be followed in real time with a setup shown schematically in Figure 2.13 [18]. The optical system provides a polychromatic radiation band. Slits in front of the sample and between sample and detector allow defining the irradiated volume.

The hydration of cement is influenced in particular by the most reactive component $Ca_3Al_2O_3$. Chemical techniques allow isolating a number of hydrated phases of cement. Time-resolved diffraction allows in addition determining the succession of crystalline phases. Figure 2.14 shows a sequence of energy-dependent diffraction patterns recorded during the rapid hydration phase. Setting aside the phases $Ca_3Al_2O_3$ and $Ca_3Al_2O_3(H_2O)_6$, one observed only one intermediate phase which can be identified as $Ca_2Al_2O_3(H_2O)_8$. This phase acts obviously as nucleation phase for the end product $Ca_3Al_2O_3(H_2O)_6$. The reaction involving crystallographic phases can thus be formulated in the following way:

$$2Ca_3Al_2O_3 + 27H_2O \rightarrow Ca_4Al_2O_3(H_2O)_{19} + Ca_2Al_2O_3(H_2O)_8 \tag{6}$$

$$2Ca_3Al_2O_3 + 21H_2O \rightarrow Ca_4Al_2O_3(H_2O)_{13} + Ca_2Al_2O_3(H_2O)_8 \tag{7}$$

$$Ca_4Al_2O_3(H_2O)_{19} + Ca_2Al_2O_3(H_2O)_8 \rightarrow 2Ca_3Al_2O_3(H_2O)_6 + 15H_2O \tag{8}$$

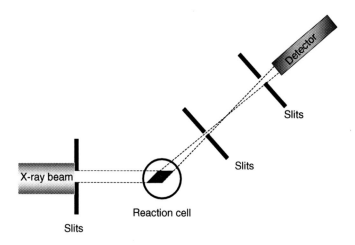

Figure 2.13 Schematic design of an experimental setup to follow the hydration of cement in real-time. The lozenge type "observation" volume in the cell is defined by the system of slits. (Adapted from [18].)

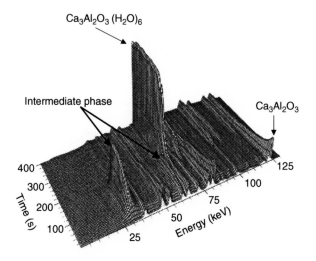

Figure 2.14 Temporal evolution of crystalline structures during the hydration of $Ca_3Al_2O_3$ as recorded by energy dispersive diffraction. (Adapted from [18].)

2.4.3   X-ray microdiffraction (μ-XRD)

X-ray microdiffraction allows us to study the local structure and orientation of materials or to investigate microcrystals [19]. Scanning μ-XRD is used to map an extended sample and to record at every position a diffraction pattern with a fast 2D-detector. This allows us to "image" the variation of the local structure and texture across the sample. Current technology employs beam sizes down to about 1 μm and submicron beam optics is under development [19–21]. A typical scanning setup is shown in Figure 2.15. In contrast to electron scattering techniques, the sample is scanned through a stationary beam. In this case a monochromatic beam, which

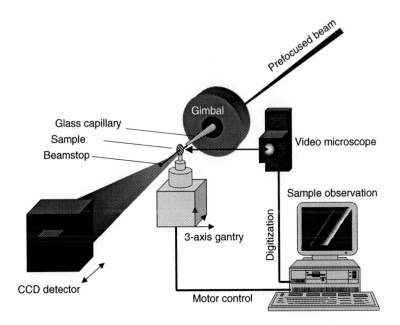

Figure 2.15 Schematic design of a scanning μ-XRD setup. The beam size can be defined by a glass capillary to about 1 μm at the sample position. The distance between the video microscope and beam position is calibrated which allows placing of the sample with micrometer accuracy in the beam. (Adapted from [19].)

has been focused to about 30 μm by a mirror, is further reduced to about 2 μm by a tapered glass capillary [19].

The scanning of a single fiber of the synthetic polymer ultrahigh molecular weight polyethylene (UHMW-PE) provides an example. A local area diffraction pattern of a UHMW-PE fiber is shown in Figure 2.16(a) [19]. The reflections correspond to the two crystalline phases of polyethylene: the orthorhombic phase and a metastable monoclinic phase. The monoclinic phase is formed by plastic deformation of the orthorhombic phase [22, 23]. The intensity ratio of the monoclinic phase (001) reflection and the orthorhombic phase (110) reflection – I_{001}^{m}/I_{110}^{o} – is shown in Figure 2.16(b). Assuming cylindrical symmetry, the monoclinic phase appears to form a band extending along the fiber axis. This information could be obtained without sectioning the sample as required for electron diffraction techniques.

Scanning μ-XRD at high energies ($E > 50$ keV) with depth selection by conical slits has been developed for texture and strain mapping of bulk materials in volume with a spatial resolution of $5 \times 5 \times 50\,\mu$m^3 [24]. Spatial resolution of about 0.1 μm and smaller can be obtained in one direction by X-ray waveguide structures. This has been used to study buried strain fields in semiconductor devices [25].

2.4.4   X-ray microfluorescence (μ-XRF)

X-ray fluorescence (XRF) is based on X-ray photoionization of elements and the detection of their characteristic fluorescent radiation. Excitation can be done both by white or monochromatic radiation. In contrast to electron probe microanalysis, XRF allows

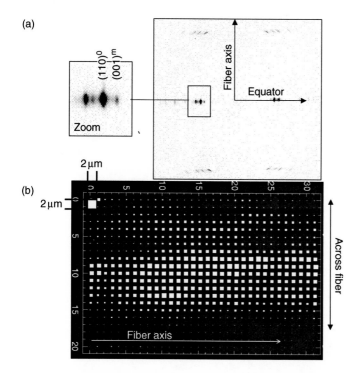

Figure 2.16 (a) Fiber diffraction pattern of a UHMW-PE fiber. The pattern shows reflections of the monoclinic and orthorhombic phases. (b) Distribution of the ratio monoclinic/orthorhombic phase calculated from the intensity ratio of a Bragg reflection of each phase across a single fiber on a mesh of $2 \times 2 \, \mu m^2$.

quantitative elemental bulk analysis down to the sub-ppm level [26, 27]. In exceptional cases, absolute detection limits in the range of $\leq 100 \, ag(10^{-18} \, g)$ can be reached [28]. The same type of scanning system as used in scanning μ-XRD (Figure 2.15) in combination with an energy dispersive X-ray detector can be used for scanning μ-XRF [26, 29]. One obtains thus an "image" of the elemental distribution across the sample. Three-dimensional image reconstruction based on μ-XRF data obtained at different sample rotation angles has been demonstrated [26].

Figure 2.17 shows the result of a 2D scan of spruce wood cells through a $2 \times 2 \, \mu m^2$ monochromatic beam. X-ray microfluorescence images were obtained with traces of Mn and Fe in the cell walls at the 100 ppm level [30]. This information is of great interest for products derived from wood such as the production of pulp where the bleaching process by H_2O_2 can be influenced by the presence of transition metals. In a related experiment, the orientation of cellulose fibrils in wood cell walls has been determined by scanning μ-XRD [31]. Combined μ-XRF and μ-XRD experiments are, in principle, possible [32].

2.4.5 Extended X-ray absorption fine structure (EXAFS) [7]

X-ray absorption is dominated by the photoelectric effect up to about 0.03 nm. The absorption of a photon with a specific energy by an atom (Figure 2.18(a)) can liberate a

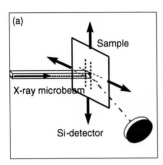
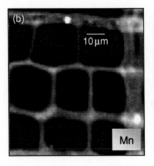
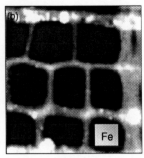

Figure 2.17 Imaging of wood cell walls for the Mn- and Fe-fluorescence on a mesh of $2 \times 2\,\mu m^2$ at $\lambda = 0.0984\,nm$. The cell walls of wood can be clearly distinguished. Bright spots correspond to larger particles containing Fe or Mn [30].

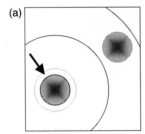
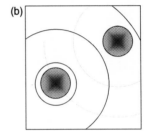

Figure 2.18 Principle of X-ray absorption spectroscopy: (a) after the absorption of a photon by the atom at the center, a photoelectric wave is created; (b) the reflection of this wave by the environment of the central atom is at the origin of the observed interference phenomenon.

photoelectron with a characteristic energy. The neighbouring atoms reflect the photoelectron wave (Figure 2.18(b)). The interference of the two waves creates an absorption spectrum, which allows extraction of the EXAFS function (Figure 2.19). The oscillations of this function extend up to $1000\,eV$ beyond the absorption edge (K, L, M, ...). In practice, one can transform the EXAFS function into real space by a Fourier transformation (Figure 2.19) [33]. This function is close to the pair correlation function of the central atom with its environment. A structural analysis allows us in principle to obtain information on the first coordination sphere of an atom (number of atoms, geometry and distances). The interest in such an analysis is due to the fact that one can analyze in principle for every atom its sphere of coordination.

Figure 2.20 shows the EXAFS-spectra at the L_3 edge of uranium in a uranium silicate glass [34]. The patterns were obtained at different incidence angles. For a large incidence angle the spectrum (a) reflects the bulk environment of uranium. Spectrum (b) was obtained for grazing incidence and reflects the local structure at the surface. The other spectra correspond to different degrees of corrosion and can be related to the formation of a gel at the surface. The increase in U–U correlation at the surface suggests the formation of clusters containing uranium atoms.

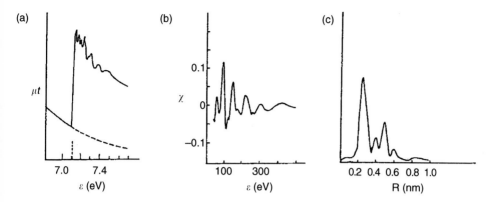

Figure 2.19 (a) Absorption spectrum at the K-edge of iron (μ, linear absorption coefficient; t, thickness of sample). (b) EXAFS function – $\chi(\varepsilon)$ – extracted from the total spectrum. (c) Fourier transformed EXAFS function. (Adapted from [33].)

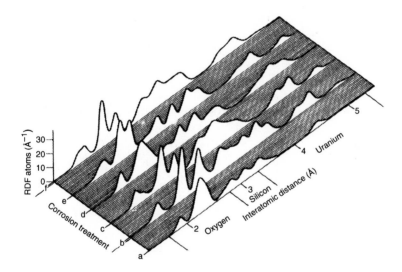

Figure 2.20 Fourier transformed EXAFS spectra of uranium silicate glasses: (a) bulk structure; (b) surface structure; (c–f) surface structures at different corrosion times [34].

Appendix

Synchrotron radiation applications in materials research

Diffraction [35, 36]

Crystal structures of single crystals, powders and fibers
Time-resolved and *in situ* diffraction
Structures at high pressures
Distinction of neighbouring elements by exploiting anomalous dispersion
Microdiffraction with beam sizes extending into the submicron range

Structures of surfaces and interfaces by glancing incidence diffraction
Analysis of textures and strain distribution including experiments with microbeams [37]

Diffuse scattering [38–40]

Small-angle scattering
Defects in metals and alloys
Polymer crystallization
Structures of colloidal systems and emulsions

Extended X-ray absorption Fine structure (EXAFS) [33, 41]

Short-range correlation
Kinetics of phase formation
Local structure of catalysts

Topography [42, 43]

Defects in crystals
Analysis of dislocations and stress

Tomography [13, 14]

Three-dimensional reconstruction of materials based on absorption or phase contrast.

Microfluorescence [26, 44]

Quantitative analysis of elements
Elemental mapping with beam sizes extending into the submicrometer range

Lithography [45]

Production of artificial structures in three dimensions (LIGA process)
Microlithography applications in semiconductor industry

References

[1] Kunz C., Properties of synchrotron radiation, in *Synchrotron Radiation – Techniques and Applications*, C. Kunz (ed.), 1979, Springer Verlag: Berlin. pp. 1–24.

[2] Winick H., Properties of synchrotron radiation, in *Synchrotron Radiation Research*, H. Winick and S. Doniach (eds), 1980, Plenum Press: New York. pp. 11–26.

[3] Krinski S., Perlman M. L. and Watson R. E. (eds), Characteristics of synchrotron radiation and its sources, in *Handbook on Synchrotron Radiation*, E. E. Koch (ed.), Vol. 1b, 1983, North-Holland: Amsterdam, pp. 65–172.

[4] Webpage: http://www.esrf.fr/machine/support/ids/Public/ID6Movies/ID6movies.html

[5] Tatchyn R. *et al.*, X-ray optics design studies for the 1.5 to 15 A Linac coherent light source (LCLS) at the Stanford Linear Accelerator Center (SLAC), in *Coherent Electron-Beam X-Ray Sources: Techniques and Applications*, 1997, SPIE Vol. 3154.

[6] Freund A., X-ray optics for synchrotron radiation, in *Neutron and Synchrotron Radiation for Condensed Matter Studies*, J. Baruchel *et al.* (eds), 1993, Springer Verlag: Berlin. pp. 79–94.

[7] Greaves G. N. and Catlow C. R. A., Synchrotron radiation instrumentation, in *Applications of Synchrotron Radiation*, C. R. A. Catlow and G. N. Greaves (eds), 1990, Blackie: Glasgow. pp. 1–38.

[8] Matsushita, T. and Hashizume, H. O., X-ray Monochromators, in *Handbook of Synchrotron Radiation*, E. E. Koch (ed.), Vol. 1b, 1983, North-Holland: Amsterdam. pp. 261–314.

[9]   Webpage: http://www-cxro.lbl.gov/optical_constants/mirror2.html
[10] Rosenbaum G. and Holmes K. C., Small-angle diffraction of X-rays and the study of biological structures, in *Synchrotron Radiation Research*, H. Winick and S. Doniach (eds), 1980, Plenum Press: New York. pp. 533–564.
[11] Glover I. D. and Helliwell J. R., Protein crystallography, in *Applications of Synchrotron Radiation*, C. R. A. Catlow and G. N. Greaves (eds), 1990, Blackie: Glasgow. pp. 241–267.
[12] Elsner G., Riekel C. and Zachmann, H. G., Synchrotron radiation in polymer science, in *Advances in Polymer Science*, H. H. Kausch and H. G. Zachmann (eds), 1985, Springer: Berlin.
[13] Hmelo A. B., *J. X-ray Sci. Technol.* **4**, 290–300 (1994).
[14] Flannery B. F. *et al.*, *Science* **237**, 1439–1444 (1987).
[15] Spanne P. *et al.*, *Phys. Rev. Lett.* **73**, 2001–2004 (1994).
[16] Pannetier J., Time resolved experiments, in *Neutron and Synchrotron Radiation for Condensed Matter Studies*, J. Baruchel *et al.* (eds), 1993, Springer: Berlin. pp. 425–436.
[17] Srajer V. *et al.*, *Science* **274**, 1726–1729 (1996).
[18] Jupe A. C. *et al.*, *Phys. Rev. B* **53**, 14697–14700 (1996).
[19] Riekel C., New avenues in X-ray microbeam experiments, *Rep. Prog. Phys.* **63**, 233–262 (2000).
[20] *X-ray Microbeam Technology and Applications*, 1995, San Diego, CA: SPIE Vol. 2516.
[21] *X-ray Microfocusing: Applications and Techniques*, 1998, San Diego, CA: SPIE Vol. 3449.
[22] Turner-Jones A., *Science* **62**, S53–S56 (1962).
[23] Seto T., Hara, T. and Tanaka, K., *Jpn. J. Appl. Phys.* **7**, 31 (1968).
[24] Lienert U. *et al.*, A high energy X-ray microscope for the local structural characterization of bulk materials, 40th Conference of AIAA on Structures, "Structural Dynamics of Bulk Materials," 1999, St Louis (USA), 2067–2075.
[25] Di Fonzo S. *et al.*, *Nature* **403**, 638–640 (2000).
[26] Janssens K. H. A., Adams, F. C. V. and Rindby, A. (eds), *Microscopic X-ray Fluorescence Analysis*, 2000, John Wiley & Sons: Chichester.
[27] Vis R. D., Synchrotron radiation trace element analysis, in *Applications of Synchrotron Radiation*, C. R. A. Catlow and G. N. Greaves (eds) 1990, Blackie: Glasgow. pp. 311–332.
[28] Vekemans B., *et al.*, *Book of Abstracts*, EDXRS-2000, Krakow, Poland, 2000, p. 106.
[29] Rindby A., X-ray Spectometry, **18**(3), 187–191 (1993).
[30] Berlund A. *et al.*, *Holzforschung* **53**, 474–480 (1999).
[31] Lichtenegger H. *et al.*, *J. Appl. Crystallogr.* **32**, 1127–1133 (1999).
[32] Rindby A., Engström P. and Janssens K., *J. Synchrotron Radiat.* **4**, 228–235 (1997).
[33] Gurman S. G., EXAFS and structural studies of glasses, in *Applications of Synchrotron Radiation*, C. R. A. Catlow and G. N. Greaves (eds), 1990, Blackie: Glasgow. pp. 140–170.
[34] Greaves G. N., *J. Am. Chem. Soc.* **111**, 4313 (1998).
[35] Catlow C. R. A., X-ray diffraction from powders and crystallites, in *Applications of Synchrotron Radiation*, C. R. A. Catlow and G. N. Greaves (eds), 1990, Blackie: Glasgow. pp. 1–38.
[36] Coppens P., *Synchrotron Radiation Crystallography*, 1992, Academic Press: London.
[37] Heidelbach F., Riekel, C. and Wenk, H. R., *J. Appl. Crystallogr.* **32**, 841–849 (1999).
[38] Lambert M., Diffuse scattering, in *Neutron and Synchrotron Radiation for Condensed Matter Studies*, J. Baruchel *et al.* (eds), 1993, Springer Verlag: Berlin. pp. 223–234.
[39] Williams C., Small angle scattering from solids and solutions, in *Neutron and Synchrotron Radiation for Condensed Matter Studies*, J. Baruchel *et al.* (eds), 1993, Springer Verlag: Berlin. pp. 235–245.
[40] Russell T. P., Small-angle scattering, in *Handbook of Synchrotron Radiation*, G. S. Brown and D. E. Moncton (eds), Vol. 3, 1991, North-Holland: Amsterdam. pp. 379–470.
[41] Stern E. A. and Heald S. M., Basic principles and applications of EXAFS, in *Handbook of Synchrotron Radiation*, E. E. Koch (ed.), Vol. 1b, 1983, North-Holland: Amsterdam. pp. 955–1014.
[42] Miltat J. and Dudley M., X-ray topography, in *Applications of Synchrotron Radiation*, C. R. A. Catlow and G. N. Greaves (eds.), 1990, Blackie: Glasgow. pp. 65–99.

[43] Baruchel J., Neutron and synchrotron radiation topography, in *Neutron and Synchrotron Radiation for Condensed Matter Studies*, J. Baruchel *et al.* (eds), 1993, Springer Verlag: Berlin. pp. 79–94.

[44] Baryshev V., Kulipanov G. and Skrinsky A., X-ray fluorescent elemental analysis, in *Handbook of Synchrotron Radiation*, G. S. Brown and D. E. Moncton (eds), Vol. 3, 1991, North-Holland: Amsterdam. pp. 639–688.

[45] Grobman W. D., Synchrotron radiation X-ray lithography, in *Handbook of Synchrotron Radiation*, E. E. Koch (ed.), Vol. 1b, 1983, North-Holland: Amsterdam. pp. 1131–1165.

Part 2

Methods and problems in residual stress determination by diffraction

3 Calculation of residual stress from measured strain

A. Lodini

3.1 General overview

Residual stress evaluation is an important stage in the improvement of the performance of materials, the control of the deformation of components and the understanding of industrial processes. In general, residual stress has various origins: mechanical (machining, shot peening), thermal (heat treatment, laser treatment), thermomechanical (forging, welding) or thermochemical (carburizing, nitriding).

According to the standard definition [1], the residual stresses are the auto-balancing stresses that exist in a material that is submitted either to no forces or external stresses and that is in a constant temperature condition.

In general, these residual stresses are caused by an inhomogeneity in deformation, from some source of local incompatibility which is generated from one or more of three fundamental physical origins: plastic flow, volume change, and thermal dilatation. Incomplete relaxation of the elastic deformation associated with these phenomena leads to residual stress.

In the solid there are additionally local incompatibilities caused by crystal defects, dislocations, grain boundaries, second phase particles, etc. The exact origin of stress is therefore going to depend on the scale of observation.

A classification of the residual stresses into three orders or types, related to the scale on which one considers the material, was proposed by Macherauch and Kloss [2].

- Residual stresses of the first order, or type I residual stresses, are homogeneous over a very large number of crystal domains of the material. Such stresses are also termed macrostresses $\left(\sigma_R^I\right)$. The internal forces related to this stress are balanced on all planes. The moments related to these forces are equal to zero around all axes.
- Residual stresses of the second order, or type II residual stress, are homogeneous within small crystal domains of the material (a single grain or phase). The internal forces related to these stresses are in balance between the different grains or phases.
- Residual stresses of the third order, or type III residual stress, are homogeneous on the smallest crystal domains of the material (over a few interatomic distances). The internal forces coupled to these stresses are in balance in very small domains (such as around dislocations or point defects). Type II and III residual stresses are collectively termed microstresses $\left(\sigma_R^{II} \text{ and } \sigma_R^{III}\right)$.

In the case of real materials, the actual residual stress state at a point comes from the superposition of stresses of type I, II and III stresses [3–5, 7], as is illustrated in Figure 3.1.

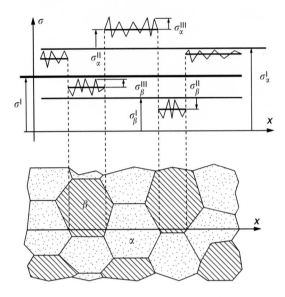

Figure 3.1 Three orders of stresses in two-phase materials.

3.2   Mechanical definition of stress and strain

In engineering science, the mechanical stress state at a point in a material and in a reference system is represented by a second-order tensor σ_{ij}:

$$\sigma_{ij} = \frac{\partial F_i}{\partial S_j}$$

where F_i represents the component of force in a direction x_i that acts on the element of volume characterized by surface dS_j, whose normal is in direction x_j.

In its complete notation σ_{ij} can be written in the referential system (1, 2, 3):

$$\sigma_{ij} = \begin{pmatrix} \sigma_{11} & \sigma_{12} & \sigma_{13} \\ \sigma_{21} & \sigma_{22} & \sigma_{23} \\ \sigma_{31} & \sigma_{32} & \sigma_{33} \end{pmatrix}$$

The diagonal elements represent normal stresses on planes normal to the direction of the chosen axes, the other elements being shear stresses.

The stress tensor is symmetrical ($\sigma_{ij} = \sigma_{ji}$) and can be diagonalized, which means that for any point we can choose a system of particular axes called *principal axes* of the tensor so that only elements of the diagonal are not equal to zero: that is, an axis system where there are only normal stresses along the axes, and no shear stresses between them. These components are the principal components of the stress tensor.

The stress tensor can be transposed to other reference axes using the law of transformation:

$$\sigma'_{ij} = a_{ik} a_{jl} \sigma_{kl}$$

The stress in direction ij of the initial set of axes is equal to the stress in direction kl of the new set of axes after the transformation $a_{ik}a_{jl}$.

While developing the equation in the direction of the plane normal, the stress is given by the ellipsoid of stresses:

$$\sigma'_{11} = a^2_{11}\sigma_{11} + a^2_{12}\sigma_{22} + a^2_{13}\sigma_{33} + 2a_{12}a_{11}\sigma_{12} + 2a_{13}a_{11}\sigma_{13} + 2a_{13}a_{12}\sigma_{23}$$

The stress tensor can also be expressed in the matrix notation of Voigt

$$\sigma = \begin{bmatrix} \sigma_{11} \\ \sigma_{22} \\ \sigma_{33} \\ 2\sigma_{12} \\ 2\sigma_{13} \\ 2\sigma_{23} \end{bmatrix}$$

The advantage of this notation is that it is sometimes convenient to express the elastic tensor as a matrix.

In practice, it is not stress that is obtained by a diffraction measurement: it is imprecise to speak of stress measurement. It is necessary to transform strain in stress.

In a solid, applied external forces deform the structure. It is then necessary to define a strain field characteristic of the resulting displacements. In any reference sample the state of strain of a small element of volume is defined by the second-order tensor ε_{ij}. For small strains the expression is written:

$$\varepsilon_{ij} = \frac{1}{2}\left[\frac{\partial u_i}{\partial x_j} + \frac{\partial u_j}{\partial x_i}\right]$$

in which $u(1, 2, 3)$ is the vector of displacements.

In its complete notation and for the standard reference system, it can be written as:

$$\varepsilon_{ij} = \begin{bmatrix} \varepsilon_{11} & \varepsilon_{12} & \varepsilon_{13} \\ \varepsilon_{21} & \varepsilon_{22} & \varepsilon_{23} \\ \varepsilon_{31} & \varepsilon_{32} & \varepsilon_{33} \end{bmatrix}$$

The diagonal elements represent strains along the direction of the chosen axes and the other elements are the shear strains between the axes.

For tensor ε_{ij}, the same rules as the stress tensor are applied: it is symmetrical,

$$\varepsilon_{ij} = \varepsilon_{ji}$$

it obeys the law of transformation,

$$\varepsilon'_{ij} = a_{ik}a_{jl}\varepsilon_{kl}$$

and, as for the stress tensor, matrix notation can be used for the strain tensor.

Table 3.1 Stress and strain notations

	Tensor	Matrix	Engineering notation
Normal stress	σ_{11}	σ_1	$\sigma_{xx}\sigma_x$
	σ_{22}	σ_2	$\sigma_{yy}\sigma_y$
	σ_{33}	σ_3	$\sigma_{zz}\sigma_z$
Shear sress	σ_{23}	σ_4	$\sigma_{yz}\tau_{yz}$
	σ_{13}	σ_5	$\sigma_{xz}\tau_{xz}$
	σ_{12}	σ_6	$\sigma_{xy}\tau_{xy}$
Longitudinal strain	ε_{11}	ε_1	$\varepsilon_{xx}\varepsilon_x$
	ε_{22}	ε_2	$\varepsilon_{yy}\varepsilon_y$
	ε_{33}	ε_3	$\varepsilon_{zz}\varepsilon_z$
Shear strain	ε_{23}	$(1/2)\varepsilon_4$	$\varepsilon_{yz}(1/2)\gamma_{yz}$
	ε_{13}	$(1/2)\varepsilon_5$	$\varepsilon_{xz}(1/2)\gamma_{xz}$
	ε_{12}	$(1/2)\varepsilon_6$	$\varepsilon_{xy}(1/2)\gamma_{xy}$

In engineering science, stress and/or strain analysis is done generally by choosing a Cartesian reference system (o, x, y, z). The relations between the different notations are given in Table 3.1. However, these definitions are not sufficient to describe the origins of the stress field in the material. It is necessary to separate clearly the macrostress and macrostrain from microstress and microstrain.

3.3   Macroscopic and microscopic relations

When a material is submitted to a applied load, it develops a field of microstress σ and microstrain ε. By definition, the macrostress $\bar{\sigma}$ is equal to the average value of the field of microstress σ in the volume V

$$\bar{\sigma} = \frac{1}{V} \int_V \sigma \, dV$$

In the same way, the macrostrain $\bar{\varepsilon}$ is written as:

$$\bar{\varepsilon} = \frac{1}{V} \int_V \varepsilon \, dV$$

If the sample is homogeneous and elastic, there is a linear correlation, on a macroscopic scale between the two terms $\bar{\varepsilon}$ and $\bar{\sigma}$:

$$\bar{\sigma} = \overline{C}\bar{\varepsilon}$$

where $\overline{C}$ represents the macroscopic elastic constant matrix.
  In the same way, on the microscopic scale:

$$\sigma = C\varepsilon$$

where C is the microscopic elastic coefficients matrix.
  In most materials, the polycrystalline values of $\overline{C}$ cannot generally be used interchangeably with the average single crystal C because there are very strong interactions between single crystals (grains).

3.3.1   *Case of a monocrystal*

The microscopic relation between the strain and the stress tensor is given by the generalized Hooke's law:

$$\varepsilon_{ij} = S_{ijkl}\sigma_{kl}$$

or by expressing the stress according to the strain:

$$\sigma_{ij} = C_{ijkl}\varepsilon_{kl}$$

C_{ijkl} and S_{ijkl} are coefficients of stiffness and compliance of the material, respectively, and are fourth-order tensors.

For any transformation a_{ij}, tensors of compliance and stiffness can be expressed as:

$$S_{ijkl} = a_{im}a_{jn}a_{ko}a_{lp}S_{mnop}$$
$$C_{ijkl} = a_{im}a_{jn}a_{ko}a_{lp}C_{mnop}$$

These formulas represent nine equations each of nine terms. That gives 81 constants in the tensors of compliance and stiffness.

However, relations of symmetry of the stress and strain tensor decreases the number of constants to 36.

We can also demonstrate that:

$$C_{ijkl} = C_{klij}$$
$$S_{ijkl} = S_{klij}$$

This decreases the number of independent constants to 21.

In general, tensors of stiffness and compliance are expressed with the help of the matrix notation of Voigt.

$$S_{ijkl} = S_{mn} \quad \text{for } m \text{ and } n = 1, 2 \text{ or } 3$$
$$S_{ijkl} = \tfrac{1}{2}S_{mn} \quad \text{for } m \text{ or } n = 4, 5 \text{ or } 6$$
$$S_{ijkl} = \tfrac{1}{4}S_{mn} \quad \text{for } m \text{ or } n = 7, 8 \text{ or } 9$$

In this case, the generalized Hooke's law is simplified as:

$$\sigma_i = C_{ij}\varepsilon_j$$
$$\varepsilon_i = S_{ij}\sigma_j$$

The number of independent constants of the matrix of compliance or stiffness can be reduced according to these symmetries.

For a cubic monocrystal it is, for example, necessary to know only three independent constants to deduce its elastic behaviour.

Matrices of compliance and stiffness are written in the following way:

$$S_{ij} = \begin{bmatrix} S_{11} & S_{12} & S_{12} & 0 & 0 & 0 \\ S_{12} & S_{11} & S_{12} & 0 & 0 & 0 \\ S_{12} & S_{12} & S_{11} & 0 & 0 & 0 \\ 0 & 0 & 0 & S_{44} & 0 & 0 \\ 0 & 0 & 0 & 0 & S_{44} & 0 \\ 0 & 0 & 0 & 0 & 0 & S_{44} \end{bmatrix}$$

$$C_{ij} = \begin{bmatrix} C_{11} & C_{12} & C_{12} & 0 & 0 & 0 \\ C_{12} & C_{11} & C_{12} & 0 & 0 & 0 \\ C_{12} & C_{12} & C_{11} & 0 & 0 & 0 \\ 0 & 0 & 0 & C_{44} & 0 & 0 \\ 0 & 0 & 0 & 0 & C_{44} & 0 \\ 0 & 0 & 0 & 0 & 0 & C_{44} \end{bmatrix}$$

If we consider the cubic monocrystal system as isotropic, the elastic constants must be the same in all directions.

So, the independent constants are S_{11} and S_{12}, with $S_{44} = 2(S_{11} - S_{12})$. In the same way, C_{11} and C_{12} with $C_{44} = (C_{11} - C_{12})/2$; by introducing Lamé constants of $\lambda = C_{12}$ and $\mu = C_{44}$, the matrix can be expressed thus:

$$C_{ij} = \begin{bmatrix} (\lambda + 2\mu) & \lambda & \lambda & 0 & 0 & 0 \\ \lambda & (\lambda + 2\mu) & \lambda & 0 & 0 & 0 \\ \lambda & \lambda & (\lambda + 2\mu) & 0 & 0 & 0 \\ 0 & 0 & 0 & \mu & 0 & 0 \\ 0 & 0 & 0 & 0 & \mu & 0 \\ 0 & 0 & 0 & 0 & 0 & \mu \end{bmatrix}$$

Hooke's law is simplified by the following expression:

$$\varepsilon_{ij} = \frac{1 + v}{E}\sigma_{ij} - \delta_{ij}\frac{v}{E}\sigma_{kk}$$

$$\sigma_{ij} = \frac{E}{1 + v}\varepsilon_{ij} - \frac{vE}{(1 + v)(1 - 2v)}\delta_{ij}\varepsilon_{kk}$$

with δ being the Kronecker delta-function, E the Young's modulus and v the Poisson's ratio of the material or of the cubic crystal variety considered:

$$E = \frac{1}{S_{11}} \quad \text{and} \quad v = -\frac{S_{12}}{S_{11}}$$

For a cubic monocrystal, the microscopic relations can generally be used. It is only necessary to consider the anisotropy of the single crystal.

3.3.2   Elastic constants of the single crystal (cubic system)

A single crystal is an anisotropic system and its elastic properties depend on the direction which is considered. To determine the Young's modulus a uniaxial stress σ_{app}, is applied to the cubic crystal in a direction [hkl].

In this direction, the ε_{11} strain is given by:

$$\varepsilon_{11} = \frac{1}{(E)_{hkl}}\sigma_{appl}$$

Since the material is anisotropic, the equation can be written in the generalized form:

$$\varepsilon_{11} = S'_{1111}\sigma_{11}$$

It is supposed that the direction $[hkl]$ can also be expressed in terms of the compliance constants S, defined in the reference system of the crystal, with the use of the transformation law:

$$\frac{1}{(E)_{hkl}} = S'_{1111} = a_{1m}a_{1n}a_{1o}a_{1p}S_{mnop}$$

in which a_{ij} are the direction cosines of the direction $[hkl]$.

By developing the equation:

$$\frac{1}{(E)_{hkl}} = S_{11}\left(a_{11}^4 + a_{12}^4 + a_{13}^4\right) + S_{12}\left(a_{11}^2 a_{12}^2 + a_{11}^2 a_{13}^2 + a_{11}^2 a_{23}^2\right)$$
$$+ S_{44}\left(a_{11}^2 a_{12}^2 + a_{11}^2 a_{13}^2 + a_{11}^2 a_{23}^2\right)$$

Since:

$$a_{ij}a_{jk} = 1 \quad \text{if } i = j$$
$$a_{ik}a_{jk} = 0 \quad \text{if } i \neq j$$

The equation becomes

$$\frac{1}{(E)_{hkl}} = S'_{1111} = S_{11} - 2\left\{(S_{11} - S_{12}) - \frac{S_{44}}{2}\right\}\left(a_{11}^2 a_{12}^2 + a_{11}^2 a_{13}^2 + a_{11}^2 a_{23}^2\right)$$

and according to the Miller indices [3]:

$$\frac{1}{(E)_{hkl}} = S_{1111} = S_{11} - 2A_{hkl}S$$

or

$$S = (S_{11} - S_{12}) - \frac{S_{44}}{2} \quad \text{and} \quad A_{hkl} = \frac{h^2 k^2 + h^2 l^2 + l^2 k^2}{(h^2 + k^2 + l^2)^2}$$

From the equation, we get the Young's modulus:

$$E_{hkl} = \frac{1}{S_{11} - 2A_{hkl}S}$$

The Poisson's ratio can be obtained by considering the direction perpendicular to the loading direction:

$$\varepsilon_\perp = \frac{1}{(E_\perp)_{hkl}} \sigma_{appl}$$

With the help of the transformation law:

$$S'_{2111} = a_{2i}a_{2j}a_{1k}a_{1l}S_{ijkl} = \frac{1}{(E_\perp)_{hkl}} = S_{12} + 2A_{hkl}$$

and therefore:

$$(E_\perp)_{hkl} = \frac{1}{S_{12} + 2A_{hkl}}$$

So, the Poisson's ratio for the plane (hkl) is:

$$(v)_{hkl} = -\frac{(E_\parallel)_{hkl}}{(E_\perp)_{hkl}} = \frac{S_{12} + 2A_{hkl}}{S_{11} - 2SA_{hkl}}$$

3.3.3 Case of one phase in an isotropic polycrystalline material

For a cubic and isotropic polycrystalline material, Hooke's law is simplified by the following expression where the micro–macro transformation takes place by knowledge of $E(hkl)$ and $v(hkl)$:

$$\bar{\sigma}_{ij} = \frac{E(hkl)}{1 + v(hkl)}\varepsilon_{ij} + \frac{v(hkl)E(hkl)}{1 + v(hkl)(1 - 2v(hkl))}\delta\varepsilon_{kk}$$

δ being the Kronecker delta-function.

3.3.4 Elastic constants of a polycrystalline material (cubic system)

The behaviour of a polycrystal plane (hkl) in a non-textured sample is generally considered to be isotropic. However, the material is composed of aggregated monocrystals whose properties are anisotropic. The elastic behaviour of the polycrystal depends not only on the elastic constants of the single crystal, but also on the interaction between grains. A rigorous theoretical calculation of the elastic constants of the polycrystal requires a complete theoretical solution of the influence of the elastic anisotropy of every crystallite and the interaction between crystallites on the elastic response of polycrystalline material. Since such a precise solution has not yet been obtained, it is necessary to use coupling models for the interaction between crystallites. Historically several models have been proposed, such as the models of Voigt, Reuss and Kroner [8–13].

3.3.4.1 Model of Voigt

In the approximation of Voigt the field of distortions (strain) is uniform in the polycrystal, and therefore each grain experiences the same strain. There is a continuity strain on grain boundaries. In this hypothesis the elastic constants of the material are calculated from five stiffness constants of the single crystal and are independent of the crystallographic direction. For example, for a non-textured cubic material:

$$E_{hkl} = \frac{2S_{44} + 6(S_{11} - S_{12})}{2S_{11}(S_{11} + S_{12} + 2S_{44}) - 2S_{12}(2S_{12} + S_{44})}$$

$$\nu_{hkl} = -\frac{S_{11}(2S_{11} + 2S_{12} - S_{44}) + S_{12}(3S_{44} - 4S_{12})}{2S_{11}(S_{11} + S_{12} + 2S_{44}) - 2S_{12}(2S_{12} + S_{44})}$$

3.3.4.2 Model of Reuss

In the approximation of Reuss, the stress field in the material is considered to be uniform, and therefore each crystallite develops a strain that is proportional to the anisotropic modulus in a particular direction. In this hypothesis, the elastic constants therefore depend on the crystallographic orientation [*hkl*].

For example, for a non-textured cubic material:

$$E_{hkl} = \frac{1}{S_{11} - A_{hkl}\{2(S_{11} - S_{12}) - S_{44}\}}$$

$$\nu_{hkl} = -\frac{S_{12} + A_{hkl}\{(S_{11} - S_{12}) - S_{44}/2\}}{S_{11} - A_{hkl}\{2(S_{11} - S_{12}) - S_{44}\}}$$

where A_{hkl} is the orientation factor of the plane (*hkl*).

3.3.4.3 The self-consistent model of Kröner

In this extended Eshelby model, the method of Kröner concerns the calculation of the global elastic properties of a polycrystalline material, from the elastic constants of the single crystal. This model is based on the treatment of one anisotropic crystallite, spherical or ellipsoidal, that is embedded in an infinite, elastic and isotropic matrix.

For every crystallite, the distortion is directly proportional to the average stress that exists in the polycrystal:

$$\varepsilon_{ij} = (S_{ijkl} + t_{ijkl}(\Omega))\bar{\sigma}_{kl}$$

where S_{ijkl} is the compliance tensor of the single crystal of volume Ω, and t_{ijkl} is a fourth-order tensor that describes the interaction between crystallites.

After integration of the equation over all orientations, one gets the macroscopic average values:

$$\bar{\varepsilon}_{ij} = (S_{ijkl})_B \bar{\sigma}_{kl}$$

$(S_{ijkl})_B$ being the compliance tensor of the material. So

$$\int_D t_{ijkl} \, dD = 0$$

56   A. Lodini

From calculations using the Kröner model, for a spherical, anisotropic crystal and cubic symmetry, the equation becomes:

$$G^3 + \alpha G^2 + \beta G + \gamma = 0$$

with

$$\alpha = \frac{9A_1 + 4A_2}{8}$$

$$\beta = -\frac{(3A_1 + 12A_3)A_2}{8}$$

$$\gamma = -\frac{3A_1 A_2 A_3}{8}$$

$$A_1 = \frac{1}{3(S_{11} + S_{12})} \quad A_2 = \frac{1}{S_{44}} \quad A_3 = \frac{1}{2(S_{11} - S_{12})}$$

The solution of G represents the macroscopic Coulomb modulus and enables us to calculate E and v:

$$E_{hkl} = \frac{\omega}{1 - 2\omega(1 - 5A_{hkl})t_{44}}$$

$$v_{hkl} = -\frac{2G\{1 + \omega(1 - 5A_{hkl})t_{44}\} - \omega}{2G\{1 - 2\omega(1 - 5A_{hkl})t_{44}\}}$$

where the tensor is composed as:

$$t_{44} = \frac{(G - A_2)(3A_1 + 6G)}{2G\{8G^2 + G(9A_1 + 12A_2) + 6A_1 A_2\}} \quad \text{and} \quad \omega = \frac{9A_1 G}{3A_1 + G}$$

Kröner's model gives the most realistic value of the elastic constants of a polycrystal, whereas the models of Voigt and Reuss give lower and upper bounds of these values, respectively. A simpler procedure, introduced by Hill and Neerfeld, consists of taking the arithmetic average of values of Reuss and Voigt, and which yields a result which will be closer to the value of Kröner.

3.3.5   Case of a two-phase polycrystalline material

In a two-phase polycrystalline material, the grains belonging to different phases have different physical and elastic properties and hence, the second order stresses averaged over one phase are not equal to zero. It is possible to approximate this problem. The mean value of stress calculated over one phase is called the phase stress ($\bar{\sigma}^{ph}$) and it can be subdivided into the following terms with the formalism proposed by Clyne and Withers [14]:

$$\bar{\sigma}_{ij}^{ph} = \bar{\sigma}_{ij}^{I} + \sigma_{ij}^{phE} + \sigma_{ij}^{phTh} + \sigma_{ij}^{phPl}$$

where the superscript I is used for the macro (first order) stress, and the mismatch elastic thermal and plastic stresses are denoted by E, Th and Pl, respectively [6, 14, 15].

The mismatch stresses are defined as the average value of the second-order stresses over all grains belonging to one phase; for both phases of the material the second-order stresses must sum to zero over a large volume of material and we obtain:

$$\overline{\sigma_{ij}^{I}} = f\overline{\sigma_{ij}^{ph1}} + (1-f)\overline{\sigma_{ij}^{ph2}}$$

The overall macrostress is simply the summation of two macrostress terms characteristic of each phase in the material. Some theoretical model (Eshelby or autocoherent) can be used to complete this evaluation (see Chapter 16).

3.4 Experimental determination of diffraction elastic constants (DECs)

Finally, it is generally necessary to measure the diffraction elastic constants (DECs) in order to verify model results, or when the parameters are unknown for a particular peak. The simplest way to measure DECs is to apply a uniaxial stress to a sample of the material and to measure the strain evolution $\varepsilon_{\phi\psi}$ for different values of the applied stress. The macroscopic stress state is the sum of the residual stress and the applied stress.

Generally, we apply a tensile or compressive stress by a stress rig or a four-point bending device. We consider the effect of the applied stress only, taking any residual stress to be our reference state. With $\phi = 0$, the general formula is:

$$\varepsilon_{\psi} = \tfrac{1}{2}S_2\overline{\sigma}_{xx}\sin^2\psi + S_1(\overline{\sigma}_{xx} + \overline{\sigma}_{yy})$$

where $(1/2)S_2$ and $(1/2)S_1$ are the DECs with:

$$\frac{1}{2}S_2 = \frac{1+\nu}{E}; \quad S_1 = -\frac{\nu}{E}$$

where E is the Young's modulus and ν the Poisson coefficient of the plane (hkl).

3.4.1 First method: $\sin^2\psi$ method

For different values of the applied load ($\overline{\sigma}_{app} = \overline{\sigma}_{xx}$) and with different determination of $\varepsilon_{\psi} = f(\sin^2\psi)$ we have the general result given by Figure 3.2.
   If we consider for different $\overline{\sigma}_{app}$

$$\alpha = \frac{\partial\varepsilon_{\psi}}{\partial\sin^2\phi} = \frac{1}{2}S_2\overline{\sigma}_{app}$$

the slope of the graph gives S_2

$$\left(\frac{1}{2}S_2 = \frac{\partial\alpha}{\partial\overline{\sigma}_{app}}\right)$$

and also for different $\overline{\sigma}_{app}$ the origin β ($\beta = \varepsilon_{(\psi=0)}$) gives S_1 ($S_1 = \partial\beta/\partial\overline{\sigma}_{app}$).

58   A. Lodini

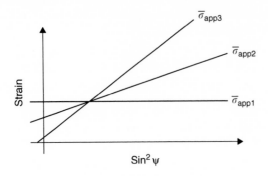

Figure 3.2 $\varepsilon_\psi = f(\sin^2 \psi)$ for different applied stresses.

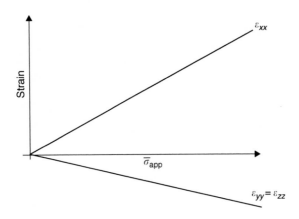

Figure 3.3 $\varepsilon_{xx}, \varepsilon_{yy}$ and ε_{zz} for different applied stresses.

3.4.2   Second method: 3D method

We measure the three principal strains for different applied stress

$$\varepsilon_{xx} = \varepsilon_{0,90}$$
$$\varepsilon_{yy} = \varepsilon_{90,90}$$
$$\varepsilon_{zz} = \varepsilon_{0,0}$$

The slope of the graph gives directly $E_{(hkl)}$ and $\nu_{(hkl)}$ (see Figure 3.3).

It is easy to calculate S_1 and $(1/2)S_2$. This method is used with the neutron diffraction technique.

Acknowledgements
The author wishes to thank Dr M. Ceretti and Prof. J. M. Sprauel. Their contributions are gratefully acknowledged.

References

[1] Mura T., *Micromechanics of Defects in Solids*, 1982, Martinus Nijhoff Publishers, The Hague, Netherlands.

[2] Macherauch E. and Kloss K. H., *Proceedings of the International Conference on Residual Stresses*, Garmisch-Partenkirchen, FRG, 1986, pp. 167–174.

[3] Lodini A., Perrin M., *Analyse des contraintes mécaniques par diffraction des rayons X et des neutrons*, 1996, Editions du CEA, Saclay.

[4] Lodini A., *Analysis of Residual Stresses from Materials to Biomaterials*, 1997, Editions IITT, Marne la Vallée.

[5] Hirschi K., *Thesis Université de Reims*, Reims, France, 1999.

[6] Levy R., *Thesis Université de Reims*, Reims, France, 1999.

[7] Ceretti M., *Thesis Université de Reims*, Reims, France, 1993.

[8] Voigt W., *Lehrbuch der Kristallphysik*, Teubner, Leipzig/Berlin, 1928.

[9] Voigt W., *Wied. Ann.* **38**, 573–587 (1889).

[10] Reuss A., *Z. Angew. Math. Mech.* **9**, 49–58 (1929).

[11] Kröner E., *J. Mech. Phys. Solids* **15**, 319–329 (1967).

[12] Hill R., *Proc. Phy. Soc.* **A65**, 349–354, (1952).

[13] Neerfeld H., *Mitt. K.W.I. Eisen Düsseldorf* **24**, 64–70 (1942).

[14] Clyne T. W. and Withers P. J., *An Introduction to Metal Matrix Composites*, 1993, Cambridge Solid State Sciences Series, Cambridge University Press, Cambridge.

[15] Baczmanski A., Braham C., Lodini A. and Wierzbanowski K., *Mater. Sci. Forum* **404–7**, 729–734 (2002).

4 Characterization of macrostresses

R. A. Winholtz

4.1 Macrostresses

Residual macrostresses are what have classically been thought of as residual stresses. Macrostresses arise in considering a macroscopic view where the material is considered homogeneous. They are the stresses that are revealed by dissection methods of residual stress measurement such as hole drilling or sectioning. Residual macrostresses are well known to affect material properties. They can be beneficial or detrimental to a material's performance. Fatigue resistance, for example, can be improved or reduced by tensile or compressive stresses, respectively, near the surface. Macrostresses also inhibit the ability to maintain dimensional control of components during manufacturing operations because, as stressed material is removed, stresses redistribute and the remaining material distorts.

For these reasons, the characterization of residual stresses is an important engineering consideration. Measurement of residual stresses is important both for directly knowing the stresses produced by various processes and for validating models of residual stress-producing processes. Diffraction methods are attractive for measuring residual stresses because they are nondestructive, precise, able to measure stresses both near the surface and in the interior of materials, and able to give the entire three-dimensional (3D) stress tensor.

4.2 Measurement of stresses with diffraction

Diffraction can be used to precisely measure a lattice spacing utilizing Bragg's law:

$$\lambda = 2d \sin \theta \qquad (1)$$

Here, λ is the wavelength of the radiation used, θ is one-half the scattering angle, and d is the average interplanar spacing for a given reflection in a crystalline material. In a stressed material, the lattice spacing can be used as a strain gauge, giving a measure of linear strain in the direction of the diffraction vector. If d_0 and θ_0 are the lattice spacing and the corresponding Bragg angle measured for the stress-free material, the strain can be computed as

$$\varepsilon = \frac{d - d_0}{d_0} = \frac{\sin \theta_0}{\sin \theta} - 1 \qquad (2)$$

The state of strain at some location in a material is a second-order tensor quantity represented by shear and normal components referenced to a given coordinate system. We can determine the state of strain in a material using diffraction by measuring the linear or normal strain in a number of directions and utilizing the rules specifying how the components of a second-order tensor transform with direction. Having determined the strain components, the stress components are then computed using Hooke's law.

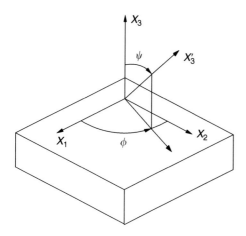

Figure 4.1 Coordinate systems used to measure stresses with diffraction. The X_i coordinate axes are the sample coordinate system. Oriented by the angles ϕ and ψ with the sample coordinate system is the X'_3 axis of the laboratory coordinate system.

First, we set up two coordinate systems illustrated in Figure 4.1. The X_i coordinate system is the sample coordinate system. We wish to determine the strain and stress components in this sample coordinate system. Although we are free to attach this coordinate system to the material in any way we desire, the geometry of the component will often suggest a natural set of sample coordinates. For near surface stress measurements using X-rays it is customary to place the X_3 axis normal to the surface. The X'_3 coordinate axis is in the laboratory system, which is the reference frame for making the diffraction measurements. The X'_3 coordinate axis is oriented with respect to the sample coordinate axes by the angles ϕ and ψ as shown in Figure 4.1. By orienting the specimen with respect to the incident and diffracted beams such that the diffraction vector is along the X'_3 axis, we can measure the strain in this direction, designated $\varepsilon_{\phi\psi}$. A diffraction peak is recorded and its Bragg angle precisely determined. The strain is then determined using equation (2).

Next, we relate the strain measured in the laboratory coordinate system to the unknown strain components in the sample coordinate system. Strain components transform according to the relation [1]

$$\varepsilon'_{ij} = \alpha_{ik}\alpha_{jl}\varepsilon_{kl} \tag{3}$$

where α_{ij} are the direction cosines between the laboratory and sample coordinate systems defined as

$$\alpha_{ij} = \cos(X'_i, X_j) \tag{4}$$

In equation (3) and in what follows, repeated indices imply summation. The direction cosines between the laboratory and sample coordinate systems are

$$\alpha_{31} = \cos\phi \sin\psi$$
$$\alpha_{32} = \sin\phi \sin\psi \tag{5}$$
$$\alpha_{33} = \cos\psi$$

Substituting these into equation (3) we obtain

$$\varepsilon_{\phi\psi} = \frac{d_{\phi\psi} - d_0}{d_0} = \varepsilon_{11}\cos^2\phi\sin^2\psi + \varepsilon_{22}\sin^2\phi\sin^2\psi + \varepsilon_{33}\cos^2\psi$$

$$+ \varepsilon_{12}\sin 2\phi\sin^2\psi + \varepsilon_{13}\cos\phi\sin 2\psi + \varepsilon_{23}\sin\phi\sin 2\psi \qquad (6)$$

This equation relates the quantities we measure with diffraction, $d_{\phi\psi}$ and d_0, to the unknown quantities we want to determine, the components of strain in the material ε_{ij}. There are six unknown strain components, thus we must measure at least six strains, $\varepsilon_{\phi\psi}$, at ϕ and ψ angles that do not form a singular set of equations. A least-squares solution of equation (6) can be obtained when more than six diffraction strain measurements are made [2]. This is good practice, minimizing the effects of a single bad measurement and allowing an assessment of the quality of the least-squares fit to the experimental data.

Having obtained the strain tensor, the stress tensor is then obtained from Hooke's law

$$\sigma_{ij} = \frac{1}{(1/2)S_2}\left[\varepsilon_{ij} - \delta_{ij}\frac{S_1}{(1/2)S_2 + 3S_1}\varepsilon_{ii}\right] \qquad (7)$$

where S_1 and $(1/2)S_2$ are the diffraction elastic constants [3, 4].

4.3   Least-squares determination of the strain tensor

To facilitate a matrix formulation for computer solution we introduce the following definitions [2]:

$$
\begin{aligned}
\varepsilon_1 &= \varepsilon_{11} & f_1(\phi,\psi) &= \cos^2\phi\sin^2\psi \\
\varepsilon_2 &= \varepsilon_{22} & f_2(\phi,\psi) &= \sin^2\phi\sin^2\psi \\
\varepsilon_3 &= \varepsilon_{33} & f_3(\phi,\psi) &= \cos^2\psi \\
\varepsilon_4 &= \varepsilon_{12} & f_4(\phi,\psi) &= \sin 2\phi\sin^2\psi \\
\varepsilon_5 &= \varepsilon_{13} & f_5(\phi,\psi) &= \cos\phi\sin 2\psi \\
\varepsilon_6 &= \varepsilon_{23} & f_6(\phi,\psi) &= \sin\phi\sin 2\psi
\end{aligned}
\qquad (8)
$$

We may then rewrite equation (6) as

$$\varepsilon_{\phi\psi} = \varepsilon' = \sum_{j=1}^{6} \varepsilon_j f_j(\phi,\psi) \qquad (9)$$

We define the weighted sum of squared error χ^2 as

$$\chi^2 = \sum_{i=1}^{n}\frac{(\varepsilon_{obs} - \varepsilon_{calc})^2}{\mathrm{var}(\varepsilon_i')} = \sum_{i=1}^{n}\frac{1}{\mathrm{var}(\varepsilon_i')}\left(\varepsilon_i' - \sum_{j=1}^{n}\varepsilon_j f_j(\phi_i,\psi_i)\right)^2 \qquad (10)$$

Here, the number of strain measurements collected is n and $\mathrm{var}(\varepsilon_i')$ is an estimate of the variance (squared standard deviation) or uncertainty in the ith strain measurement. The values of $\mathrm{var}(\varepsilon_i')$ are found by propagating the uncertainty in the peak locations to the strains.

Each var(ε_i') provides a weighting factor on each strain measurement's contribution to χ^2 so that those with a smaller standard deviation make a larger contribution. We want to find the values of the ε_j that minimize χ^2. A minimum in χ^2 exists when its partial derivative with respect to each ε_j equals zero. Taking the partial derivatives results in the set of equations

$$\frac{\partial \chi^2}{\partial \varepsilon_j} = \sum_{i=1}^{n} \left[\left(\sum_{k=1}^{6} \varepsilon_k f_k(\phi_i, \psi_i) \right) - \varepsilon_i' \right] \frac{f_j(\phi_i, \psi_i)}{\text{var}(\varepsilon_i')} = 0 \tag{11}$$

Forming the matrix **B** and the vector **E** with elements

$$B_{jk} = \sum_{i=1}^{n} f_j(\phi_i, \psi_i) f_k(\phi_i, \psi_i)/\text{var}(\varepsilon_i')$$

$$\tag{12}$$

$$E_j = \sum_{i=1}^{n} \varepsilon_i' f_j(\phi_i, \psi_i)/\text{var}(\varepsilon_i')$$

gives the matrix equation

$$\mathbf{B}\varepsilon = \mathbf{E} \tag{13}$$

This is the set of normal equations for the least-squares solution. It can be solved by matrix inversion or other methods [5] for the strain components ε_j,

$$\varepsilon = \mathbf{B}^{-1}\mathbf{E} \tag{14}$$

4.4 The $\sin^2 \psi$ method

The $\sin^2 \psi$ method is the traditional method for measuring near surface stresses with X-rays [3, 4]. There is a wealth of experience in interpreting diffraction stress data in this form and much of the anomalous behaviors seen have been studied and interpreted in this context. The $\sin^2 \psi$ method utilizes the fact that the d-spacing between any two principal directions will be linear with the square of the sine of the angle between them. This method has its origins in graphical solution methods before widespread computer usage and is still very useful for visualizing and understanding data.

We first substitute Hooke's law, equation (7), into the fundamental diffraction strain relation, equation (6), and use $\cos^2 \psi = 1 - \sin^2 \psi$ to obtain

$$\varepsilon_{\phi\psi} = \frac{d_{\phi\psi} - d_0}{d_0} = (1/2)S_2(\sigma_{11} \cos^2\phi + \sigma_{12} \sin 2\phi + \sigma_{22} \sin^2\phi - \sigma_{33}) \sin^2\psi$$

$$+ (1/2)S_2\sigma_{33} + S_1(\sigma_{11} + \sigma_{22} + \sigma_{33})$$

$$+ (1/2)S_2(\sigma_{13} \cos\phi + \sigma_{23} \sin\phi) \sin 2\psi \tag{15}$$

In the near-surface region, the stresses σ_{i3} cannot exist as macrostresses [3, 6]. In many instances microstresses will be zero and these terms will vanish. We define the stress

$$\sigma_\phi = \sigma_{11} \cos^2\phi + \sigma_{12} \sin 2\phi + \sigma_{22} \sin^2\phi \tag{16}$$

as the stress component along the ϕ direction in the sample surface. Equation (15) then becomes

$$\varepsilon_{\phi\psi} = \frac{d_{\phi\psi} - d_0}{d_0} = (1/2)S_2\sigma_\phi \sin^2\psi + S_1(\sigma_{11} + \sigma_{22}) \tag{17}$$

This equation shows that $\varepsilon_{\phi\psi}$ or $d_{\phi\psi}$ is linear with $\sin^2\psi$ for a constant ϕ. The slope of the measured d-spacings will be proportional to the stress σ_ϕ. Because we have a biaxial stress state, d_0 is just a multiplier in the slope and a precise value is not necessary to determine the stress σ_ϕ. Any uncertainty in d_0, easily determinable to a fraction of a percent, is reflected to the same proportion in the uncertainty in stress and is generally negligible in comparison to other uncertainties.

The stresses σ_{i3} can be present in the near-surface region as microstresses or they may be present in the interior or materials and measurable with neutrons or high-energy X-rays. The $\sin^2\psi$ method is also applicable for these situations where the stresses σ_{i3} are present. First, consider σ_{33} being present. Examining equations (15) and (16), we see that the slope of d-spacing (or strain $\varepsilon_{\phi\psi}$) versus $\sin^2\psi$ is proportional to the difference between stresses

$$\text{slope} \propto \sigma_\phi - \sigma_{33} \tag{18}$$

In order to determine the values of these stresses independently, we must also utilize the intercept of d versus $\sin^2\psi$. To determine the intercept with sufficient precision we must know the value of d_0 better than the precision of the strains being measured. This will be illustrated further in Section 4.8.

Now consider the case where the shear stresses σ_{13} and σ_{23} are present in the diffracting volume. Equation (15) shows that these two stress components will cause the d-spacing to be nonlinear with respect to $\sin^2\psi$. For measurements at a constant ϕ, the $\sin 2\psi$ term will have the same magnitude but opposite sign for positive and negative ψ value. The d-spacing will show two branches for positive and negative ψ forming an ellipse as shown in Figure 4.2.

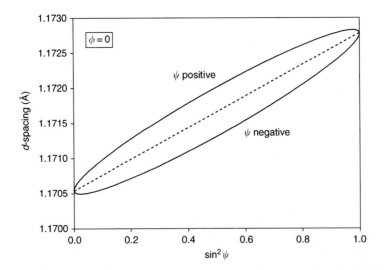

Figure 4.2 Illustration of ψ-splitting in a d versus $\sin^2\psi$ plot. The d-spacings are calculated from equation (6) with a shear strain ε_{13} present.

This is known as ψ-splitting. For near-surface measurements one cannot typically get to $\sin^2\psi = 1.0$ ($\psi = 90°$) and the ellipse does not close back up. For bulk measurements with penetrating radiations, however, one can measure out to $\psi = \pm 90°$ where one is measuring the same strain in opposite directions. From the centerline of the ellipse, or the average value between positive and negative ψ branches, one can get the normal stress components from the slope and intercept as previously done. This works because the averaging removes the shear components σ_{13} or σ_{23} since they have opposing contributions for the positive and negative ψ branches. The height of the ellipse can then be used to obtain the shear stress values. The deviation of d-spacing from the centerline versus $\sin 2\psi$ will have a slope proportional to the shear stress. By looking at the deviation from the centerline of the ellipse we have removed to the contributions of the normal stress components.

Analyzing data in this fashion is known as the Dölle–Hauk method [4, 7, 8]. It should be noted that the least-squares approach will handle all the stress components automatically and can give improved precision [2] but does not preclude one from plotting the d-spacings at constant ϕ sections versus $\sin^2\psi$ to inspect it for anomalies.

4.5 Determining the stress-free lattice spacing d_0

For the measurement of triaxial stress states a precise value of the stress-free lattice spacing d_0 is necessary to obtain the stresses and strains. Determining this value can be problematic [9]. One must measure a stress-free version of the material with the same experimental configuration used to measure the stressed material in order to remove any systematic 2θ shifts present. While such deviations from the absolute d-spacings measured can be made small, strains are also quite small and one should not try to rely on published values of lattice parameters. Another reason published values cannot be used is that, for many materials of engineering interest, the lattice parameter will vary with thermo-mechanical treatment. This can occur due to the formation of precipitates that can change the composition of phases. Differences in the stacking fault density can also change the stress-free value, in some materials [9, 10]. Another problem is that the value of d_0 can vary with location in a material due to differences in thermo-mechanical treatment with position (which may have produced the residual stresses).

Several methods can be used to determine the stress-free value at a given location in the material. First, a stress-free powder can be measured. A powder will be genuinely stress free. Any internal stresses in a powder particle must balance with each other giving an average stress of zero in all the particles. Care must be taken in using this method to be sure that the powder has the stress-free value of interest. The powder may not have the same thermo-mechanical treatment as the stressed material and hence may be unusable. This presents a difficulty because the method to produce a powder must be inherently different than the method used to produce the component of interest. For example, precipitation hardening aluminum alloys will have a different stress-free lattice spacing when used as the matrix for a composite than when they are a bulk alloy even with the same heat treatments. The presence of the reinforcing phase changes the precipitation kinetics, changing the stress-free lattice spacing [11–13]. Thus, filings from a bulk alloy may be unsuitable for measuring d_0.

Second, one may go to an unstressed region of a component to measure the stress-free lattice spacing. This method works nicely when appropriate because it does not involve any extra experimental set up. Here, one must be concerned that there is truly an unstressed region in the component. One must also be sure that d_0 is independent of position in the material. For a multiphase material, this method gives a macrostress-free lattice spacing that may still contain effects from microstresses. If one wants to determine only the macrostresses and the microstresses do not vary with position, this is appropriate.

Another method is to take an identical companion component and cut it up into small stress-free coupons to measure [14]. One must ensure that the cutting operation relaxes the residual macrostresses without introducing new residual stresses. Metals can often be electro-discharge machined to produce the coupons. Concerns in using this method include being sure that the coupons are cut to a small enough size so that macrostresses are substantially relieved and that the d_0 variations are captured sufficiently. It is also a disadvantage that a second component must be sacrificed, giving up the nondestructive nature of the diffraction stress measurement. This method will again give only macrostress-free values.

If one is confident that it does not vary with location in the component, equilibrium can be used to determine d_0. Any component of stress must integrate to zero over a cross-section of the component

$$\int_A \sigma_{ij} dA = 0 \tag{19}$$

One may iteratively try d_0 values until equation (19) is satisfied and then use this value.

Finally, if none of these methods is amenable, one may use an inaccurate value of d_0 and report the estimated error. It will be shown in Section 4.8 that an inaccurate value of d_0 leads to an error in the hydrostatic part of the stress. In many instances accurate values of the deviatoric stresses may provide the information necessary for the problem at hand.

4.6   Assuming principal directions

For subsurface measurements of stress with penetrating radiations, there is an incentive to save time in making measurements because intensities are often quite low, due to absorption by the sample, and source accessibility can be limited and expensive. If one knows the directions of the principal strains, time can be saved by measuring only in the three principal directions instead of trying to measure the general state of strain represented by equation (6). Geometry or loading can often suggest the principal directions.

Ignoring any shear stresses that are present and measuring three perpendicular strains will not lead to any error in the three normal components of stress. From equation (6) one can see that the shear strain components do not make any contribution to the measured strains along the sample coordinate axes, that is, for (ϕ, ψ) equal to $(0, 90°)$, $(90°, 90°)$ and $(0, 0)$. The form of Hooke's law used for polycrystalline materials, equation (7), has no coupling between shear and normal stresses and strains. This means that the three normal components of stress can be determined whether or not they are the principal directions [15]. This can be useful in simplifying measurements made to compare to a finite element computation or analytic model where components of stress are sufficient. One should be careful in neglecting the shear stress components, however. The normal stress components may not accurately reflect the complete stress state of the material. The principal stresses may be much larger, limited only by the strength of the material.

The stress equilibrium relation $\sigma_{ij,j} = 0$ can be helpful in examining assumptions about the principal stress directions [16]. If one has stress data obtained from measuring strains in three orthogonal directions in a sample, this relation may be used to prove they are not the principal directions. If any normal component of stress σ_{ii} varies with the direction X_i then the equilibrium relation tells us there must be gradients in the shear stresses as well. For example, if we see that σ_{11} has a gradient in the X_1 direction we may use

$$\frac{\partial \sigma_{11}}{\partial X_1} + \frac{\partial \sigma_{12}}{\partial X_2} + \frac{\partial \sigma_{13}}{\partial X_3} = 0 \tag{20}$$

to see that there are gradients in the shear stresses in the X_2–X_3 plane. Since the shear stresses are varying in this plane, they cannot be zero everywhere in this plane. If there are shear stress components present, the normal stress components are not the principal stresses. While the equilibrium relation may confirm that a set of stress measurements are not the principal stresses, it cannot be used to prove that they are the principal directions. This relation only deals with gradients so a constant shear stress can be present if σ_{ii} does not vary with X_i.

4.7 Error analysis

An error analysis of diffraction stress measurements is important for a number of reasons. First, elastic strains in crystalline materials are very small and it is easy to have the uncertainties in the measurements exceed the size of the strains due to the stresses. One needs an estimate of the errors to ensure that the results are meaningful. Second, one can use an error analysis to examine the feasibility of a proposed measurement. Time considerations may be important here as well, as higher precision measurements usually take more counting time. For example, after a preliminary error analysis one might conclude that a proposed set of measurements can achieve a desired level of uncertainty but the time necessary far exceeds that reasonably available. One should generally use an error analysis to design an experiment to achieve a given level of uncertainty in the final results. Third, error analysis can be used to optimize a measurement strategy. One will want to examine absorption path lengths for various potential measurement orientations to choose the ϕ and ψ values used to make the measurements. One wants to minimize the measurement uncertainties for a given amount of measurement time. Conversely, one may want to minimize the measurement time necessary to achieve a given level of measurement uncertainty. Access to neutron and synchrotron sources can be expensive and limited so the best use of the available time should be used.

For a quantity x observed repeatedly n times, the variance is defined as

$$\mathrm{var}(x) = \frac{\sum_{i=1}^{n}(x_i - \bar{x})^2}{n - 1} \tag{21}$$

where $\bar{x}$ is the average of the observations.

The objective of the error analysis is to identify and quantify the sources of error in the measurements and propagate them into the final results. Measurement uncertainties or errors are propagated as follows. Consider a result y calculated from several measured variables x_i

$$y = f(x_1, x_2, \ldots) \tag{22}$$

For random errors in the measured variables, the variance in y as a function of the variances in x_i is given by [17]

$$\mathrm{var}(y) = \left(\frac{\partial y}{\partial x_1}\right)^2 \mathrm{var}(x_1) + \left(\frac{\partial y}{\partial x_2}\right)^2 \mathrm{var}(x_2) + \cdots$$

$$+ 2\left(\frac{\partial y}{\partial x_1}\right)\left(\frac{\partial y}{\partial x_2}\right) \mathrm{cov}(x_1, x_2) + \cdots \tag{23}$$

The error in y is then often reported as one standard deviation in y

$$\text{Error in } y = \text{std}(y) = \sqrt{\text{var}(y)} \tag{24}$$

For diffraction stress measurements, the primary source of uncertainty in the stress and strain tensors is the uncertainties in the locations of the diffraction peaks. These will be both random and systematic. Random errors in the location of diffraction peaks include the counting statistics of measuring intensities with a detector, electronic noise and stability in the detector system, temperature fluctuations during the measurements, and many others. In many cases, for a properly working system, the counting statistics can dominate the random errors. We would like to be able to estimate the random errors from a single observation of a diffraction peak rather than observing it often enough to see the actual variation given by equation (21). The random error in the peak location is usually estimated by the fitting program that is used to determine the peak location. This will give an estimate of the variation expected due to counting statistics if the peak were observed many times.

Systematic errors include diffractometer misalignment, absorption effects and geometrical effects associated with a gauge volume. There can also be systematic irregularities in the specimen leading to measurement errors such as an unsuitable stress-free standard or the stress-free lattice spacing varying with location in the material. It is hoped that care is taken and that the systematic errors are removed from the measurements or at least understood and the data corrected for them. Systematic errors that are constant in both the measurement of $d_{\phi\psi}$ and d_0, such as an error in the radiation wavelength or a 2θ offset, will cancel each other in equation (2) resulting in no errors in the strain measurements.

The total variance in peak location can be estimated as the sum of the variance due to counting statistics and the variance due to other sources

$$\text{var}(2\theta_t) = \text{var}(2\theta_{cs}) + \text{var}(2\theta_{os}) \tag{25}$$

The variance due to counting statistics can be controlled by the number of counts collected in the diffraction peak. Generally, one must count four times longer to reduce the standard deviation by one half. One may estimate the variance due to other sources for a particular instrument by repeatedly observing a diffraction peak with long count times to give as equation (21) and then equation (25) is used to estimate $\text{var}(2\theta_{os})$.

To get the error in each strain measurement we apply equation (23) to equation (2) and obtain

$$\text{var}(\varepsilon_{\phi\psi}) = \text{var}(\varepsilon_i') = \left(\frac{\sin\theta_0\cos\theta}{\sin^2\theta}\right)^2 \text{var}(\theta) + \left(\frac{\cos\theta_0}{\sin\theta}\right)^2 \text{var}(\theta_0) \tag{26}$$

or

$$\text{var}(\varepsilon_{\phi\psi}) \cong \cot\theta\frac{1}{4}[\text{var}(2\theta) + \text{var}(2\theta_0)]\left(\frac{180}{\pi}\right)^2 \tag{27}$$

which relates the variance in a strain to the variances in the measured Bragg peak locations. In equation (27) we have switched from θ to 2θ, the experimentally determined variable, and added a conversion from radians to degrees because peak positions are generally determined in degrees.

The errors in each measured strain are then propagated through the least-squares solution to the strain components in the sample coordinate system

$$\text{var}(\varepsilon_k) = \sum_{i=1}^{n} \left(\frac{\partial \varepsilon_k}{\partial \varepsilon_i'} \right)^2 \text{var}(\varepsilon_i') \tag{28}$$

This result assumes that the individual strain measurements, ε_i', are independent so that the covariance terms are zero. It can be shown that applying equation (28) to equation (14) gives the result [5]

$$\text{var}(\varepsilon_k) = [\mathbf{B}]_{kk}^{-1} \tag{29}$$

That is, the diagonal elements of the inverse of the matrix $\mathbf{B}$ are the variances of the corresponding strain components. Furthermore, $\mathbf{B}^{-1}$ is the covariance matrix and the off-diagonal elements give the covariance between corresponding strain components [5]

$$\text{cov}(\varepsilon_i, \varepsilon_j) = [\mathbf{B}]_{ij}^{-1} \tag{30}$$

Care must be exercised in including the variance in the stress-free lattice spacing in equation (27). If a number of measurements of strain all use the same d_0 (or θ_0) value, they are not random and independent of each other but are perfectly correlated [18]. This makes the use of equation (28) invalid. Handling the error in d_0 in this case is shown in Section 4.8. If a separate value of d_0 is measured for each $\varepsilon_{\phi\psi}$, then the random errors are uncorrelated and its inclusion in equation (27) is the correct way to handle the error.

After obtaining the errors in the strain components, it is propagated into the stress components by applying equation (23) to equation (7) resulting in

$$\text{var}(\sigma_{11}) = \left[\frac{1}{(1/2)S_2} - \frac{S_1}{(1/2)S_2((1/2)S_2 + 3S_1)} \right]^2 \text{var}(\varepsilon_{11})$$

$$+ \left[\frac{S_1}{(1/2)S_2((1/2)S_2 + 3S_1)} \right]^2 [\text{var}(\varepsilon_{22}) + \text{var}(\varepsilon_{33})]$$

$$+ 2 \left[\frac{1}{(1/2)S_2} - \frac{S_1}{(1/2)S_2((1/2)S_2 + 3S_1)} \right] \left[\frac{S_1}{(1/2)S_2((1/2)S_2 + 3S_1)} \right]$$

$$\times [\text{cov}(\varepsilon_{11}, \varepsilon_{22}) + \text{cov}(\varepsilon_{11}, \varepsilon_{33})]$$

$$+ 2 \left[\frac{S_1}{(1/2)S_2((1/2)S_2 + 3S_1)} \right]^2 \text{cov}(\varepsilon_{22}, \varepsilon_{33}) \tag{31}$$

for the normal stress σ_{11}, the variances for σ_{22} and σ_{33} being similar in form. For the shear stress components we obtain

$$\text{var}(\sigma_{ij}) = \left[\frac{1}{(1/2)S_2} \right]^2 \text{var}(\varepsilon_{ij}); \quad i \neq j \tag{32}$$

Covariance terms exist in equation (31) because the normal stresses are coupled to all three normal strains, while they are absent in equation (32) because each shear stress component depends only on its corresponding shear strain.

If normal stress components are determined by measuring strain in three orthogonal directions, equation (31) is used to estimate the errors in the stress components by setting the strain covariance terms equal to zero. They are zero because the orthogonal strain measurements are independent of each other.

The relations for errors in diffraction stress measurements are also useful in estimating the time necessary for a set of experiments and their feasibility. Preliminary measurements of a sample of the material to be studied can be scaled to estimate the time necessary to achieve the desired precision. Such planning will help prevent unacceptably high errors in a set of measurements or running out of available time.

The time necessary to achieve a given error level in a set of measurements can be estimated from a preliminary measurement of a diffraction peak. The desired level of error in the stress and strain tensors is used to compute the necessary errors in the peak positions using the relations above. The time necessary to measure each diffraction peak for a series of measurements is then estimated from a preliminary measurement using the relation [19]

$$\text{var}(2\theta_i) = \text{var}(2\theta_0) \left(\frac{t_0}{t_i}\right) \exp[\mu(l_i - l_0)] \tag{33}$$

Here t is the time taken to collect a diffraction peak with a variance $\text{var}(2\theta)$ through a beam path length through the sample of l. The subscript '0' indicates these quantities for a preliminary measurement and the subscript 'i' designates these quantities for each of the diffraction peaks in the planned measurements. The absorption coefficient for the material is given by μ. By summing the times necessary to achieve the necessary precision in the peak positions one can estimate the time needed. Some iteration may be necessary to find a satisfactory measurement plan.

4.8   Uncertainties in the stress-free lattice spacing d_0

For triaxial stress measurements the stress-free lattice spacing d_0 must be measured to a precision better than the precision desired in the strains. We can illustrate the role of d_0 in diffraction stress measurement by substituting hydrostatic, ε_H, and deviatoric, $'\varepsilon_{ij}$, parts of the strain tensor into equation (6) and obtaining

$$d_{\phi\psi} = d_0 + d_0\varepsilon_H + d_0['\varepsilon_{11}\cos^2\phi\sin^2\psi + '\varepsilon_{22}\sin^2\phi\sin^2\psi + '\varepsilon_{33}\cos^2\psi$$
$$+ '\varepsilon_{12}\sin 2\phi\sin^2\psi + '\varepsilon_{13}\cos\phi\sin 2\psi + '\varepsilon_{23}\sin\phi\sin 2\psi] \tag{34}$$

This equation still has only six unknown strain components on the right-hand side as the relation

$$'\varepsilon_{11} + '\varepsilon_{22} + '\varepsilon_{33} = 0 \tag{35}$$

follows from the definition of the strain deviator. Using this, we obtain

$$d_{\phi\psi} = d_0 + d_0\varepsilon_H + d_0['\varepsilon_{11}(\cos^2\phi\sin^2\psi - \cos^2\psi) + '\varepsilon_{22}(\sin^2\phi\sin^2\psi - \cos^2\psi)$$
$$+ '\varepsilon_{12}\sin 2\phi\sin^2\psi + '\varepsilon_{13}\cos\phi\sin 2\psi + '\varepsilon_{23}\sin\phi\sin 2\psi] \tag{36}$$

By measuring $d_{\phi\psi}$ as a function of direction and utilizing equations (35) and (36), one can determine the deviatoric components of the strain since each has a different trigonometric dependence on the angles ϕ and ψ. An accurate value of d_0 is unnecessary since it is only a multiplier in each term. From the dependence of $d_{\phi\psi}$ with orientation one can also determine the constant component that does not vary with ϕ and ψ. However, the hydrostatic part of the strain cannot be determined without a precise value for d_0 because there are two constant terms in equation (36), which cannot be distinguished from each other with measurements of $d_{\phi\psi}$ [18, 20]. Thus, the precision to which d_0 is known determines the precision to which the hydrostatic strain can be known. Any error in d_0 results directly in an error in the hydrostatic component of strain and stress.

Having obtained the hydrostatic and deviatoric parts of the strain, the corresponding parts of the stress are obtained from

$$\sigma_H = \frac{E}{(1-2v)}\varepsilon_H = \frac{1}{(1/2)S_2 + 3S_1}\varepsilon_H \tag{37}$$

$$'\sigma_{ij} = \frac{E}{1+v}{'\varepsilon_{ij}} = \frac{1}{(1/2)S_2}{'\varepsilon_{ij}} \tag{38}$$

The error in the hydrostatic part of the stress tensor due to d_0 is then

$$\text{var}(\sigma_H) = \left(\frac{1}{(1/2)S_2 + 3S_1}\right)^2 \left(\frac{1}{d_0}\right)^2 \text{var}(d_0) \tag{39}$$

In practice, one can just fit equation (6) and perform an error analysis assuming no uncertainty in d_0. One can then use equation (39) to find the error in the hydrostatic part of the stress and add it in quadrature to the previous errors on the normal stress components.

4.9 Uncertainties in the principal stresses and directions

Having determined the components of the stress tensor one may then compute the principal stresses and directions by solving the eigenvalue problem [21]

$$\begin{bmatrix} \sigma_i - \sigma_{11} & -\sigma_{12} & -\sigma_{13} \\ -\sigma_{12} & \sigma_i - \sigma_{22} & -\sigma_{23} \\ -\sigma_{13} & -\sigma_{23} & \sigma_i - \sigma_{33} \end{bmatrix} \begin{bmatrix} \alpha_{i1} \\ \alpha_{i2} \\ \alpha_{i3} \end{bmatrix} = 0 \tag{40}$$

Three solutions for σ_i exist which are the principal stresses. The α_{ij} for each σ_i give the direction cosines describing the corresponding principal direction. Along with the principal stresses and directions, estimates of their errors should also be reported. Since equation (23) cannot be applied to the solution of equation (40), an alternate approach to estimate the errors is necessary.

A Monte Carlo approach can be used to simulate the propagation of errors through the data analysis by repeatedly generating sets of synthetic data with a random number generator and analyzing them [22]. The distribution of principal stresses and their directions are then used to estimate errors. It should be emphasized that the errors in the synthetic data sets are distributed about one particular experimental measurement and not about the true values of stress. It is assumed the distribution of errors about the true values have the same shape and variance as the distribution produced from the computer simulation and is thus useful for estimating the errors in the measurement.

To estimate the errors in the stresses with the Monte Carlo approach, we proceed as follows. Each observation of an intensity in the original data set is replaced with a new one from a random number generator. Each diffraction peak collected at an orientation of the sample will consist of the original intensities measured over a range of 2θ values. This data is fit to a function, such as a Gaussian, Lorentzian, or pseudo-Voigt, to determine the peak position. The peak fits to the original data form the basis for generating the synthetic data sets. The value of the best fit function is taken at each 2θ and then an error is added. This error is taken from a random number generated from a normal distribution with a zero mean and a standard deviation equal to the square root of the number of counts. This simulates adding counting statistics to the best fit peak function to get a synthetic data set. Figure 4.3 illustrates this process, showing the best fit Gaussian function to an original diffraction peak, a new set of synthetic data generated from this profile, and a new best fit Gaussian function to the new synthetic data. Notice that the fit to the synthetic data gives a slightly different peak position because of the new random noise added.

Once this process has been applied to all the diffraction peaks in a measurement, the new synthetic data set is analyzed in the same manner as the original data giving a stress tensor with principal stresses and directions. These values are recorded and the process repeated with a newly generated set of synthetic data. An ensemble of stress tensors and corresponding principal stresses and directions is thus generated by repeating this process. Several hundred synthetic data sets should be sufficient. The standard deviation in each stress component and principal stress can be computed from this ensemble giving an estimate of the errors in the original measurement. Errors estimated in the stress components by the Monte Carlo approach have been shown to be equivalent to those given by equations (31) and (32) [23]. The Monte Carlo method has the advantage that it can be extended to the principal stresses.

To estimate the likely errors in the principal stress directions, the ensemble of principal directions from the Monte Carlo analysis can all be plotted on an equal area projection [24]

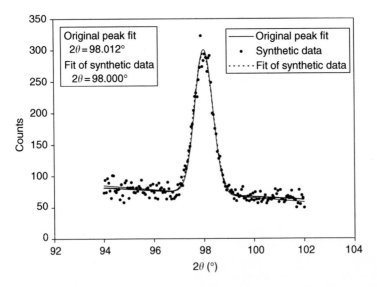

Figure 4.3 Illustration of generating a synthetic data set. The peak fit to the original data is shown along with a synthetic data set created by adding noise to the original peak fit. The best fit to the synthetic data is also shown which has a slightly different peak position.

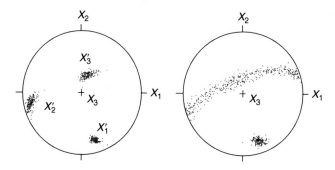

Figure 4.4 Equal area plots showing the results of a Monte Carlo calculation of the principal stress directions. These plots reveal the likely uncertainty in the principal directions.

to observe their distribution. An equal area projection is used because random directions will show up with a uniform density. Figure 4.4 shows two such plots from the same location in a weld before and after post-weld heat treatment [25]. The measured stress tensors, left and right respectively, for these two plots are

$$
\begin{bmatrix} 499 & -99 & -2 \\ -99 & 667 & -214 \\ -2 & -214 & 220 \end{bmatrix} \pm \begin{bmatrix} 150 & 34 & 36 \\ 34 & 77 & 41 \\ 36 & 41 & 107 \end{bmatrix} \quad \text{and}
$$

$$
\begin{bmatrix} 132 & -90 & 38 \\ -90 & 342 & -97 \\ 38 & -97 & 116 \end{bmatrix} \pm \begin{bmatrix} 132 & 31 & 31 \\ 31 & 67 & 44 \\ 31 & 44 & 76 \end{bmatrix} \text{MPa}
$$

The corresponding principal stresses are

$$
\begin{bmatrix} 779 & 0 & 0 \\ 0 & 438 & 0 \\ 0 & 0 & 129 \end{bmatrix} \pm \begin{bmatrix} 44 & 0 & 0 \\ 0 & 99 & 0 \\ 0 & 0 & 80 \end{bmatrix} \quad \text{and} \quad \begin{bmatrix} 412 & 0 & 0 \\ 0 & 99 & 0 \\ 0 & 0 & 80 \end{bmatrix} \pm \begin{bmatrix} 39 & 0 & 0 \\ 0 & 64 & 0 \\ 0 & 0 & 51 \end{bmatrix} \text{MPa}
$$

The principal directions were always chosen on the upper half of the projection to simplify the plots. If a principal direction was computed to be on the lower half of the projection, it was replaced with its opposite direction (also a principal direction), which is on the upper half. The first plot shows a case where the principal directions are well defined. The second plot illustrates a case where the principal directions are not well determined by the data. This has two sources. First, the errors are large in comparison to the stress values and, second, two of the principal stresses are nearly equal within the experimental error. This means that there is a plane through this point where the stress components in any direction in that plane are nearly equal. This is reflected by the band in the plot representing this plane. With the experimental errors, the measurement cannot distinguish within this plane where the principal directions are.

It should be emphasized again that the directions are distributed about the original measurement's principal directions and not about the true principal directions, which are unknown. It is assumed that the distribution of directions about the true principal directions would have the same shape in the plot.

4.10   Goodness-of-fit

In Section 4.4, examining data in a $\sin^2 \psi$ plot for anomalies was advocated. Knowing how the data should look will help one discover experimental difficulties. Here, we will develop a more quantifiable test that can be used to judge the quality of diffraction stress data [26].

The quantity χ^2, as defined in equation (10), gives a measure of the discrepancy between the best-fit model to the data and the actual data. For some unknown stress state being measured, an experiment will result in a value of χ^2_{obs}. This quantity is a random variable with statistical properties that can be tested to determine the quality of the fit to the data. We will use χ^2_{obs} to test the hypothesis that the data are a good fit to equation (6). If we reject this hypothesis, we must conclude that the strains computed by fitting the data to equation (6) are not a good measure of the true state of strain. This can be a result of systematic experimental errors or having a material where equation (6) does not accurately represent the state of strain for the different groups of grains giving rise to diffraction peaks for the different orientations of the diffraction vector.

χ^2_{obs} follows the χ^2 probability distribution given by [5, 17]

$$P(\chi^2, v) = \frac{(\chi^2)^{1/2(v-2)}e^{-\chi^2/2}}{2^{v/2}\Gamma(v/2)} \tag{41}$$

Here v is the degrees of freedom and $\Gamma(z)$ is the gamma function. The degrees of freedom are defined as the number of observations minus the number of parameters fit. If we are fitting data to equation (6) to determine six components of strain, the degrees of freedom will be the number of strain measurements minus six. Notice that if we measure only six strains, the model will fit the data exactly and χ^2 will be exactly zero because there are no degrees of freedom left.

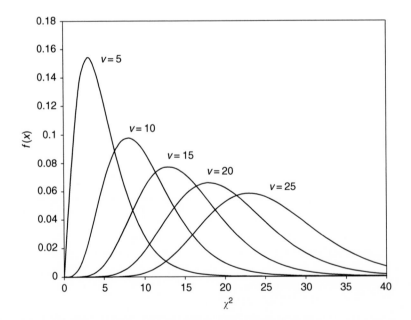

Figure 4.5 The χ^2 probability distribution for different degrees of freedom.

The χ^2 distribution for several different degrees of freedom is shown in Figure 4.5. The expected value of χ^2_{obs} is v. Notice that as χ^2_{obs} gets larger than v, the probability of experimentally realizing this χ^2_{obs} gets very small.

We can find the probability of experimentally realizing a χ^2 value of χ^2_{obs} or larger by integrating the probability distribution function from χ^2_{obs} to infinity. This gives the probability that a high or higher χ^2_{obs} could have occurred by random chance given our estimates of the experimental errors. We will call this probability the goodness-of-fit statistic Q, defined as

$$Q(\chi^2, v) = \int_{\chi^2}^{\infty} P(\chi^2, v)\mathrm{d}\chi^2 \qquad (42)$$

We can now accept or reject the following hypothesis about a set of data:

The model of equation (6) accurately estimates the state of strain in the sample and χ^2_{obs} is due to the estimated experimental uncertainties.

This hypothesis will be rejected if Q is too small, that is, if the probability of observing this much discrepancy between the model and the data, given the estimated uncertainties, is too low. Conversely, we accept the hypothesis when the probability Q indicates that χ^2_{obs} is likely to have occurred.

Figure 4.6 shows data from the near-surface region of a ground steel plate. Here, we observe a good fit to the data and the probability Q shows that the discrepancy between the

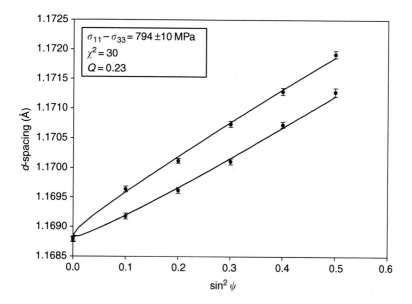

Figure 4.6 Measured *d*-spacings from a ground steel plate compared to the best-fit model, equation (6). This data illustrates a good fit to the strain transformation equation. The probability Q indicates that the observed variations of the data from the model are most likely due primarily to the counting errors in the measurement.

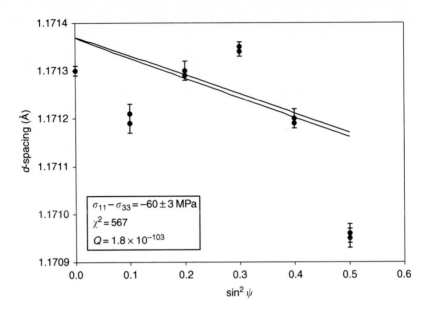

Figure 4.7 Measured *d*-spacings from a rolled steel plate compared to the best-fit model, equation (6). This data illustrates a poor fit to the strain transformation equation. The probability Q indicates that the data do not fit the model.

data and the model is likely to be accounted for by the estimated experimental uncertainties. In contrast, Figure 4.7 shows data from the near surface region of a cold-rolled steel sheet. This data shows oscillations with $\sin^2 \psi$ that cannot be fit with the model. The extremely low probability Q gives us reason to reject our hypothesis and not accept the fitted strains as a valid measurement. Cold-rolled materials are well known to develop these oscillations due to the inhomogeneous partitioning of stress among grains of different orientations in the material. The fitting of the data to equation (6) is inappropriate, which is revealed by our goodness-of-fit test.

The following criteria have been offered for judging the quality of data with the goodness-of-fit test and accepting or rejecting the stress and strain results [26]:

$1.0 < Q < 0.01$:   Accept the hypothesis. The experimental data fits the model and the observed discrepancy is accounted for by the estimated uncertainties.

$10^{-5} < Q < 10^{-2}$:   The quality of the fit is questionable. Results may still be valid because some small experimental errors have not been accounted for. Take a closer look.

$Q < 10^{-5}$:   Reject the hypothesis. The data do not fit the model.

References

[1] Nye J. F., *Physical Properties of Crystals: Their Representation by Tensors and Matrices*, 1985, Clarendon Press, Oxford.

[2] Winholtz R. A. and Cohen J. B., *Aust. J. Phys.*, **41**, 189–199 (1988).

[3] Noyan I. C. and Cohen J. B., *Residual Stress: Measurement by Diffraction and Interpretation*, 1987, Springer-Verlag, New York.
[4] Hauk V., *Structural and Residual Stress Analysis by Nondestructive Methods*, 1997, Elsevier, Amsterdam.
[5] Press W. H., Teukolsky S. A., Vetterling W. T. and Flannery B. P., *Numerical Recipes in C: The Art of Scientific Computing*, 1992, Cambridge University Press, New York.
[6] Noyan I. C., *Metall. Trans. A*, **14A**, 1907–1914 (1983).
[7] Dölle H. and Hauk V., *Härterei-Tech. Mitt.*, **31**, 165–169 (1976).
[8] Dölle H., *J. Appl. Crystallogr.*, **12**, 489–501 (1980).
[9] Noyan I. C., *Adv. X-ray Anal.*, **28**, 281–288 (1985).
[10] Warren B. E., *X-ray Diffraction*, 1990, Dover Publications Inc., New York.
[11] Shi N., Arsenault R. J., Krawitz A. D. and Smith L. F., *Metall. Trans. A*, **24A**, 187–196 (1993).
[12] Papazian J., *Metall. Trans. A*, **19A**, 2945–2953 (1988).
[13] Dutta I. and Bourell D. L., *Mater. Sci. Eng. A*, **112**, 67–77 (1989).
[14] Krawitz A. D. and Winholtz R. A., *Mater. Sci. Eng. A*, **185**, 123–130 (1994).
[15] Winholtz R. A. and Krawitz A. D., *Mater. Sci. Eng. A*, **205**, 257–258 (1996).
[16] Winholtz R. A. and Krawitz A. D., *Mater. Sci. Eng. A*, **221**, 33–37 (1996).
[17] Bevington P. R., *Data Reduction and Error Analysis for the Physical Sciences*, 1969, McGraw-Hill, New York.
[18] Winholtz R. A. and Cohen J. B., *Mater. Sci. Eng. A*, **154**, 155–163 (1992).
[19] Winholtz R. A. and Krawitz A. D., *Proceedings of the Fifth International Congress on Residual Stresses*, 1997, Linköping, Sweden, 598–603.
[20] Winholtz R. A. and Cohen J. B., *Adv. X-ray Anal.*, **32**, 341–353 (1989).
[21] P. Karasudhi, *Foundations of Solid Mechanics*, 1991, Kluwer Academic Publishers, Dordrecht.
[22] Witte D. A., Winholtz R. A. and Neal S. P., *Adv. X-ray Anal.*, **37**, 265–278 (1994).
[23] Winholtz R. A., *J. Appl. Crystallogr.*, **28**, 590–593 (1995).
[24] Wenk H. R., in *Preferred Orientation in Deformed Metals and Rocks: An Introduction to Modern Texture Analysis*, H. R. Wenk (ed.) 1985, Academic Press, Orlando, FL, pp. 11–47.
[25] Winholtz R. A. and Krawitz A. D., *Metall. Trans. A*, **26A**, 1287–1295 (1995).
[26] Lohkamp T. A. and Winholtz R. A., *Adv. X-ray Anal.*, **39**, 281–289 (1997).

5   Study of second- and third-order stresses

J. M. Sprauel

5.1   Introduction

It has long been known, from the use of X-rays to analyse polycrystals, that the shape of diffraction peaks greatly depends on the microstructure of the material being studied. The first method which allowed quantification of this information, was proposed by Warren and Averbach [1, 2] and is based on a fine Fourier analysis of the diffraction peaks. This theory allows estimation of the mean size $\langle L \rangle$ of the crystallites in the diffracting volume and the mean square of their micro-strains $\langle \varepsilon^2 \rangle$. In the case of unstressed powders, the value $\langle L \rangle$ obtained by this method is generally in good agreement with the true particle size as characterized by other experimental techniques (electron microscopy, for example). For polycrystalline aggregates, however, such a correlation is often difficult to establish. In the case of plastically deformed metals, for example, the crystallite size estimated by X-ray or neutron diffraction is several orders of magnitude smaller than the dimension of the dislocation cells generated by the cold working. This is linked to a great sensitivity of the diffraction techniques to any mobile crystal defect. Such defects are eliminated when cutting slides and cannot be observed by electron microscopy. Nevertheless, mobile defects play an important role in the damage of the material. The X-ray and neutron diffraction techniques are thus complementary to electron microscopy. However, Warren and Averbach's theory only applies to powders, since it does not account for type II internal stresses that may exist in polycrystalline aggregates. A physical and micro-mechanical model of the material has therefore to be developed to improve the method. This will be the main aim of this chapter.

5.2   Physical and micro-mechanical model of the material

The definition of the three stress scales has been made in Chapter 3; [3, 4], Figure 3.1:

- stresses of type I or macro-stresses: across a large area (several grains) of a material;
- stresses of type II across a grain or parts of a grain;
- stresses of type II across several atomic distances within a grain.

Type II and III stresses collectively are also called micro-stresses. These definitions are, however, not necessarily useful in analysis of diffraction peaks. A diffraction peak results from the coherent scattering of the incident beam from the periodic structure of a crystal. This means that only periodic domains of the material can be analysed by the diffraction techniques (X-rays or neutrons). However, the crystal structure is altered by any discontinuity in the material (vacancy, interstitial, substitution, dislocation, grain or sub-grain boundary, phase boundary, hole, crack, etc.). An accurate model of a true heterogeneous material needs therefore to define a small reference volume for which both the theories of diffraction and the concepts of continuous medium mechanics can be applied.

To be considered as continuous, each elementary volume should contain at least a few hundred atoms. The equations of compatibility should also be valid within it. This requires that the discontinuities (crystal defects, precipitates, holes, cracks) contained in each volume are small enough to be mechanically negligible. Depending on its microstructure, the material can be divided thus into a great number of such elements which will be called *elementary volumes for mechanics* (EVM). Micro-mechanical models can be used then to predict the macroscopic behaviour of the aggregate from the local characteristics of each elementary volume.

An EVM can contain small precipitates, but it is always composed of one major phase. If this phase is crystalline, the EVM will be called a *crystallite*. Each crystallite is characterized by its position P in the sample and by a given orientation Ω defined by three Euler angles (in relation to a coordinate system fixed to the specimen). It also has specific mechanical characteristics such as strain, stress state and elastic constants.

The crystallite is the base element used to make the connection between diffraction and mechanics. In the case of a recovered metal, the crystallite will generally correspond to a sub-grain of the material (Figure 5.1). If the metal is plastically deformed, dislocation cells will be formed usually inside the grains or sub-grains. These domains are separated by dislocation walls which induce a strong misorientation between neighbouring cells. A grain or sub-grain cannot thus be considered as a continuous medium. In this case, the crystallite will finally fit to a dislocation cell.

A crystallite may contain minor discontinuities (crystal defects, small precipitates, etc.). These discontinuities will not participate in the diffraction phenomenon, but will lead to a diffuse scattering of X-rays or neutrons. For this reason, only a part of each crystallite, called the *coherently diffracting domain*, will contribute to diffraction. The evaluation of macro- and micro-strains by X-ray or neutron diffraction assumes, however, that the mean lattice spacing of the coherently diffracting domain corresponds to the mechanical state of the crystallite in which it is included.

A crystallite behaves like a single crystal. It can thus be described by a stacking of lattice planes defined by three integers h, k, and l (the Miller indices). If the crystallite and the corresponding coherently diffracting domain are correctly oriented to obey Laue's conditions [5],

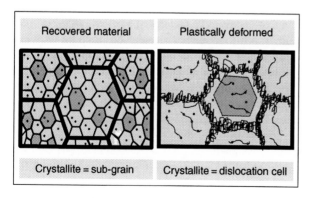

Figure 5.1 Definition of a crystallite.

then the scattering of the X-rays or the neutrons will lead to a constructive interference which builds a diffraction spot. The mean angular position $\theta_{hkl}(P, \Omega)$ of this diffraction spot is linked to the lattice spacing $d_{hkl}(P, \Omega)$ by Bragg's law:

$$\lambda = 2d_{hkl}(P, \Omega)\sin(\theta_{hkl}(P, \Omega)) \tag{1}$$

where λ is the wavelength of the incident beam.

The length $d_{hkl}(P, \Omega)$ is related to the lattice strain $\varepsilon_{hkl}(P, \Omega)$ in the direction normal to the diffracting planes and to the spacing $d_{0\,hkl}(P, \Omega)$ corresponding to the stress-free condition:

$$d_{hkl}(P, \Omega) = d_{0\,hkl}(P, \Omega)[1 + \varepsilon_{hkl}(P, \Omega)] \tag{2}$$

The spacing $d_{0\,hkl}(P, \Omega)$ depends on the temperature and on the local composition of the coherently diffracting domain. This composition may vary inside the material owing to chemical segregation. Concerning the strain component $\varepsilon_{hkl}(P, \Omega)$, it has to be pointed out that diffraction is not directly sensitive to plastic deformation. In fact, plastic deformation is produced by the movement and/or multiplication of crystal defects (vacancies, dislocations). These defects do not belong to the coherently diffracting domain and therefore do not participate directly in the diffraction. Nevertheless local plastic strains (or other "stress-free strains" like thermal strains or volume changes owing to phase transformations) are generally incompatible and are thus compensated by elastic strains which create local residual stresses. This last effect is that which is actually detected by X-ray or neutron diffraction.

Each coherently diffracting domain leads to an elementary peak which is characterized by three parameters:

- the peak *position*, which depends on the current *temperature*, on the chemical *composition* of the diffracting domain and on its *elastic strain*;
- the peak *intensity*, which is linked to the number of elementary sources contributing to the diffraction phenomenon. It is thus related to the *volume* of the coherently diffracting domain;
- the peak *shape*, which is related to the *size* of the coherently diffracting domain and to its local *micro-strain distribution* (type III strains).

A polycrystalline aggregate generally contains a great number of crystallites that have the same orientation Ω. Moreover, for a given incidence of the primary beam, Laue's conditions will be satisfied for different particular orientations Ω. Generally, however, all the crystallites of a grain do not diffract simultaneously. A diffraction peak results thus from the contribution of a great number of coherently diffracting domains, scattered in the grains of the irradiated volume; [6], Figure 5.2. Hence the grain structure of the material cannot be characterized by X-ray or neutron diffraction. The discontinuous volume formed by all these coherently diffracting domains is called the *diffracting volume*.

The shape of a diffraction peak depends on the convolution of four effects:

- The *instrumental broadening* which depends on the geometrical characteristics of the incident and diffracted beams (size of the primary and secondary slits) and on the distribution of wavelength across the incident beam.
- The *compositional heterogeneity* of the coherently diffracting domains which contribute to the diffraction peak. No model has yet been proposed to quantify this effect.

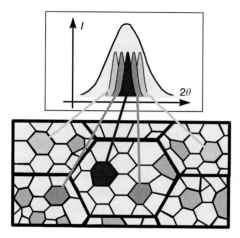

Figure 5.2 Physical generation of a diffraction peak.

- The *elastic strain heterogeneity* of the domains belonging to the diffracting volume. This effect is linked to the heterogeneity of the elastic behaviour of the crystallites (elastic anisotropy) and to local strain incompatibilities (thermal, phase transformation and/or plastic misfits). It can be characterized by the mean square of the type II micro-strains $\langle(\varepsilon^{II})^2\rangle$, which can be predicted by micro-mechanical models [7, 8].
- The mean *size* of the coherently diffracting domains $\langle L \rangle$ and the local elastic *micro-strain distribution* (type III strain). This last effect can be characterized by the mean square of the type III elastic micro-strains $\langle(\varepsilon^{III})^2\rangle$.

As detailed in Warren and Averbach's method [1, 2], these different effects can be separated through a precise Fourier analysis of the diffraction peaks recorded for several orders of reflection. This will allow quantification of the three parameters $\langle L \rangle$, $\langle(\varepsilon^{II})^2\rangle$ and $\langle(\varepsilon^{III})^2\rangle$.

5.3 Peak treatment and removal of instrumental broadening

An accurate analysis of a diffraction peak requires the profile to be plotted as a function of the reciprocal lattice unit $s = 2\sin\theta/\lambda$. The normalized diffracted intensity $i(s)$ can thus be represented by a Fourier integral:

$$i(s) = \int_{l=-\infty}^{l=+\infty} C(l, s_n) \exp(-2\pi i l(s - s_n)) \, dl \tag{3}$$

where s_n is the diffraction peak centroid and l is a distance in the crystal.

The Fourier coefficients $C(l, s_n)$ are a dual representation of the peak shape and may be computed through a Fourier transform of the intensity function $i(s)$:

$$C(l, s_n) = \int_{s=-\infty}^{s=+\infty} i(s) \exp(2\pi i l(s - s_n)) \, ds \tag{4}$$

5.3.1   Determination of the Fourier coefficients of the diffraction peak

The method that has been used classically to determine the Fourier coefficients of a diffraction peak is based on direct processing of the acquired data. After background subtraction and different corrections (like the Lorentz, polarization and absorption corrections), the diffraction profile is smoothed and interpolated, to build the function $i(s)$ with a constant step. The Fourier coefficients $C(l, s_n)$ are finally computed through a fast Fourier transform [9]. This method does not require any assumption about the peak shape. However, owing to experimental restrictions it is often impossible to achieve this computation with sufficient accuracy.

The first restriction is imposed by the statistical fluctuations of the diffracted intensity. If the Fourier coefficients are computed directly from experimental values, they will exhibit spurious oscillations for the higher l values. To remove these oscillations, an optimal filtering can be applied to the raw data [10]. However, since optimal filtering is equivalent to a damping of the highest frequencies, it assigns an upper limit to the number of significant Fourier coefficients.

The second and most fundamental restriction is imposed by the truncation of the experimental profile. According to the Warren–Averbach method, several reflection orders should be used and this often requires working with diffraction peaks at very high diffraction angles. For cold-worked materials, the broadening of these lines is so great that they must be truncated either owing to instrumental limitations or because of their overlap with adjacent peaks. This truncation affects all the Fourier coefficients and leads to an error called the "hook effect". The only way to avoid this error is to model the shape of the diffraction line with a parametric function. This function is then adjusted to the experimental profile by a least-squares optimization.

Among the various proposed shape functions, it is recognized generally that the best candidates are the Pearson-VII, the pseudo-Voigt [11] and the Voigt [12]. Of these three functions, the most satisfactory from a physical viewpoint is the Voigt function.

The Voigt function is defined as the convolution of a Lorentzian and a Gaussian. This shape has not been used often owing to the complexity of its algebraic expression. The expression of its Fourier transform is however rather simple:

$$C(l, s_n) = \exp\left(-2\beta_L|l| - \pi\beta_G^2 l^2\right) \tag{5}$$

where β_L and β_G are the integral breadths of the Lorentz and Gauss components, respectively.

The integral breadth is the mean width of the diffraction peak. It is the ratio of the peak-area and peak-height.

5.3.2   Correction for instrumental broadening

The method most commonly used to remove the instrumental broadening is the Stokes correction [13]. In this procedure, a sample without any defects (generally a recovered powder) is analysed in the same experimental condition as the studied specimen. This sample is not necessarily of the same material as the studied specimen, but its diffraction pattern should show a peak in the same angular range as the data to be corrected. This reference peak is assumed then to be representative of all the instrumental aberrations: wavelength distribution, divergence of the incident and diffracted beam, size of the primary and secondary slits etc. [2, 5].

The normalized peak $i(s)$ obtained on the studied specimen is the convolution of the reference data $i_R(s)$ recorded on the calibration sample and of the corrected profile $i_D(s)$ to be computed:

$$i(s) = i_R(s) * i_D(s) \qquad (6)$$

After Fourier transform, this convolution product becomes a simple product of the corresponding complex Fourier coefficients $C_R(l, s_R)$ and $C_D(l, s_n)$:

$$C(l, s_n) = C_R(l, s_R)C_D(l, s_n) \qquad (7)$$

with

$$C_R(l, s_R) = \int_{s=-\infty}^{s=+\infty} i_R(s) \exp(2\pi i l(s - s_R)) \, ds$$

where s_R is the centroid of $i_R(s)$.

The corrected profile $i_D(s)$ is thus computed through an inverse Fourier transform:

$$i_D(s) = \int_{l=-\infty}^{l=+\infty} C_D(l, s_n) \exp(-2\pi i l(s - s_n)) \, dl \quad \text{with} \quad C_D(l, s_n) = C(l, s_n)/C_R(l, s_R)$$

$$(8)$$

If the diffraction peak-shapes $i(s)$ and $i_R(s)$ are both approximated by a Voigt function, then the profile $i_D(s)$ becomes itself Voigtian. The integral breadths (β_{DL} and β_{DG}) of the Lorentz and Gauss components of the corrected profile are thus easily defined:

$$\beta_{DL} = \beta_L - \beta_{RL} \quad \text{and} \quad \beta_{DG}^2 = \beta_G^2 - \beta_{RG}^2 \qquad (9)$$

where β_{RL} and β_{RG} are, respectively, the integral breadths of the Lorentz and Gauss components of the reference peak $i_R(s)$.

5.4 Effect of the elastic strain heterogeneities of the crystallites

5.4.1 Theoretical approach

As has already been pointed out, the lattice spacing $d_{hkl}(P, \Omega)$ of a given crystallite depends on the temperature, on the local chemical composition and on the elastic lattice strain component $\varepsilon_{hkl}(P, \Omega)$. Since no simple way exists actually to characterize the chemical segregation inside a grain of the material, this effect is always assumed to be negligible. Concerning the temperature, two different cases have to be considered:

For cubic materials, the thermal expansion coefficient is independent of the crystallographic direction. For that reason, the temperature will lead to the same change of lattice spacing for all the crystallites of the diffracting volume. This will result in a global shift of the diffraction peak, but will not induce any broadening of this profile.

For non-cubic materials, the thermal expansion coefficient is anisotropic. Nevertheless, even in this case, the lattice spacing $d_{0\,hkl}(P, \Omega)$ corresponding to the stress-free condition remains constant in the diffracting volume. This is due to the fact that diffraction only analyses a fixed direction of the crystal: that is, the direction normal to the diffracting planes. Temperature will thus lead to a global shift of the diffraction peaks. However, the thermal expansion of the crystallites depends on the orientation and varies inside the material. This results in thermal strain misfits that create local residual stresses. This effect will of course produce a broadening of the diffraction peak, but this phenomenon is already accounted for by the elastic lattice strain component $\varepsilon_{hkl}(P, \Omega)$.

The heterogeneity of the crystallites belonging to the diffraction volume may finally be characterized completely by the distribution of the lattice strain component $\varepsilon_{hkl}(P, \Omega)$. This strain component $\varepsilon_{hkl}(P, \Omega)$ not only depends on the macroscopic stresses within the diffracting volume $\hat{\sigma}_{ij}$, but is also affected by the incompatibility between the stress-free strain $\varepsilon_{f_{ij}}(P, \Omega)$ (thermal strains, plastic strains and volume changes owing to phase transformations) of the crystallite and the stress-free strain $\hat{\varepsilon}_{f_{ij}}$ of the diffracting volume:

$$\varepsilon_{hkl}(P, \Omega) = p_{ij}(P, \Omega)\hat{\sigma}_{ij} + q_{ij}(P, \Omega)[\varepsilon_{f_{ij}}(P, \Omega) - \hat{\varepsilon}_{f_{ij}}] \tag{10}$$

The polarization matrix coefficients $p_{ij}(P, \Omega)$ depend on the single crystal compliances of the crystallite, on its orientation and shape, and on the macroscopic elastic constants of the material. These factors can be calculated through micro-mechanical models of the polycrystalline aggregate [8, 14–19]. Owing to the anisotropy of the crystallite's elastic constants, the coefficients $p_{ij}(P, \Omega)$ may vary inside the diffracting volume. This *elastic heterogeneity* will lead to a non-uniform distribution of the strains $\varepsilon_{hkl}(P, \Omega)$ and will thus result in a broadening of the diffraction peak that is proportional to the level of the macro-stresses $\hat{\sigma}_{ij}$.

The same kind of micro-mechanical approach may also be used to define the elastic–plastic accommodation factors $q_{ij}(P, \Omega)$. In this case, however, a knowledge of the whole history of the material is required to predict the *local strain incompatibilities* which are not directly related to the macro-stresses.

As has already been pointed out, not all the crystallites within the irradiated volume will contribute to diffraction, but only particular orientations Ω_d which satisfy Laue's conditions will be taken into account. To define these orientations Ω_d more precisely, let us consider a coordinate system $(\vec{e}_1, \vec{e}_2, \vec{e}_3)$ linked to the studied specimen (where $\vec{e}_3$ is usually selected as the normal to the sample surface). A given measurement direction $\vec{l}_3$ (bisector between the primary and secondary beam) may be characterized by two angles ϕ and ψ which define a coordinate system $(\vec{l}_1, \vec{l}_2, \vec{l}_3)$ called the laboratory coordinate system (where $\vec{l}_2$ belongs to the plane tangent to the surface) (Figure 5.3a). The orientation Ω_d can be characterized by a set of three Euler angles (α, β, γ) which links the macroscopic coordinate system $(\vec{e}_1, \vec{e}_2, \vec{e}_3)$ to the crystallographic coordinate system $(\vec{c}_1, \vec{c}_2, \vec{c}_3)$ fixed by the lattice vectors of the crystallite (Figure 5.3b). At the microscopic scale, another auxiliary crystallographic coordinate system $(\vec{a}_1, \vec{a}_2, \vec{a}_3)$ may also be defined so that its direction $\vec{a}_3$ matches the normal to the diffracting lattice planes (hkl). This coordinate system $(\vec{a}_1, \vec{a}_2, \vec{a}_3)$ is related to the main crystallographic coordinate system by two rotations r_1 and r_2 (Figure 5.3c). The angles r_1 and r_2 do not depend on the orientation Ω_d, but are fixed by the choice of the diffracting planes (hkl). Finally, to satisfy Laue's conditions, both the measurement direction $\vec{l}_3$ and the normal to the diffracting planes $\vec{a}_3$ should be identical. The crystallographic coordinate system $(\vec{a}_1, \vec{a}_2, \vec{a}_3)$ is then related to the laboratory coordinate system $(\vec{l}_1, \vec{l}_2, \vec{l}_3)$ by just one angle δ (Figure 5.3d). The diffracting volume will thus correspond to such particular orientations Ω_d for δ varying from 0 to 2π. The relation between the different coordinate system is summarized in Figure 5.4.

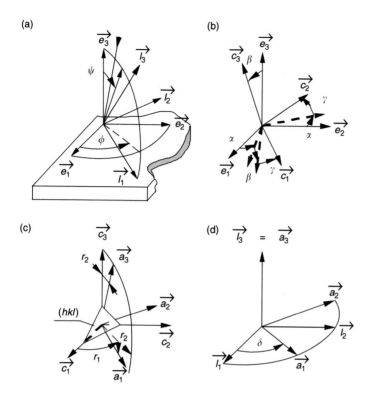

Figure 5.3 Definition of the coordinate systems.

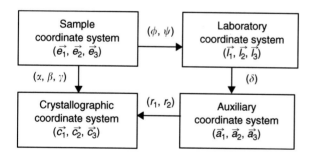

Figure 5.4 Relation between the different coordinate systems.

For every crystallite satisfying Laue's conditions, diffraction will lead to a simple peak which can be characterized by a normalized shape function $i(P, \Omega, s)$. This contribution to the diffracted intensity is called a *type III effect*. The shape of each elementary peak depends on the size of the coherently diffracting domain and on the distribution of the type III micro-strains. Its position $s(P, \Omega)$ (centroid) is directly related to the elastic strain of the crystallite:

$$s(P, \Omega) = s_0[1 - \varepsilon_{hkl}(P, \Omega)] \quad \text{where } s_0 = 1/d_{0hkl}(P, \Omega) \tag{11}$$

The chemical composition has been considered as homogeneous inside the diffracting volume. With this assumption, the reciprocal lattice spacing s_0 becomes constant and the evolution of the peak position $s(P, \Omega)$ only depends on the lattice strain $\varepsilon_{hkl}(P, \Omega)$. This strain varies inside the diffracting volume; thus leading to a broadening of the diffraction peak called a *type II effect*. Its distribution can be characterized, however, by a function $F_{hkl}(\varepsilon)$ which defines the density of crystallites owing to the fixed strain value $\varepsilon_{hkl}(P, \Omega) = \varepsilon$. The type II effect is usually assumed to be independent of the type III effect. With this condition, the global diffraction peak $i_D(s)$ obtained from the studied specimen after applying the instrumental broadening correction, becomes the convolution of two profiles $i^{II}(s)$ and $i^{III}(s)$ which characterize each phenomenon:

$$i_D(s) = i^{II}(s) * i^{III}(s) \tag{12}$$

where $i^{II}(s)$ characterizes the effect of the type II elastic strains $\varepsilon_{hkl}(P, \Omega)$ and $i^{III}(s)$ characterizes the crystallite's size and the type III micro-strains.

After Fourier transform, this convolution product becomes a simple product of complex Fourier coefficients:

$$C_D(l, s_n) = C^{II}(l, s_n) C^{III}(l, s_n) \tag{13}$$

where $C^{II}(l, s_n)$ is the Fourier transform of the type II diffraction profile $i^{II}(s)$ and $C^{III}(l, s_n)$ is the Fourier transform of the type III diffraction profile $i^{III}(s)$.

The type II diffraction profile $i^{II}(s)$ is directly related to the strain distribution function $F_{hkl}(\varepsilon)$ by the following equation:

$$i^{II}(s) = \frac{1}{s_0} F_{hkl}(\varepsilon) \quad \text{with } \varepsilon = 1 - \frac{s}{s_0} \tag{14}$$

Its Fourier transform therefore becomes:

$$C^{II}(l, s_n) = \int_{\varepsilon=-\infty}^{\varepsilon=+\infty} F_{hkl}(\varepsilon) \exp(ui(\varepsilon - \varepsilon_n)) \, d\varepsilon \quad \text{with } u = -2\pi l s_n \tag{15}$$

where $\varepsilon_n = 1 - s_n/s_0$ is the mean elastic lattice strain within the diffracting volume.

For small values of the parameter l or for low strain levels, this expression can be approximated by a Gauss function which is fully defined by the mean square of the type II micro-strains, $\langle (\varepsilon^{II})^2 \rangle$:

$$C^{II}(l, s_n) = \exp(-2\pi^2 l^2 s_n^2 \langle (\varepsilon^{II})^2 \rangle) \quad \text{where } \langle \varepsilon^{II^2} \rangle = \int_{\varepsilon=-\infty}^{\varepsilon=+\infty} F_{hkl}(\varepsilon)(\varepsilon - \varepsilon_n)^2 \, d\varepsilon \tag{16}$$

This simplification is also mathematically exact when the strain distribution $F_{hkl}(\varepsilon)$ is Gaussian.

5.4.2   *Effect of the elastic heterogeneities*

As has already been pointed out (equation 10), the elastic heterogeneity of the diffracting volume is fully characterized by the polarization coefficients $P_{ij}(P, \Omega)$, which depend on the single crystal compliances of the crystallites, on their orientation and shape, and on the macroscopic elastic constants of the material. These factors can be calculated through micro-mechanical models of the polycrystalline aggregate [8].

The self-consistent model [17–19] has been used to define the strain distribution function $F_{hkl}(\varepsilon)$ induced by pure elastic loading of cubic isotropic materials [8]. These calculations show that the shape of this function $F_{hkl}(\varepsilon)$ depends on the choice of the diffracting planes (hkl). In the case of $(h00)$ and (hhh) reflections, the type II micro-strains $\varepsilon_{hkl}(P, \Omega)$ are uniform inside the diffracting volume. The distribution $F_{hkl}(\varepsilon)$ therefore becomes a Dirac function. It means that for these particular lattice planes, no broadening of the diffraction peaks are induced by the macro-stresses. For any other reflection, unusual shapes are obtained which show the greatest density of crystallites at the most opposed strain states (Figure 5.5). The strain distribution $F_{hkl}(\varepsilon)$ is thus very far from a Gaussian. However, since the elastic heterogeneity induced by the macro-stresses remains always low, the approximation of equation (16) is still valid. This heterogeneity can therefore be characterized by the mean square of the type II micro-strains, $\langle(\varepsilon^{II})^2\rangle$. The width of the distribution $F_{hkl}(\varepsilon)$ not only depends on the single crystal's anisotropy, but also varies up to a factor of five with the measurement direction (ϕ, φ).

The diffraction peak broadening induced by the elastic heterogeneity remains low in comparison with the instrumental effects. For this reason, the whole shape of the distribution $F_{hkl}(\varepsilon)$ cannot be characterized experimentally. The theoretical model shows, however, that the width of this distribution is proportional to the stress components. In the case of pure elastic loading, this leads to a quasi-linear dependency of the diffraction peak broadening versus the applied stress. Such behaviour is clearly illustrated by the theoretical results obtained in the case of the elastic torsion loading of α iron. Figure 5.6 shows the simulation of measurements carried out on the {220} reflection with a wave length of 0.198 nm. The plotted peak widths correspond to raw data including the instrumental broadening.

To validate these calculations, different materials were elastically deformed by torsion up to a stress level of 75% of the yield strength [8]. In order to improve the reliability of the X-ray measurements, all these experiments were carried out using synchrotron radiation (LURE, Orsay, France). In fact, the high intensity of such X-ray sources allows us to greatly reduce the statistical fluctuations of the diffraction peaks. Moreover, the small divergence of the incident beam and the precise wavelength selection lead to very low instrumental broadening. The effects induced by the applied load are thus better detected. The results obtained for a recovered alloy steel (0.35%C, 1.0%Cr) are plotted in Figure 5.7, as an example. This graph shows a reversible increase of the diffraction peak width versus the applied stress. It

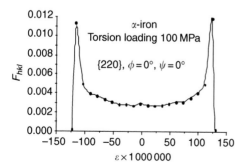

Figure 5.5 Strain distribution function calculated in the case of the elastic torsion loading of α iron.

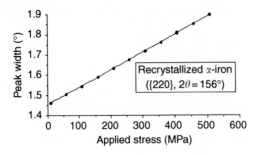

Figure 5.6 Peak broadening induced by the elastic torsion loading of iron (theoretical prediction).

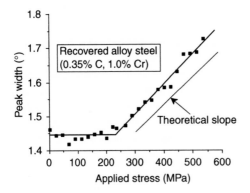

Figure 5.7 Diffraction peak broadening induced by the elastic torsion loading (simulated data).

consists of two straight lines, which are probably related to two different deformation modes. The slope of the second line is in good agreement with the value derived from the self-consistent micro-mechanical approach. This means that this domain corresponds to a pure elastic accommodation of the crystallites. The good agreement between the measured and the predicted results demonstrates the validity of the theoretical model. However, in the first stage of deformation, the slope determined experimentally is much smaller than the calculated one. This might be due to reversible micro-rotations of the crystallites or micro-slips, which are linked to the movements of mobile crystal defects. These micro-displacements lead to a lowering of the local strain heterogeneities. The transition between the two deformation modes would be related then to the locking of the mobile crystal defects when the stress level exceeds a fixed threshold which depends on the microstructure of the sample. For all our experiments, a good correlation has been established between this threshold and the fatigue limit of the studied material.

5.4.3   *Effect of the stress-free-strain incompatibilities*

Contrary to the previous case, the elastic deformations induced by local stress-free-strain incompatibilities are not related directly to the external loads, but depend on the whole history of the studied specimen. Therefore, it will not be easy, in general, to predict the diffraction peak broadening caused by these strains. In fact, the manufacturing of an industrial component consists usually of a great number of complex operations which are difficult to model. However, when the whole history of the material is well known, micro-mechanical models can be used to describe these different operations. For that purpose, the material is divided usually into a great number of elementary crystallites. The orientations and shapes of these crystallites are selected so that they represent statistically the true studied sample. Such an aggregate is called a "pseudo-material". It allows us to define the accommodation of the crystallites during the plastic deformation and thus to predict the type II residual stresses (or strains) induced by any external loading. Such a model has already been developed to predict the texture induced by different manufacturing processes [20, 21].

The same kind of method can also be used to define the distribution of the elastic strains inside the diffracting volume. To model the broadening of the diffraction peak, it is however necessary to compute this distribution very precisely. This requires description of the diffracting volume with a sufficient number of elementary crystallites. Nevertheless, with classical measurement conditions (divergence of the incident beam less than $2°$), only a small portion of the total crystallites (less than 3%) will contribute to diffraction. A pseudo-material of more than 10 000 elementary orientations is therefore required to obtain accurate results.

Such calculations have been performed to predict the diffraction peak broadening induced by elastic-plastic tension loading of an austenitic stainless steel (316L grade). Figure 5.8 shows an example of the strain distribution function $F_{hkl}(\varepsilon)$ obtained for a plastic deformation of about 3%. Contrary to the case of a pure elastic loading, the shape of this function $F_{hkl}(\varepsilon)$ is now much closer to a Gaussian. All our theoretical results show that this agreement is improved when the plastic deformation increases. The approximation of equation (16) is thus always valid. For that reason, even in the case of large plastic deformations, the elastic-strain heterogeneity of the diffracting volume can be characterized accurately by the mean square of the type II micro-strains, $\langle(\varepsilon^{II})^2\rangle$.

To check the contribution of the plastic-strain incompatibilities to the diffraction peak broadening, X-ray measurements were also carried out during a step-by-step tension loading of annealed 316L stainless steel specimens. At each step of the uniaxial tension test, the (311)

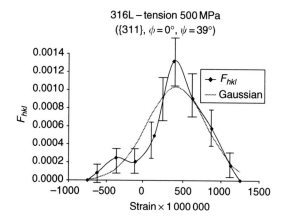

Figure 5.8 Strain distribution function calculated in the case of an elastic–plastic tension loading.

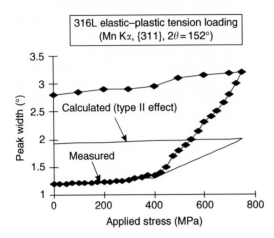

Figure 5.9 Diffraction peak broadening induced by the tension loading of a stainless steel.

diffraction peak of the austenite was recorded using a mobile goniometer (SET-X diffractome-ter manufactured by ELPHYSE under ENSAM license) with Mn Kα X-ray radiation. These diffraction profiles were then fitted by a Voigt function to define their width. The results obtained for this material (Figure 5.9) show that more than 40% of the peak broadening induced by the plastic deformation arises from the type II elastic strain heterogeneities.

5.5   Effect of the size of the coherently diffracting domains on the distribution of the type III micro-strains

5.5.1   *Theoretical approach*

As described earlier, this effect is defined by the elementary peak obtained for each diffract-ing crystallite and can be characterized by a mean normalized profile $i^{III}(s)$. It has been demonstrated by Warren and Averbach [1, 2] that the shape of this elementary peak depends on the size of the coherently diffracting domains and on the distribution of the local elastic micro-strains (type III strains).

The Fourier coefficients of the profile $i^{III}(s)$, $C^{III}(l, s_n)$, can thus be written as the product of two factors:

$$C^{III}(l, s_n) = A^{III}(l, s_n) + iB^{III}(l, s_n) = A^{\langle L \rangle}(l)C^\varepsilon(l, s_n) \qquad (17)$$

The real function $A^{\langle L \rangle}(l)$ characterizes the effect of the size of the coherently diffract-ing domains. It is independent of the reflection order. The complex coefficients $C^\varepsilon(l, s_n)$ characterize the effect of the type III micro-strains.

To define these two types of Fourier coefficients, each coherently diffracting domain is described as a set of orthorhombic cells which form diffracting columns normal to the reflect-ing lattice planes. Each diffracting column is characterized by its length L. The size effect is defined then by the column-length distribution function $p(L)$ and by the mean length $\langle L \rangle$, called the mean size of the coherently diffracting domains. The distribution function $p(L)$ is always positive. Since columns of zero length do not exist it should tend to zero for $L = 0$. The parameter $\langle L \rangle$ represents the mean distance, in the scattering direction, between the discontinuities of the crystallites. The function $A^{\langle L \rangle}(l)$ is proportional to the distribution of pairs of elementary cells belonging to the same column and separated by the fixed distance l. Its expression has been derived by Bertaut [22]:

$$A^{\langle L \rangle}(l) = \frac{1}{\langle L \rangle} \int_{x=l}^{x=+\infty} (x-l)p(x)\,dx \tag{18}$$

$$\frac{dA^{\langle L \rangle}}{dl}(l) = -\frac{1}{\langle L \rangle} \int_{x=l}^{x=+\infty} p(x)\,dx \quad \text{and} \quad \frac{d^2 A^{\langle L \rangle}}{dl^2} = \frac{1}{\langle L \rangle} p(l)$$

with

$$A^{\langle L \rangle}(l=0) = 1; \quad \frac{dA^{\langle L \rangle}}{dl}(l=0) = -\frac{1}{\langle L \rangle}; \quad p(l=0) = 0$$

The plot of $A^{\langle L \rangle}(l)$ versus l gives a decreasing curve without any inflection point. The tangent to the origin of this curve intercepts the horizontal axis at the point $l = \langle L \rangle$. This property thus allows definition of the mean size of the coherently diffracting domains $\langle L \rangle$.

The determination of the tangent to a curve is however a difficult and unreliable operation. It is therefore necessary to define the shape of the function $A^{\langle L \rangle}(l)$ for a large range around the origin $l = 0$. An exponential approximation of the size effect $A^{\langle L \rangle}(l)$ can be used for that purpose. Its mathematical expression is derived from a power series expansion of its logarithm:

$$A^{\langle L \rangle}(l) = \exp(-F^{\langle L \rangle}(l)) \tag{19}$$

with

$$F^{\langle L \rangle}(l) = \left(\frac{l}{\langle L \rangle}\right) + \frac{1}{2}\left(\frac{1}{\langle L \rangle}\right)^2 + \cdots$$

The column-length distribution function $p(L)$ corresponding to this approximation is plotted in Figure 5.10. To define the strain effect $C^\varepsilon(l, s_n)$, continuous elements of length l are built

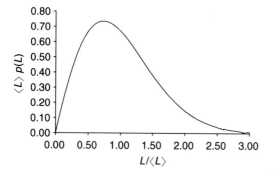

Figure 5.10 Distribution of the size of the diffracting columns.

in each diffracting column. The mean type III lattice strain $\varepsilon_x(l)$ of a given element located at the distance x to the column base is also considered. The Fourier coefficients $C^\varepsilon(l, s_n)$ are finally written as follows [1, 2]:

$$C^\varepsilon(l, s_n) = \langle \exp(2\pi i l s_n \varepsilon_x(l)) \rangle \tag{20}$$

This mean value integrates all the columns of height L greater or equal to the gauge length l. In each column it also accounts for all the distances x between 0 and $L - l$. In most cases, the distribution of the strains $\varepsilon_x(l)$ can be assumed to be Gaussian. The strain effect simplifies then to real coefficients which only depend on the mean square of the type III lattice strain $\varepsilon_x(l)$:

$$C^\varepsilon(l, s_n) = A^\varepsilon(l, s_n) = \exp(-2\pi^2 l^2 s_n^2 \langle \varepsilon_x(l)^2 \rangle) \tag{21}$$

The following relation is finally obtained:

$$\ln(A^{III}(l, s_n)) = \ln(A^{\langle L \rangle}(l)) - 2\pi^2 l^2 s_n^2 \langle \varepsilon_x(l)^2 \rangle = F^{\langle L \rangle}(l) - 2\pi^2 l^2 s_n^2 \langle \varepsilon_x(l)^2 \rangle \tag{22}$$

This equation is the basis of the separation between the size effect $A^{\langle L \rangle}(l)$ and the strain effect $\langle \varepsilon_x(l)^2 \rangle$. In fact, the graph $\ln(A^{III}(l, s_n))$ versus s_n^2 obtained for a fixed gauge length l is a straight line (Figure 5.11). The function $\langle \varepsilon_x(l)^2 \rangle$ is thus easily defined through the slope of this line, while $A^{\langle L \rangle}(l)$ is derived from its intercept. This procedure requires, however, a determination of the Fourier coefficients $A^{III}(l, s_n)$ for at least two different values of the reciprocal lattice spacing s_n, thus analysing several diffraction peaks. Owing to the elastic anisotropy of the crystallites, it will not be possible in practice to select these diffraction peaks in an arbitrary way, but the same lattice planes have to be analysed at successive orders of reflection. In the case of cubic materials, only the two first orders of reflection are available because higher-order peaks always overlap with other diffraction profiles.

The function $\langle \varepsilon_x(l)^2 \rangle$ is always positive. It should also decrease and tend to zero for large gauge lengths because the strains $\varepsilon_x(l)$ correspond to a smoothing of the local lattice displacements over this increasing distance. A hyperbolic approximation of the strain effect $\langle \varepsilon_x(l)^2 \rangle$ is therefore often proposed [23, 24]:

$$\langle \varepsilon_x(l)^2 \rangle = \frac{\langle (\varepsilon^{III})^2 \rangle}{l} \tag{23}$$

5.5.2   *Experimental validation*

The separation of the size and strain effects is based on equation (22). In order to verify this linear relationship, measurements have been carried out on cold-worked tungsten [9]. In fact, the elastic behaviour of this material is isotropic. For this reason, any diffraction peak can be used to check the linear dependency of $A^{III}(l, s_n)$ versus s_n^2. The sample in this case was composed of fine particles filed off a cold rolled plate. The type II elastic heterogeneities are thus very low and can be neglected. The instrumental broadening was characterized through measurements carried out on a recovered tungsten powder. The first 12 reflections of the α phase of this material were recorded using synchrotron radiation (LURE, Orsay, France). As has already been pointed out, such X-ray sources allow great improvement of the reliability of the results. This is due to the high intensity of the synchrotron radiation, to the small divergence of the incident beam and to a precise wavelength selection. The Fourier coefficients of the diffraction peaks were determined by a direct processing of the acquired

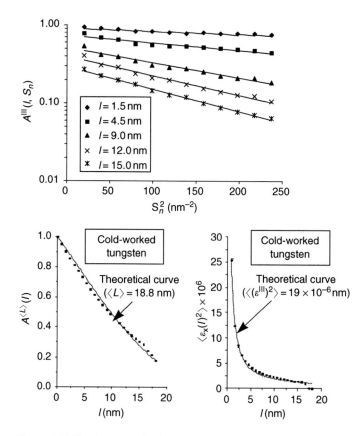

Figure 5.11 Fourier analysis of a cold-worked tungsten specimen.

data (background subtraction, Lorentz polarization–absorption correction, smoothing and interpolation to the reciprocal lattice units, fast Fourier transform).

The results (Figure 5.11) show excellent linearity of the curves $A^{III}(l, s_n)$ versus s_n^2. The Gaussian distribution of the strains $\varepsilon_x(l)$ is thus confirmed. The two components of the diffraction peak broadening, that is, the size effect $A^{\langle L \rangle}(l)$ and the strain effect $\langle \varepsilon_x(l)^2 \rangle$ could therefore be defined very precisely. These curves were fitted then by the corresponding exponential and hyperbolic approximations (of equations (19) and (23)). The experimental data and the theoretical approximations are in very good agreement.

5.6 Global expression of the diffraction peak broadening

5.6.1 *Theoretical approach*

As demonstrated above, the diffraction peak obtained from the specimen after instrumental broadening correction is a convolution of the following effects:

- The elastic strain heterogeneity of the diffracting volume. This effect is linked to the heterogeneity of the elastic behaviour of the crystallites (elastic anisotropy) and to the local strain incompatibilities (thermal, phase transformation and/or plastic mismatches). It can be characterized by the mean square of the type II micro-strains $\langle (\varepsilon^{II})^2 \rangle$.

• The mean size of the coherently diffracting domains $\langle L \rangle$ and the distribution of the local elastic micro-strains (type III strains). This last effect can be characterized by a coefficient $\langle (\varepsilon^{III})^2 \rangle$, which defines the mean square of this micro-strains.

Exponential approximations have been proposed for all these effects (equations 16, 19 and 21).

The Fourier coefficients of the corrected diffraction profile can be written finally as follows:

$$C_D(l, s_n) = \exp(-2\beta_{DL}|l| - \pi\beta_{DG}^2 l^2) \tag{24}$$

with

$$\beta_{DL} = \frac{1}{2\langle L \rangle} + \pi^2 s_n^2 \langle (\varepsilon^{III})^2 \rangle$$

and

$$\beta_{DG} = \frac{1}{\sqrt{2\pi}\langle L \rangle} + s_n \sqrt{2\pi \langle (\varepsilon^{II})^2 \rangle}$$

This expression demonstrates clearly that the diffraction peak can be approximated by a Voigt function. This property has already been found in the very pragmatic and useful work of Langford [25]. In this work it is assumed that the Gauss component of the Voigt profile is linked to micro-strains. The Lorentz component is used to define the crystallite size.

In our model, the integral breadths of the Gauss and Lorentz components of the diffraction profile both depend on the size of the coherently diffracting domains and on the mean square of the second and third kind micro-strains. These peak widths depend also on the reciprocal lattice spacing s_n. As has already been pointed out, this property allows separation of the effects of size and strain. For that purpose, the same diffracting planes have to be analysed at successive orders of reflection.

5.6.2   Experimental application

The whole Fourier analysis method has been applied to the characterization of the diffraction peak broadening induced by the plastic deformation of a 316L austenitic stainless steel. For that purpose, the (220) and (440) reflections of the austenitic phase were recorded using synchrotron radiation (LURE, Orsay, France). The instrumental broadening has been defined through measurements carried out on different annealed powders (copper, nickel and cerium oxide). All the acquired peaks were fitted by a Voigt profile. The mean size of the coherently diffracting domains $\langle L \rangle$, the mean square of the type II micro-strains $\langle (\varepsilon^{II})^2 \rangle$, and the type III micro-strains coefficient $\langle (\varepsilon^{III})^2 \rangle$ have been computed finally through our improved procedure (equations 9 and 24). The error bars on these parameters have been estimated through an inverse Monte Carlo simulation method.

Figure 5.12 presents the evolution of these different parameters with the plastic deformation induced by uniaxial stretching of the material. On these graphs, the experimental results have been smoothed as guides to the eye.

The sample has been analysed by transmission electron microscopy to obtain a direct observation of the microstructure. The 316L stainless steel used in our study is very stable.

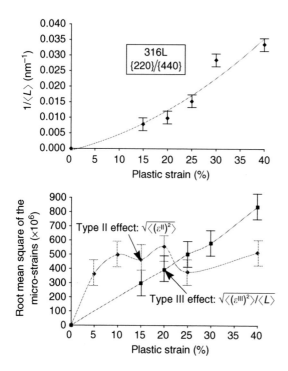

Figure 5.12 Results of the Fourier analysis of a plastically stretched 316L stainless steel.

For this reason, no phase transformation is induced by the plastic deformation. For low plastic deformations, the dislocations remain coplanar, but over 10% elongation, these defects reorganize into cells. This leads to a saturation of the type II internal stresses and micro-strains which is clearly observed in the X-ray results. The size of the coherently diffracting domains decreases with increasing plastic deformation and comes to about 25 nm for 40% of elongation. This is one order of magnitude smaller than the dimension of the dislocation cells generated by the cold working. This difference can be explained by the existence of mobile dislocations inside the cells. These defects are eliminated when cutting slides and cannot be observed by electron microscopy. Nevertheless, mobile dislocations play an important role in the fatigue and corrosion cracking behaviour the material. The X-ray diffraction technique, which is sensitive to any crystal defect, is thus complementary to electron microscopy.

5.7   Conclusion

In this chapter, a physical and micro-mechanical approach has been used to define the main factors which influence the width of a diffraction peak. It has been demonstrated, thus, that the shape of a diffraction profile is not only linked to the instrumental broadening, to the size of the coherently diffracting domains and to their local lattice strains (as is usually assumed), but is also affected by the elastic-strain heterogeneities of the crystallites belonging to the diffracting volume. Theoretical models have been developed to account for all these effects. The different theories and their corresponding assumptions have also been verified by experiments carried out using synchrotron radiation.

References

[1] Warren B. E., Averbach, B. L., *Journal of Applied Physics*, **21**, 595 (1950).
[2] Warren B. E., *X-ray Diffraction*, (1969) Addison Wesley, London.
[3] Macherauch E., Müller P., *Arch. Eisenhüttenwesen*, **29**, 257 (1958).
[4] Maeder G., Lebrun J. L., Sprauel J. M., *Non-Destructive Testing International*, **10**, 235 (1981).
[5] Klug H. P., Alexander L. E., *X-ray Diffraction Procedures*, (1974) Wiley, London.
[6] Sprauel J. M., Castex L., *Materials Science Forum*, **79–82**, 143 (1991).
[7] Mabelly P., Hadmar P., Desvignes M., Sprauel J. M., Role of the internal stresses on the diffraction peak's broadening, *ECRS4*, ENSAM Cluny, (1996).
[8] Sprauel J. M., Castex L., *Analyse des contraintes résiduelles par diffraction des rayons et des neutrons, Etude des contraintes d'ordre 2 et 3, influence sur la forme des pics de diffraction*, Ed. A. Lodini et M. Perrin, CEA (1996), p. 45.
[9] Maeder G., Sprauel J. M., Lebrun J. L., Line profile analysis with position sensitive detector, *5th Risø International Symposium*, (1984) Roskilde.
[10] Bourniquel B., Feron J., *Journal of Applied Crystallography*, **18**, 248 (1985).
[11] Ji N., Lebrun J. L., Sprauel J. M., *Materials Science and Engineering A*, **127**, 71 (1990).
[12] Bourniquel B., Sprauel J. M., Feron J., Lebrun J. L., *ICRS2* (1989) Elsevier Applied Science, Amsterdam, p. 184.
[13] Stokes A. R., *Proceedings of Physics Society London*, **61**, 382 (1948).
[14] Voigt W., *Lehrbuch für Kristallphysik*, Teubner, Berlin, 962–967 (1928).
[15] Reuss A., *Z. Angew. Math. Mech.*, **9**, 49 (1929).
[16] Eshelby J. D., *Proceedings of the Royal Society*, **A241**, 376 (1957).
[17] Kröner E., *Zeitschrift für Physik*, **151**, 504 (1958).
[18] Kneer G., *Zur Elastizität vielkristalliner Aggregate mit und ohne Textur*, 1964, Dissertation, Clausthal, Germany.
[19] Morris P. R., *International Journal for Engineering Science*, **8**, 49 (1971).
[20] Lipinski P., Berveiller M., *International Journal of Plasticity*, **5**, 149 (1989).
[21] Lipinski P., Naddari A., Berveiller M., *International Journal of Solids*, **29**, 1873 (1992).
[22] Bertaut F., *C.R. Académie des Sciences*, **228**, 187 (1949).
[23] Rothman, R. L., Cohen J. B., *Advances in X-ray Analysis*, **12**, 208 (1969).
[24] Mignot J., Rondot D., *Acta Crystallographica*, **A33**, 327 (1977).
[25] Langford J. I., Wilson A. J. C., *Journal of Applied Crystallography*, **11**, 102 (1978).

6 The effect of texture on residual stress measurement and interpretation

T. M. Holden

The problem areas associated with the measurement of, and the interpretation of, residual elastic strains in textured material are reviewed. Practical methods of determining the most significant strain tensors for the materials are discussed in the context of highly textured zirconium alloys. An example is given of a practical approach to determining the strain corresponding to the macroscopic stress field in over-rolled pressure tube material.

6.1 Introduction

Residual stresses are stresses which exist in a component even when it is not under load. Residual stresses occur on three length scales. The first is the stress which has the length scale of the component itself, the type-I or macroscopic stress field. A familiar example of a macroscopic stress field is provided by a linear weld in a plate. There is likely to be a tensile stress, along the direction of the weld, because of the constraint offered by the cooler material far from the melted region to unrestrained cooling of the weld metal. The longitudinal stress near the weld must be balanced by a compressive distribution of longitudinal stress extending out to the boundary of the component. This is the stress of greatest concern to the engineer, since it clearly affects the strength, and susceptibility to cracking, of the part.

Type-II stresses have the length scale of the size of the grains making up the material. The stresses that exist in grains of different crystallographic orientations in single-phase materials are also known as intergranular stresses. Residual stresses also exist in the grains of composite materials, such as ferrite and cementite in steel, or in SiC in an Al metal matrix composite and these are sometimes termed interphase constraints. The origins of the intergranular stresses most often lie in the interplay between the plastic and elastic anisotropy of the grains under a load which exceeds the yield point. The plastic anisotropy comes from the fact that dislocations move through the lattice in preferred directions in certain planes. For example, dislocation motion generates slip in face-centred cubic metals only in $\langle 110 \rangle$ directions in $\{111\}$ planes. Thus, grains with different crystallographic directions aligned with an applied stress will deform plastically by different amounts.

Conceptually, if a small volume element of a component is considered, sufficient to contain many grains of different crystallographic orientations, integration over the grain stresses in any given direction must give the type-I stress. Type-I and type-II stresses, and the corresponding strains, are superposed, and a diffraction measurement alone on a component does not distinguish between the two different length scales. If a small coupon is cut from a large sample that exhibits a residual stress, the macroscopic stress will be destroyed in the coupon but the intergranular stresses will be retained and may be measured. This provides a basic method for distinguishing between the two types of stress.

Type-II stresses give rise to shifts and to peak broadening in strain measurements. The elastic anisotropy alone will give rise to different strains in [002] and [111] grains aligned parallel to an applied load. This is a straightforward effect, which can be accounted for by using the appropriate diffraction elastic constants. Intergranular residual strains arising from the interplay of plastic and elastic anisotropy will give further differences, so that the strain measured in a polycrystalline sample with the (111) reflection will generally be different from the strain measured with the (002) reflection. Perpendicular to a particular plane normal [hkl], there is a distribution of crystal orientations. Since the response of each of these crystal orientations is different because of plastic and elastic anisotropy, the resulting distribution of lattice spacings will also give an intrinsic width to the diffraction peak known as strain broadening.

The third type of stress is that surrounding the core of a defect such as a dislocation in the crystal lattice which is on the scale of thousands of Ångstroms. Type-III stresses contribute to the widths of the diffraction peaks and are not discussed further in this chapter.

If the grains making up the polycrystalline aggregate are oriented randomly in crystallographic direction, there is said to be random texture. A practical example would be a sample of steel made to a near net shape by powder metallurgy. However, industrial processing usually generates a non-random distribution of crystallite orientations in industrial components and most components do exhibit texture. For example, rolled steel plate exhibits a preferential alignment of the grains with $\langle 110 \rangle$ directions in the rolling direction and $\langle 111 \rangle$ directions normal to the rolling plane. A Zircaloy-2 (Zr-2) plate will exhibit preferential alignment with $\langle 10\bar{1}0 \rangle$ grains in the rolling direction. An extruded Zr-2 rod exhibits very strong alignment, primarily with $\langle \bar{1}2\bar{1}0 \rangle$ directions along the rod axis. An austenitic stainless steel weld will show very strong texture with $\langle 100 \rangle$ directions aligned with the direction of heat flow out of the weld.

The origin of a great deal of crystallographic texture is in the rotation of the grains into particular directions dictated by how the grains slip as they undergo plastic deformation. All the forming operations mentioned above involve strong plastic deformation. Both the degree of texture and the magnitude of the type-II strains provide measures of the plastic deformation in the sample. In the cases considered in this chapter, texture was generated by rolling and extrusion processes. Annealing at an elevated temperature removed the associated residual stresses but not the texture. However, the nature of the texture determined the form of the residual strains that resulted when the samples cooled to room temperature.

The purpose of this chapter is to suggest means of analysing strain data from textured samples, especially how to take advantage of the texture. The philosophy adopted is that strong texture defines the ideal orientations. It is then possible to measure the complete strain tensor for these orientations by making use of the one-to-one relationship between crystal orientation and direction in the sample. In a strongly textured sample, the ideal grain orientation strongly influences the mechanical behaviour. The strain tensor associated with this ideal texture is likely to be the most significant strain tensor. From a diffraction standpoint, measuring along sample directions corresponding to the ideal orientation maximizes the diffracted intensity. These simplifications do not appear when the texture is weak.

Good progress in understanding these problems has been made by comparing the diffraction experiments with self-consistent models of polycrystalline aggregates which permit samples of varying texture to be simulated. A number of experimental situations are examined in this chapter where strong texture is present, such as rod-textured and plate-textured Zr-2 and NbZr alloys. There have been a number of advances in the area of understanding the generation of type-II stresses under load, which have a bearing on the approach to highly textured materials taken in this chapter. The physical origins of the effects are emphasized in this chapter and how our understanding of them leads to methods of resolving problem areas. Complex mathematical formulations are avoided. Many difficulties in the interpretation of strains in textured engineering components are really the problems of the superposition of type-I and type-II strains.

In Section 6.2, we review the principles of the measurement of residual strain, making clear the assumptions in the methods. The experimental methods are described in Section 6.3, and Section 6.4 is devoted to a discussion of results on highly textured materials.

6.2   The measurement of strain by diffraction

A clear description of the principles of measurement of an elastic, type-I, strain field was given by Noyan and Cohen [1] for X-ray diffraction and is also covered in the chapters by A. Lodini and R. A. Winholtz in this book.

In the cases discussed here, instead of choosing directions related to the geometrical shape of the sample, axes are chosen to be aligned with crystal axes defined by the strong texture. These are, of course, related by a simple rotation of axes. For this reason, the general expression for strain in terms of the elements of a strain tensor forms the starting point of the analysis,

$$\frac{\Delta d}{d} = \sum_{i,j} l_i l_j \varepsilon_{ij} \tag{1}$$

Here l_i are direction cosines of the vector along which the strain measurement is made with respect to the chosen axes. Because of the high penetration of the neutron beam all elements of the strain tensor may be found.

For an applied stress which is below the yield point of an initially strain-free sample the assumptions for obtaining the macroscopic stress field are correct because only the elastic response matters. The strains measured with different reflections (*hkl*) give the identical value of the macroscopic stress when used with the appropriate plane-specific elastic constants. However, this does not correspond to the real situation for most residual stresses encountered where inhomogeneous plastic deformation has occurred and stresses have exceeded the yield point. Sets of grains in the sample, with plane normals [*hkl*] making different tilt angles to the original 'applied' stress imposed during manufacture, will have deformed plastically and elastically by different amounts under load. After the load is relaxed, the strains in these grains carry information about the original stress field but also about the individual strain response at the particular angle of tilt. The crux of the matter is that every grain orientation is characterized by its own stress tensor and the task in hand is to distinguish which is the common part associated with the stress field and which is the part corresponding to the intergranular response. This complex problem can be unravelled if a statistical model of the grains is available which incorporates the individual elastic and plastic response of each grain but considers the aggregate behaviour of a set of many thousand grains. Then the appropriate averages over spatial direction and crystallographic orientations can be made to simulate diffraction from the aggregate. However, if all the grains have a common orientation an essential simplification occurs, since they will have all deformed in the same way. Then there is only one strain tensor involved rather than a multitude of strain tensors. This simplification can be made use of in the case of highly textured materials.

The subject of stress measurement in textured samples was reviewed by Van Houtte and De Buyser [2] and their article gave a balanced view of the field up to 1994. Several useful expressions were written down for the average values of the strain in the limit of certain assumptions about the distributions of strains in the grains. The Reuss picture assumes that the stresses in all the grains are identical and the Voigt picture assumes that the strains are identical in every grain. Both of these assumptions, while useful, are inaccurate. A self-consistent treatment of strain response to stress in the elastic regime was provided by Kröner [3]. This was based on the concept of considering the elastic behaviour of a single elliptical grain embedded in a matrix that has the average isotropic properties of all the grains. This permits the strain response to stress to be calculated for any crystallographic direction, as well as the macroscopic response, once the single crystal elastic constants are specified. The method leads to calculations of the diffraction elastic constants, the slopes of the linear region of the response for a particular lattice plane to applied stress, that are in satisfactory agreement with experiment for untextured samples.

Van Houtte and De Buyser noted that, when texture was present, there were very often strong departures from linearity in $\sin^2\psi$ plots for different reflections. They showed that, in principle, texture can lead to differences in the diffraction elastic constants, as well as the macroscopic elastic constants, with sample orientation. This was thought to be a major contributor to the non-linearity of $\sin^2\psi$ plots. However, Van Houtte and De Buyser also pointed out that plasticity can also lead to deviations of the $\sin^2\psi$ plots from linearity although there was no way of demonstrating this theoretically at the time. The implementation of the elasto-plastic self-consistent (EPSC) model by Turner and Tomé [4], and later Clausen *et al.* [5], which is built on the formulation of Hill [6] and Hutchinson [7], has been a primary driver in modelling the generation of intergranular stress and strain under the action of plasticity. The self-consistent models also permit the calculation of the diffraction elastic constants for textured materials and are in exact agreement with the Kröner model in the case of zero texture. The finite element approach of Dawson *et al.* [8], which involves modelling individual grains in terms of many sub-grain elements has also been applied to the calculation of residual stresses resulting from elastic and plastic anisotropy. The theoretical methodology is, however, not quite complete since a calculation of both the development of texture and residual strain within a single code has not yet been perfected [9]. At present, given an array of 2000 or more atoms, the residual stress has been calculated for different *(hkl)* reflections for textured samples with no adjustable parameters, once the flow curve has been parameterized, with an accuracy of around 30%. The method does not permit explicit formulae to be written down for the strains and stresses but does lead to an accurate conceptual picture of the behaviour.

6.3   Experimental methods

6.3.1   Constant wavelength diffraction

The cases discussed in this article primarily deal with intergranular effects. Accordingly, the discussion of experimental methods is based on experiments to measure intergranular strains. The basic requirement is a measurement of lattice spacing, d_{hkl}, of planes characterized by Miller indices *(hkl)* with accuracy approaching ±0.0001 Å or better. The experimental arrangement is therefore similar to any high-resolution neutron diffraction experiment. For constant wavelength diffraction, high accuracy is achieved by use of a high monochromator scattering angle, of order 100° for example, and collimation of order 0.3° between the monochromator and sample, and sample and detector. The collimation is necessary because experiments are carried out as a function of the tilt of the cylindrical sample away from the azimuth. Otherwise the instrumental linewidth would vary with tilt. Typically, the linewidth is about $0.4°$ at a sample scattering angle of $50°$ and $1.2°$ at a scattering angle of $100°$.

The relationship between the wavelength of the neutron beam, λ, the true scattering angle $2\theta_{hkl}$, and the lattice spacing, d_{hkl}, is given by Bragg's law,

$$\lambda = 2d_{hkl} \sin \theta_{hkl} \tag{2}$$

The true angle is related to the setting angle of the spectrometer, $2\theta_{hkl}^{s}$, by

$$2\theta_{hkl}^{s} = 2\theta_{hkl}^{0} + 2\theta_{hkl}, \tag{3}$$

where $2\theta_{hkl}^{0}$ is the scale zero of the spectrometer. This has to be determined for each sample since it depends on the precise positioning of each sample over the centre of rotation. This is readily obtained by measuring the setting angles of two orders of reflection from a sample, for example the $(10\bar{1}0)$ and $(20\bar{2}0)$ reflections of hexagonal Zr-2, and solving equations (2) and (3) holding λ fixed. There should be no systematic variation of $2\theta_{hkl}^{0}$ with tilt if the samples are well centred and the beam is uniform in brightness, as discussed below.

The wavelength must be calibrated during the course of each experiment with a standard powder sample, such as Si or Al_2O_3 powder. This is done by measuring perhaps five diffraction lines from the standard over the range of scattering angles encountered during the experiment and solving equations (2) and (3) to find λ. The samples may be of order 1000 mm^3 in volume, so that there is no lack of scattered intensity. The sample is usually fixed securely on a goniometer, such as an Eulerian cradle, to permit any orientation in tilt and azimuth to be selected. For textured samples, this allows selection of angles corresponding to the 'ideal orientations' of the texture.

Centering the sample is important, as mentioned above, and this is achieved by aligning it by means of a sighting telescope. The uniformity of the neutron beam over its cross-section is an important requirement. For tilted samples, a left–right asymmetry in the beam intensity will generate specious systematic shifts with respect to the vertical alignment of the sample. For textured samples, the measurements should be made in a step-scan mode. The counter is set to an angle 2θ, the sample at θ, and the diffracted intensity is collected for a fixed number of incident neutrons as determined by a low efficiency counter in the incident beam. The counter is moved through a small angular increment, the sample is moved by half this amount, and the count repeated until the complete diffraction peak is collected. If this is not done the variation of texture with scattering angle at a fixed sample angle can generate a spurious shift.

6.3.2 Time-of-flight diffraction

In conventional neutron diffraction, the wavelength is fixed and the diffraction angle varies. For time-of-flight diffraction, the diffraction angle is fixed and a distribution of wavelengths falls on the sample. The initial flight from the source to the sample spreads out the neutrons in velocity and hence wavelength. Incident flight paths are typically between 20 and 100 m long; the greater the length, the higher the accuracy, but the lower the intensity. Banks of counters are arranged at constant (1–2 m) scattered flight paths from the sample, held in an evacuated sample chamber, on either side of the incident beam. The detectors record both the neutron count and the arrival time. The elapsed time from the source pulse to the arrival at the detector is used to determine the wavelength precisely. Information about interplanar spacings is obtained by rewriting equation (2) in terms of t, the time taken to traverse the flight path, L,

$$\lambda = \frac{ht}{mL} = 2d \sin \theta \tag{4}$$

where h is Planck's constant and m the neutron mass. The exact value of L and the angular position of the counters are determined by calibration experiments with standard powders. A feature of time-of-flight diffraction is that the instrumental resolution, $\Delta d/d$, is almost independent of d, so that all lattice spacings can be obtained with equal precision. The precise positioning of the sample on the centre of the spectrometer, and the uniformity of the neutron beam across its width, are as important for time-of-flight studies as for constant wavelength diffraction.

6.4   Experimental measurements of highly textured material

6.4.1   Zircaloy-2 rod

The intergranular residual stress measurements of MacEwen *et al.* [10, 11] on Zr-2 rod were the first to make use of rod texture in order to extract the full strain tensor of the sample. The variation of the scattered intensity for a Zr-2 rod as a function of angle, χ, from the rod axis for several reflections is shown [12] in Figure 6.1. The pole figure in Figure 6.1 for the (0002) reflection demonstrates the axisymmetrical nature of the texture. The results show that the texture may be summarized in terms of two ideal grain orientations. The first corresponds to a $[\bar{1}2\bar{1}0]$ crystallographic axis along the rod axis and accounts for about 67% of the grains. The second orientation corresponds to a $[10\bar{1}0]$ crystal orientation along the rod axis and this accounts for the remaining 33% of the grains. Angles which other crystallographic directions make with the rod axis may easily be obtained by imagining rotating the reciprocal lattice of zirconium about the $[\bar{1}2\bar{1}0]$ axis. Perpendicular to the rod axis, all crystallographic plane normals perpendicular to $[\bar{1}2\bar{1}0]$ are present, namely, the $\langle 10\bar{1}n \rangle$ and $\langle 0002 \rangle$ directions. Similarly, for the $[10\bar{1}0]$ ideal orientation, there are $\langle \bar{1}2\bar{1}n \rangle$ and $\langle 0002 \rangle$ directions perpendicular to the rod axis. It is clear that the strains associated with both these orientations may be found by examining $\langle 10\bar{1}n \rangle$ and $\langle \bar{1}2\bar{1}n \rangle$ reflections, respectively, when the rod axis is in a vertical orientation on the spectrometer. Because of the random distribution of plane normals perpendicular to the rod axis, all the corresponding planes may be measured in one azimuthal direction.

MacEwen *et al.* [10] first measured the thermal strains associated with cooling Zr-2 from its annealing temperature of 650°C. The thermal strains are caused by the fact that the coefficients of expansion of zirconium (and by assumption Zr-2) are different along the a- and c-axes, $\alpha_c = 10.3 \times 10^{-6} \, \text{K}^{-1}$ and $\alpha_a = 5.8 \times 10^{-6} \, \text{K}^{-1}$, respectively [13]. Grains with c-axis perpendicular to the rod axis are constrained by grains of other orientations from shrinking as much as they would if they were unconstrained, and then end up in tension. In turn, grains with $[10\bar{1}0]$ axes along the radial direction, associated with the first ideal orientation, are left in a state of residual compression. The principal axes were chosen to be the $[10\bar{1}0]$, the $[\bar{1}2\bar{1}0]$ and the $[0002]$ directions. The first ideal orientation in the Zr-2 rod therefore has ε_{22} aligned along the rod axis and ε_{11} and ε_{33} aligned perpendicular to the rod axis. The EPSC model, applied to a calculation of the thermal stresses in strongly textured

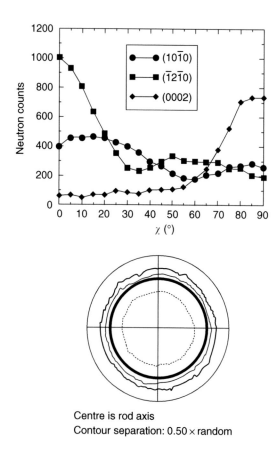

Centre is rod axis
Contour separation: 0.50 × random

Figure 6.1 Neutron intensities for the (10$\bar{1}$0), ($\bar{1}$2$\bar{1}$0) and (0002) reflections as a function of the angle, χ, to the axis of a Zr-2 rod. The pole figure for the (0002) reflection displays the rotational symmetry about the rod axis. The bold line in the pole figure indicates random texture and contours above and below random are represented by solid and dashed lines, respectively.

Zr-2 on cooling from 650°C, predicted that the elements of the stress tensor would be (−49, −7, +66) MPa. The calculation took into account the measured texture of the Zr-2 sample. This result compared well with experiment (−41 ± 7, −3 ± 7, 67 ± 7) MPa. The residual stresses were shown to be completely elastic in origin: there was no plasticity involved in cooling from 650°C. (Initially it was thought that plasticity was involved but the Taylor-like model used in the earlier work [10] was too stiff and was later shown to be incorrect in detail.) It is interesting to note that the strains and stresses along the rod axis are small. This is because, in this direction, the grains have either [10$\bar{1}$0] or [$\bar{1}$2$\bar{1}$0] directions which have the same coefficient of expansion. There is very little constraint offered to the thermal shrinkage of these grains and hence very low residual strain. This is a situation where the strong extrusion texture has considerably influenced the development of the residual strains on cooling.

MacEwen *et al.* [11] showed that the residual thermal strain state is changed strongly by plastically deforming rod textured Zr-2. Pang *et al.* [12] extended this work by studying

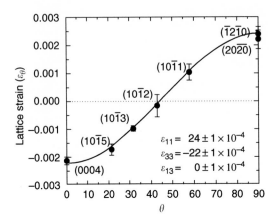

Figure 6.2 Residual elastic lattice strains determined for the $\{10\bar{1}n\}$ and (0002) reflections perpendicular to the rod axis for rod-textured Zr-2. The strains are plotted as a function of the angle, θ, between the $[10\bar{1}n]$ and $[0002]$ axes and are fitted well by a tensor form with elements given on the figure. The measurements were made at an applied stress of 2 MPa at the end of a load cycle during *in situ* measurements on rod-textured Zr-2.

the strain response of Zr-2 rod to applied stress *in situ*, examining both the strain response parallel and perpendicular to the axis of the applied stress which was directed along the rod axis. It was shown that, for an applied stress below the yield point, the variation of the residual strains was completely described by a stress in the material directed along the rod axis and with a magnitude equal, within the experimental uncertainties, to the applied stress. Above the yield point, the plastic deformation induced large changes in the intergranular strains which were followed as a function of applied stress. Figure 6.2 shows the residual strains for the $\{10\bar{1}n\}$ and $\{0002\}$ reflections of Zr-2 rod [12]. Also shown is the least squares fit to a strain tensor form, equation (1), where the principal axes were the $[10\bar{1}0]$, the $[1\bar{2}\bar{1}0]$ and the $[0002]$ directions. Measurements on the second ideal orientation gave the elements of the strain tensor to be the same, to within the experimental uncertainty, as the first orientation. ε_{33} is common to the two ideal orientations.

After the *in situ* experiments, the residual strains were measured at various specific angles to the rod as dictated by the rod texture. For example, the $[11\bar{2}0]$ direction corresponding to the first ideal orientation may be investigated at 60° to the rod axis. The results showed that the strains at various angles to the rod axis were consistent with the ε_{12} and ε_{23} elements of the strain tensor being zero. The residual strains agreed well with the earlier results of MacEwen *et al.* [11] parallel and perpendicular to the rod axis and with the EPSC model calculation.

Finally, it is interesting to consider how stress balance is achieved in this case. From the axisymmetric texture, there is a random distribution of all grains perpendicular to the rod axis and the compressive and tensile stresses balance each other along every radial direction.

6.4.2   Zircaloy-2 plate

Rolled Zr-2 plate has extremely strong texture, as shown by representative $\{10\bar{1}0\}$ and $\{0002\}$ pole figures [13] in Figure 6.3. These indicate high rolling texture with the $\{10\bar{1}0\}$ poles enhanced in the rolling direction five times more than a random distribution. Figure 6.3

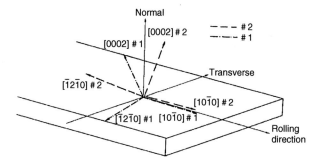

Representation of the two ideal textures in rolled Zircaloy plate

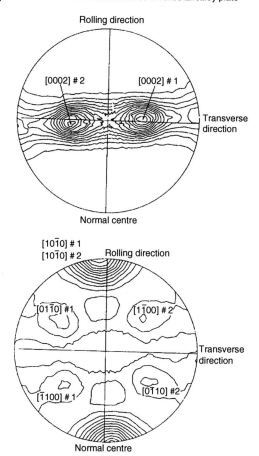

Figure 6.3 A schematic representation of the two ideal orientations of Zr-2 rolled plate with a common [10$\bar{1}$0] crystallographic orientation along the rolling direction and [0002] orientations situated at ±40° to the plate normal in the normal–transverse plane. The experimental (10$\bar{1}$0) and (0002) pole figures for the Zr-2 plate are also shown in which regions of high intensity are labelled with particular crystallographic plane normals of the ideal orientations.

also shows a schematic representation of the ideal orientations in the plate in which two orientations are identified, corresponding to the two maxima in the {0002} pole figures about ±40° from the normal direction, with a common [10$\bar{1}$0] direction along the plate rolling direction. The ideal orientations have a spread of about ±20° and are referred to as orientation 1 and 2. In the {10$\bar{1}$0} pole figure, for example, every region of high intensity may be identified with a specific crystal orientation for either orientation 1 or 2. Because there is a one-to-one correspondence between direction in the sample and crystallographic direction for 1 and 2, it is straightforward to orient the sample on an Eulerian cradle, arrange to diffract from particular reflections and hence measure the strains for each orientation in turn. It is true that other grain orientations, which are not as prominent as grains 1 and 2, will also contribute to these particular reflections but they are weighted far less than the ideal orientations. The schematic diagram of Figure 6.3 is vital for visualizing how to plan experiments on highly textured samples.

Measurements were made [13] of the {10$\bar{1}$0}, {20$\bar{2}$0}, {0002}, {0004}, {10$\bar{1}$1}, {10$\bar{1}$2} and {$\bar{1}$2$\bar{1}$0} reflections at specific angles of tilt and orientation determined from the pole figures as corresponding to grains 1 and 2. Measurements were also made in sets of directions based on crystal orientations with an [0002] axis directed along the plate normal and a [$\bar{1}$2$\bar{1}$0] axis along the transverse direction, and with an [0002] axis along the transverse direction and a [$\bar{1}$2$\bar{1}$0] axis directed along the plate normal. Both these sets have a common [10$\bar{1}$0] direction along the rolling direction. These latter sets cannot be thought of as ideal grain orientations, unlike grains 1 and 2, because many other discrete crystal orientations will also contribute to the average strain at any given sample orientation with no strong weighting from any one element of the texture. For these cases it does not make sense to describe the strains in terms of a set of axes based on crystal orientations and a sample-based coordinate system suffices.

The principal axes were taken to be [10$\bar{1}$0], [$\bar{1}$2$\bar{1}$0] and [0002] in order to describe the strain tensors associated with the ideal grain orientations. Figure 6.4(a) shows the angular coordinates (θ, ϕ) which define directions in the crystal coordinate system. Figure 6.4(b) shows the coordinates (χ, η), which describe angles in the sample system with principal axes along the rolling direction, **r**, the transverse direction, **t**, and the normal direction, **n**. The orientations of grains 1 and 2 in the sample coordinate system are shown in Figure 6.4(c).

The residual strains observed [14] for the ideal orientation 1 in the as-cooled state are shown in Figure 6.5 plotted versus crystal coordinates (θ, ϕ) for the {10$\bar{1}$0}, {0002}, {10$\bar{1}$1}, {10$\bar{1}$2} and {$\bar{1}$2$\bar{1}$0} planes. Each point is also labelled by the particular reflection in the principal axis system adopted for grain 1. Solid symbols denote the experimental results and open symbols represent the EPSC model calculation with the full rolling texture determined from the pole figures. There is very good agreement for these thermal strains where the EPSC model works with high accuracy. As in the case of the Zr-2 rod, no plasticity needs to be invoked in order to calculate the thermal strains. The curves in this figure represent the best least squares fit to the strain tensor, equation (1), which is written out in crystal coordinates as

$$\varepsilon(\theta, \phi) = (\varepsilon_{11} \cos^2 \phi + \varepsilon_{12} \sin 2\phi + \varepsilon_{22} \sin^2 \phi) \sin^2 \theta + \varepsilon_{33} \cos^2 \theta + \varepsilon_{23} \sin 2\theta \sin \phi$$

$$+ \varepsilon_{31} \sin 2\theta \cos \phi \tag{5}$$

Within the limits of experimental accuracy, different $\langle 10\bar{1}0 \rangle$ and $\langle \bar{1}2\bar{1}0 \rangle$ orientations give the strains every 30° around the basal plane equally well, since one set of reflections is not offset from the other. This is expected since the elastic and plastic properties are the same in these directions. The strains from orientation 2 were measured, and agreed with the EPSC model and fitted to the tensor form just as well as orientation 1. The elements of the strain tensors

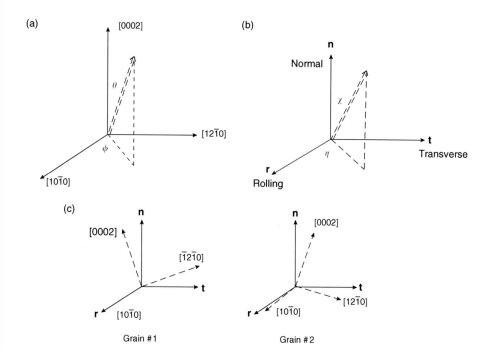

Figure 6.4 (a) Coordinate axis system for describing crystallographic directions in hexagonal close-packed Zr-2. (b) Coordinate axis system for describing directions in the sample with respect to the rolling direction, **r**, the transverse axis, **t** and the plate normal, **n**. (c) Schematic representation of the two ideal orientations within the sample axes.

are given in Table 6.1. The diagonal elements of the strain tensors for 1 and 2, each referred to their own principal axes, agreed closely. The ε_{11} elements, along the rolling direction, are small. The rolling texture is responsible for this result since the very small volume fraction of grains with *c*-axis parallel to the rolling direction provide little constraint to the predominant volume fraction of grains with [10$\bar{1}$0] orientation. These then contract as dictated by the *a*-axis coefficient of expansion. Perpendicular to the rolling direction there are $\langle 0002 \rangle$, $\langle \bar{1}2\bar{1}0 \rangle$ as well as $\langle \bar{1}2\bar{1}n \rangle$ orientations which constrain each other and give rise to the large thermally induced residual strains. The off-diagonal elements, ε_{23}, were found to be non-zero and to change sign between orientation 1 and 2. This result is a consequence of the symmetry of the rolled plate, since it is only when there is a change of sign between the two orientations, that the strains resolved along the plate normal and plate transverse directions are equal. The fact that these changes of sign are observed gives confidence that the assignment of crystal orientation was correct and that the accuracy of the measurement is high.

Sufficient data were obtained to map the strains onto a quarter pole figure for the {10$\bar{1}$0}, {$\bar{1}$2$\bar{1}$0}, and {0002} reflections in order to include directions not corresponding to the ideal orientations. Measurements made in completely different sample directions (χ, η), which were related by symmetry, were found to agree within the assigned errors and so were averaged and are plotted in Figure 6.6. The results (Figure 6.6a) are shown as a function of angular deviation from the rolling direction ($\chi = 90, \eta = 0$), to the transverse direction ($\chi = 90, \eta = 90$), thence to the normal direction ($\chi = 0, \eta = 0$) and back to the rolling

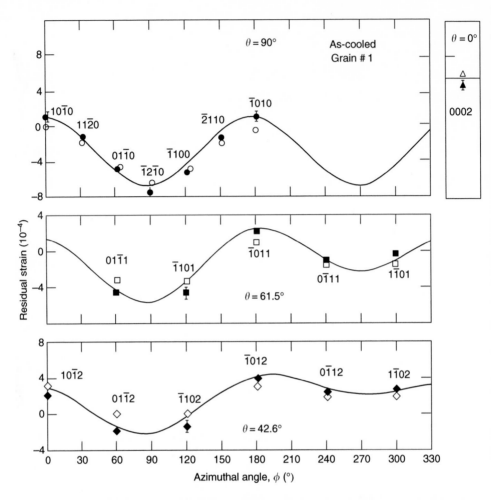

Figure 6.5 Residual strains as a function of the crystal coordinates (θ, ϕ) for the first ideal orientation of as-cooled Zr-2 plate. The filled symbols denote experimental data, open symbols represent the EPSC calculations and the curves are the least-squares fit of the experiment to a tensor form of strain. The results are plotted as a function of azimuthal angle at particular values of θ. The measurements are also labelled by the corresponding $(hkil)$ planes in the crystal coordinate system.

Table 6.1 Residual strain tensor elements (10^{-4}) for the ideal orientations in highly textured Zircaloy-2 plate

	ε_{11}	ε_{22}	ε_{33}	ε_{12}	ε_{23}	ε_{31}
Grain 1	1.1 ± 0.4	-6.6 ± 0.5	5.6 ± 0.6	-0.5 ± 0.4	-2.1 ± 0.4	-0.7 ± 0.3
Grain 2	1.4 ± 0.6	-7.4 ± 0.5	5.8 ± 0.6	-0.6 ± 0.4	2.5 ± 0.4	-0.4 ± 0.4

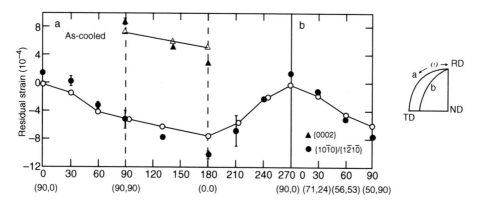

Figure 6.6 Angular variation of the residual strain in sample coordinates (χ, η) plotted as a function of angular deviation from the rolling direction for a locus (a) around the rim of the pole figure and (b) for a locus within the pole figure as shown in the accompanying schematic. The Zr-2 rolled plate sample was in the as-cooled state. The coordinates (χ, η) are identical to pole-figure coordinates. The experimental results are designated by filled circles for the ($10\bar{1}0$) and ($\bar{1}2\bar{1}0$) reflections and filled triangles for the (0002) reflection. The corresponding EPSC calculations are shown by open circles and triangles joined by lines.

direction around the perimeter of a quarter-pole figure. The results along a locus within the quarter pole figure are shown in Figure 6.6b. The experimental results and the EPSC calculations are shown by solid symbols and open symbols joined by lines, respectively. There is as good agreement between model and experiment for orientations corresponding to the ideal orientations as for those that do not. However, it was not possible to fit the results in Figure 6.6 as well with a strain tensor of the form of equation (1) since there is no strong weighting of the texture by an ideal orientation in many cases.

Figure 6.7 shows the strain tensors for grains 1 and 2, with elements tabulated in Table 6.1, plotted in a pole figure representation for the as-cooled sample. Note that the tensile [$\bar{1}2\bar{1}0$] strain for orientation 2 occurs in nearly the same direction as the compressive strain [0002] for orientation 1. Since zirconium is fairly isotropic elastically, this implies that intergranular stress balance is achieved for Zr-2 plate by the balance of grain 1 against grain 2. It is also interesting to note that the strains along the sample principal axes (r, t, n) are not large. If the strains had only been measured along the sample principal axes, the major intergranular effects would have been missed. The principal intergranular strains tend to be close to the principal crystal axes as was also found for rod-textured Zr-2.

As a general rule, in highly textured material it is likely to be important to measure and analyse the data within a coordinate axis system which recognises the strong texture, in this case the crystal axis system, rather than the principal axes of the sample. In cases of orientations that do not correspond to peaks in the pole figures, where there are contributions from many grain orientations and no strong weighting, it does not appear to be possible to deduce tensor information from the data directly. One would then have to rely on a statistical model of the grain assembly such as the EPSC model.

6.4.3 Residual strains in a Zr2.5wt% Nb over-rolled pressure tube

With the understanding of type-II residual strains developed from measurements of highly textured Zr-2 a method of determining the type-I strains in highly textured components can

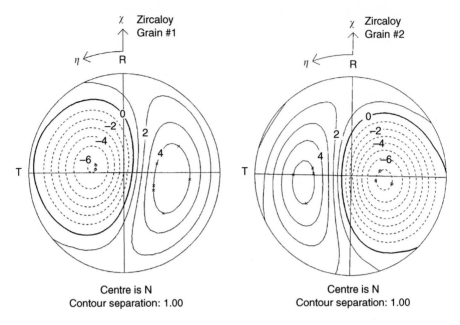

Figure 6.7 Strain tensors for the two ideal orientations of rolled Zr-2 plate displayed in a pole figure presentation in sample coordinates. The elements of the strain tensor were derived from fits to the ideal orientation data such as those shown in Figure 6.5.

be established. It is fair to say, in retrospect, that when the measurements described below were made the interpretation of the results in terms of macroscopic stresses was probably inadequate.

Zr2.5%Nb pressure tubes constitute the pressure boundary [15] in Canadian CANDU reactors. To make leak-tight seals with the end-fitting on the reactor face, a tube is rolled into grooves in the end-fitting by inserting a rolling tool from the outboard side. If the rolling tool is inserted too far, where the tube is unsupported, then strong plastic deformation can occur in the tube and high tangential stresses are generated near the boundary between the supported and unsupported regions. Zr2.5%Nb tubes are strongly textured and a schematic of the texture is shown in Figure 6.8. Grain 1 in this figure is the ideal orientation, with [10$\bar{1}$0] plane normals aligned along the tube axis, [$\bar{1}$2$\bar{1}$0] plane normals aligned along tube radii and [0002] normals aligned in hoop directions.

Figure 6.9 shows the residual hoop strain [16], measured with the (0002) reflection, as a function of distance from the end fitting. The open circles represent the results of measurements on an intact pressure tube with part of the end-fitting cut away but still restrained at distances less than zero on the abscissa in Figure 6.9, by the remainder of the end-fitting. The 50 mm thickness of the end-fitting precluded measurements in the region below zero. The reference lattice spacing was taken to be the value far from the end-fitting and far from the mechanical disturbance. This will not be a stress-free value in the sense that the last thermomechanical treatment of the component, cooling from the annealing temperature, would have generated thermal stresses in the material. However, it provides a sensible reference for assessing the perturbation caused by over-rolling.

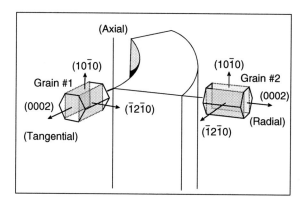

Figure 6.8 Schematic representation of the ideal orientation in a Zr2.5wt%Nb pressure tube designated grain number 1. The $[10\bar{1}0]$ direction is aligned with the axis of the pressure tube, the [0002] orientation lies in the hoop direction in the tube, and the $[\bar{1}2\bar{1}0]$ orientation lies in the radial direction. A second grain, not strongly weighted in the texture, is designated grain number 2.

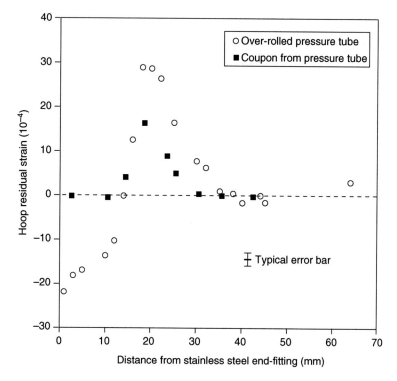

Figure 6.9 Variation of the residual hoop strain (open circles) in an over-rolled pressure tube as a function of distance from a restraining stainless steel end-fitting. The solid squares represent the residual hoop strains in a long coupon from the pressure tube cut along the axis of the tube. The squares represent the intergranular strains generated by plastic deformation during the over-rolling process.

There is clearly a very large tensile hoop strain in the over-rolled region about 20 mm from the end-fitting and a strong compressive constraint as the end fitting is approached. The solid squares in Figure 6.9 represent the hoop stresses, again measured with the (0002) reflection, but on a narrow coupon cut axially from the pressure tube. The reference lattice spacing was taken at the end far from the mechanical disturbance as previously. The hoop strain has decreased by only 40% in the neighbourhood of 20 mm but the constraint provided by the end fitting, which corresponds to a balancing stress, has disappeared. The hoop strain remaining in the coupon is however not of type-I character but is intergranular in nature, generated by the plastic deformation in the grains with [0002] along the hoop direction. There was no plastic deformation in the region of the end-fitting constraint and hence no intergranular strains were generated. The type-I strain corresponding to over-rolling is given by the difference between the intact pressure tube and the coupon.

It would appear that the only method, at present available in this case, of determining the strains corresponding to the macroscopic stress field is destructive. Measurements would be made on the intact sample and then a small coupon would be cut from the component and remeasured to obtain the intergranular contributions. Note that the intergranular strain varies with position along the coupon because the amount of plastic deformation varies along the tube (the spatial resolution of the experiment in the axial direction was about ±1 mm). This requires the use of a position-dependent reference rather than a single reference to obtain the corresponding type-I strains.

It is worthwhile making some general points based on these results. Because of the strong texture, the mechanical properties of an intact pressure tube are anisotropic. For example, the properties in the axial direction are strongly influenced by the elastic and plastic behaviour of $[10\bar{1}0]$ grains and hardly at all by the properties of [0002] grains. It would not make sense, either mechanically, or in diffraction since the signal will be very weak, to measure axial strains with the (0002) reflection. The sign of the strain might even be reversed since the role of the few [0002] grains along the axial direction may be to balance the effect of the $[10\bar{1}0]$ grains. The crystallographic alignment, the texture, will tend to influence the orientation of the principal axes of the strain tensor more than the geometry of the piece. The high texture offers the possibility of measuring the full strain tensor, type-I and type-II together in the intact component, for the ideal orientation. Subsequently the measurements can be made on the coupon and the type-I strains obtained by subtraction. Measurements of the full strain tensor were made on the coupon, which was easier to orient than the massive pressure-tube-end-fitting component.

Acknowledgements

I wish to acknowledge many productive collaborations with R. A. Holt on the topic of strain measurement in highly textured materials. I am grateful to J. W. L. Pang and C. N. Tomé for many discussions and for carrying out the EPSC model calculations.

References

[1] Noyan C. V. and Cohen J. B., *Residual Stress Measurement and Interpretation by Diffraction*, 1987, Springer-Verlag: New York.
[2] Van Houtte P. and De Buyser L., *Acta Metall. Mater.*, **41**, 323 (1993).
[3] Kröner E., *Z. Phys.*, **151**, 620 (1958).
[4] Turner P. A. and Tomé C. N., *Acta Metall. Mater.*, **42**, 4043 (1994).
[5] Clausen B., Lorentzen T. and Leffers T., *Acta Mater.*, **46**, 3087 (1998).

[6] Hill R., *J. Mech. Phys. Solids*, **13**, 89 (1965).

[7] Hutchinson J. W., *Proc. Roy. Soc.*, **A139**, 247 (1970).

[8] Dawson P., Boyce D., MacEwen S. and Rogge R., *Simulation of Materials Processing: Theory, Methods and Application, Proceedings of NUMIFORM 98*, 1998, eds. J. Huétink and F. T. P. Baaijens, Balkema: Rotterdam, p. 323.

[9] Tomé C. N., Holden T. M., Turner P. A. and Lebensohn R., *Proceedings of the International Conference on Residual Stresses, ICRS-5*, 1998, eds. T. Ericsson, M. Odén and A. Andersson, Institute of Technology: Linköpings Universitet, Vol. 1, p. 40.

[10] MacEwen S. R., Tomé C. N. and Fabe J., Jr., *Acta Metall.*, **37**, 979 (1989).

[11] MacEwen S. R., Christdoulou N. and Salinas-Rodriguez A., *Metal. Trans.*, **21A**, 1083 (1990).

[12] Pang J. W. L., Holden T. M., Turner P. A. and Mason T. E., *Acta Mater.*, **47**, 373 (1999).

[13] Holden T. M., Root J. H., Fidleris V., Holt R. A. and Roy G., *Mater. Sci. Forum*, **27/28**, 359 (1988).

[14] Holden T. M., Root J. H., Holt R. A. and Turner P. A. *J. Nucl. Mater.*, **304**, 73 (2002).

[15] Perryman E. C. W., *Nucl. Energy*, **17**, 9 (1978).

[16] Holden T. M., Powell B. M., Dolling G. and MacEwen S. R., *Proceedings of the 5th Risø International Symposium on Metallurgy and Materials Science*, 1984, eds. N. Hessel Andersen, M. Eldrup, N. Hansen, D. Juul Jensen, T. Leffers, H. Lilholt, O. B. Pedersen and B. N. Singh, Risø National Laboratory: Roskilde, p. 291.

7 Anisotropy of lattice strain response

T. Lorentzen

7.1 Introduction

The practical application of neutron diffraction was pioneered by the early experiments in 1981 by Allen *et al.* [1], Pintschovius *et al.* [2] and Krawitz *et al.* [3] and was soon followed by contributions from Schmank and Krawitz [4], Pintschovius *et al.* [5, 6], MacEwen *et al.* [7], Allen *et al.* [8], Holden *et al.* [10] and Holt *et al.* [11]. Throughout the years since then, the utilization of neutron diffraction for stress/strain characterization has been based on a continuum mechanics conception of materials deformation. Diffraction has merely been considered as an experimental technique which probes the stresses and strains at a continuum mechanics scale.

Most investigations have been conducted based on the assumption that selected crystallographic reflections could be used as simple monitors of the macroscopic strain much like a strain gauge: just using the ever-present crystallographic lattice as an embedded atomic strain gauge. Furthermore, most investigations have been based on the assumption that strains measured on this microstructurally relevant scale represented the tensorial quantity of strains in the continuum mechanics sense. As a consequence stresses have been deduced simply by a conversion of the lattice strains through a continuum mechanics formalism into stress tensors. Both stress and strain tensors have been taken to behave like ordinary second rank tensors with the associated characteristic transformation rules. Consequently, experimenters conducting diffraction-based investigations have adopted the concept of principal stresses and strains and the associated definition of principal directions along which no shear components are present. In the general sense, at least six independent measures of strain are required for a complete characterization of a strain tensor; however, when considering the inherent measurement errors, it requires an over-determination with more like 8–12 independent measures in order for errors on the stresses not to grow unacceptably large [12]. Typically it is, however, not feasible either for technical or for economic reasons to perform enough measurements to derive the full tensor, and most experimental characterizations have in fact been based on assumptions of principal directions, and strain measurements have been limited to three orthogonal directions. In short the basic concepts of continuum mechanics have been adopted and stress/strain relations have been based on the generalized Hooke's law.

It has become increasingly clear in recent years that the complexity of polycrystal deformation gives rise to criticism of some of the above-mentioned assumptions. A firm understanding of the mechanics, at the scale at which our experimental technique probes the deformation, is necessary in refining our view of the diffraction technique and in refining our data analysis. Only through this approach may the experimental technique evolve to a state where crystallographic lattice strains can be used safely in supplying engineers with valid measures of strains and stresses, which can readily be utilized in traditional engineering analysis of materials and structures.

In numerous previous investigations, the simplistic assumption of adopting a continuum mechanics view, has been a valid one as many investigations on components and structures with known stress distributions have proven, and in most cases diffraction experiments are likely to have produced relevant measures of stresses and strains for engineering and scientific purposes. However, in refining our conception of the mechanics of deformation and in realizing the inherent differences in the scale between the level at which the diffraction techniques probe the stresses and strains and the scale at which the continuum mechanics and traditional engineering analysis operate, it is necessary to engage in more general studies of polycrystal deformation; in itself a scientific discipline, which only in recent years has been brought into play in the field of studying the anisotropy of lattice strain evolution during deformation of polycrystalline aggregates. But more so, in the practical utilization of the diffraction technique and in using the atomic lattice as a microscopic strain gauge for engineering purposes, a firm insight into the micromechanics is necessary in order to bring about valid measures of macroscopic stresses and strains.

The anisotropy in lattice strain response observed when performing diffraction measurements has caught the interest of the entire neutron diffraction community as numerous recent publications indicate; see for instance Refs. [13–34]. This anisotropy was quickly realized in the early days of the technique [35, 36], but especially from the mid-1990s it has drawn much attention both from a modelling point of view and from the point of view of performing proper measurements and interpretation. It is, however, not a new discovery, and not a finding which can be attributed to the neutron diffraction community. Already in 1938, Glocker [37] addressed the issue of elastic anisotropy in diffraction experiments and several publications by Bollenrath *et al.* [38–40] and Smith *et al.* [41, 42] observed anisotropy effects following plastic deformation during bending, tension and compression. The first direct demonstration of the existence of residual intergranular stresses caused by interactions between grains of different orientation came in 1946 by Greenough [43] followed by a number of subsequent publications [44–46]. In fact, much of this work was probably inspired by Heyn [47] who already in 1921, in studying work hardening of metals, arrived at the conclusion that following plastic deformation some kind of internal self-equilibrating stress/strain state pertained. Despite Heyn himself naming these stresses and strains "hidden elastic strains" and "hidden (latent) stresses", they were more commonly called "Heyn stresses".

It is the aim of the present chapter to bring about an insight into the micro-mechanics of polycrystal deformation with specific relevance to experimental measures of lattice strains by diffraction, and to illustrate by state-of-the-art modelling techniques how we may attack these issues theoretically. I will discuss how these theoretical findings may guide us both in performing valid measurements and in performing valid data analysis.

7.2   Elastic anisotropy in single crystals

As the diffraction techniques probe the elastic strain field in the material, using the atomic lattice as a strain gauge monitoring elastic changes in the lattice plane spacings in selected crystallographic directions, the single crystal anisotropy (here more or less being used in the context of single grain elastic anisotropy) plays an important role in the interpretation of measured lattice strains and their relation to the overall macroscopic strain field and in modelling of aggregate response.

For cubic systems, the inverse Young's modulus as a function of direction (orientation) may be expressed as [48]:

$$\frac{1}{E} = S^*_{1111} = S_{11} - 2\left[(S_{11} - S_{12}) - \frac{1}{2}S_{44}\right](a_{11}a_{12} + a_{12}a_{13} + a_{11}a_{13}) \tag{1}$$

with the orientation dependency governed by the latter term of direction cosines:

$$(a_{11}a_{12} + a_{12}a_{13} + a_{11}a_{13}) \tag{2}$$

which varies between 0 and 0.33. The degree of anisotropy is governed by the term preceding this orientation term:

$$2\left[(S_{11} - S_{12}) - \frac{1}{2}S_{44}\right] \tag{3}$$

The degree of anisotropy is often expressed in terms of the ratio given below, which will be equal to unity for isotropic crystal properties and deviating from unity with increasing degree of anisotropy:

$$\frac{2(S_{11} - S_{12})}{S_{44}} \tag{4}$$

With more specific relevance to performing diffraction experiments, the orientation dependency of the elastic constants is more adequately expressed in terms of the Miller indices of the crystallographic planes:

$$\Gamma_{hkl} = \frac{h^2k^2 + k^2l^2 + l^2h^2}{(h^2 + k^2 + l^2)^2} \tag{5}$$

This quantity is seen to vary from a value of 0 for the $\langle 100 \rangle$ family of crystal directions to 0.33 for the $\langle 111 \rangle$ family of directions. The most common situation for metals is that the term which governs the degree of anisotropy in equation (1) is positive, in which case the $\langle 111 \rangle$ family of directions is the most stiff and the $\langle 100 \rangle$ family of directions is the most compliant. For special combinations of compliance constants this term will, however, be negative in which case the stiff and compliant directions are interchanged.

Typical tabulated values of modulus for selected crystallographic directions (*hkl*) as well as values of the ratio given in equation (4), as a measure of the anisotropy, are found in Table 7.1.

The materials selection of Table 7.1 represents a broad range in the degree of anisotropy from the nearly isotropic aluminum to the highly anisotropic copper and stainless steel. For aluminum, we note that the maximum deviation in stiffness from the typical macroscopic value of 70 GPa merely amounts to 8–9%, whereas for stainless steel, with a typical macroscopic modulus near 200 GPa, the deviation in stiffness exceeds 50%. Note also that for all the four materials selected here, the parameter describing the degree of anisotropy is larger than unity, and all materials show the $\langle 111 \rangle$ to be the most stiff direction, and the $\langle 200 \rangle$ to be the most compliant.

In Table 7.2 similar data are given for common bcc materials. Notice here that the quantity displaying the degree of anisotropy is less than unity for materials like molybdenum (Mo) and chromium (Cr) giving rise to the previously mentioned situation where the $\langle 111 \rangle$ family of directions becomes the most compliant one, and the $\langle 200 \rangle$ family of directions the stiff one.

Table 7.1 Values of single crystal elastic modulus for selected *hkl* reflections of common fcc materials; the materials data which form the basis for these calculations are given in Ref. [49]

	Al	Cu	Ni	Fe
$(2(S_{11} - S_{12}))/S_{44}$	1.22	3.20	2.37	3.80
E_{200} (GPa)	63.7	66.7	120.5	93.5
E_{311} (GPa)	69.0	96.2	161.4	138.3
E_{420} (GPa)	69.1	97.0	162.4	139.6
E_{531} (GPa)	71.1	113.6	182.9	165.9
E_{220} (GPa)	72.6	130.3	202.0	193.2
E_{331} (GPa)	73.6	143.6	216.2	215.5
E_{111} (GPa)	76.1	191.1	260.9	300.0

Table 7.2 Values of single crystal elastic modulus for selected *hkl* reflections of common bcc materials; the materials data which form the basis for these calculations are given in Ref. [49]

	Stainless steel	V	Mo	Cr
$(2(S_{11} - S_{12}))/S_{44}$	2.51	2.13	0.79	0.71
E_{200} (GPa)	125.0	80.5	357.1	333.3
E_{310} (GPa)	146.4	102.3	336.6	306.7
E_{411} (GPa)	149.8	104.4	334.1	303.5
E_{321} (GPa)	210.5	141.3	305.3	268.5
E_{112} (GPa)	210.5	141.3	305.3	268.5
E_{110} (GPa)	210.5	141.3	305.3	268.5
E_{222} (GPa)	272.7	176.5	291.3	252.1

For a material like ferritic iron, the maximum deviation from a typical macroscopic modulus of 210 GPa amounts to 30–40%; hence ferritic iron displays a smaller degree of elastic anisotropy than does austenitic. Note that not all the bcc materials show the $\langle 111 \rangle$ family of directions to be the most compliant one; hence this is not characteristic of the bcc structure. This difference between having the $\langle 111 \rangle$ family of directions being the most stiff or the most compliant ones is simply attributed to specific combinations of elastic constants, and the reason is related to the atomic number and the electron cloud configuration rather than a consequence of the atomic arrangement.

When referring to even the simplest of all loading conditions, uni-axial tension, naturally focus is on the stiffness of different crystallographic directions aligned with the loading direction. With the large variations in single crystal elastic stiffness it obviously makes a difference which *hkl* reflection is used as the strain monitor. However, in interpreting the effects of elastic anisotropy it is rather interesting also to study the transverse properties, not only in terms of the single crystal Poisson's ratio, but also the elastic stiffness in directions transverse to a specific *hkl* direction. One may imagine walking around a specific lattice direction and monitoring the lattice stiffness perpendicular to this direction as a function of angle. Several examples of such calculations including a description of the procedure are given in further detail in Ref. [49]; here in Figure 7.1 we only show examples for two selected *hkl* directions, $\langle 113 \rangle$ and $\langle 531 \rangle$, of stainless steel, which is a highly elastically anisotropic material.

Nature presents itself rather strikingly with dramatic variations in stiffness as a function of angle. The high symmetry of the $\langle 113 \rangle$ direction is noticed with amplitudes in the normalized

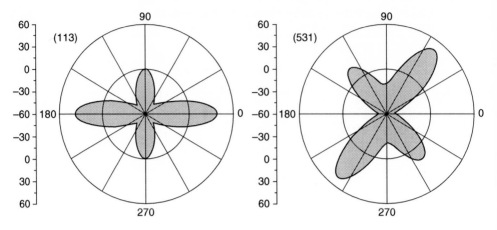

Figure 7.1 Elastic stiffness of stainless steel perpendicular to the $\langle 113 \rangle$ and $\langle 531 \rangle$ directions normalized to the calculated homogeneous effective aggregate modulus; from [49].

modulus exceeding 40%. The $\langle 531 \rangle$ direction displays similar amplitude in the variations, but furthermore shows an unsymmetric nature. Clearly, in a diffraction experiment with many grains in the diffraction gauge volume, these dramatic variations are smeared out and will in general not be noticed by the experimenter. However, the more we move away from the lattice strain measurements being an adequate monitor of the true volume average macroscopic strain field, which will be the case for highly textured materials or materials with a large grain size relative to the spatial resolution of the experimental configuration, the more diffraction results will be influenced by these matters. As we move away from this regime of good volume average measurements with a statistically significant number of grains sampled, and the experimental observations become more prone to microstructural effects, such as the grain-to-grain variation in stiffness, the more the grain rotation around the lattice plane normal will become an issue. With transverse stiffness varying more than 40%, the transverse contraction will vary greatly with angle of rotation; hence measurements of transverse strains will have to be performed averaging over the angle of rotation. It is particular noticeable that when calculations are done for the transverse properties transverse to the $\langle 111 \rangle$ direction, we find that the stiffness is completely isotropic and does not vary with angle. A similar situation is found in simple laminate theory, where it is well known that three highly anisotropic layers placed at 120° angular intervals, produce a laminate which is completely isotropic in its in-plane elastic properties; a similar situation with regards to symmetry is observed transverse to the $\langle 111 \rangle$ direction of a cubic crystal.

As mentioned, many more examples and a more detailed treatment of the implications for stress strain interpretation in diffraction-based experiments may be found in Ref. [49].

The above treatment of single crystal elastic anisotropy has focused on cubic systems only. Similarly, the elastic anisotropy of hexagonal systems can be expressed in terms of the inverse Young's modulus as a function of direction (orientation) and is given by [50]:

$$\frac{1}{E} = \left(1 - a_{13}^2\right)^2 S_{11} + a_{13}^4 S_{33} + a_{13}^2 \left(1 - a_{13}^2\right)(2S_{12} + S_{44}) \qquad (6)$$

It is left to the reader to explore the range of anisotropy and orientation dependency in hexagonal systems.

7.3   Elastic anisotropy in polycrystals

In the previous section, the large variations in elastic stiffness were exemplified through tabulated values of *hkl*-specific Young's modulus for selected directions in a single crystal. Clearly, this inherent anisotropy affects the lattice strains monitored in a polycrystalline material. Some *hkl* reflections are very compliant (small elastic stiffness) and some are very stiff (large elastic stiffness), and we may expect that for a given macroscopic strain level, say an imposed external uni-axial deformation, lattice strains vary from reflection to reflection. They will do so, but not to the extent observed on the single crystal elastic properties. Naturally some averaging procedure comes into play, as grains are not merely free to deform according to their own stiffness. Grains of different stiffnesses are restricted in their deformation by their surroundings, and in some self-consistent manner, a state of equilibrium is brought about though still rendering large variations in the lattice strains as a function of orientation relative to the imposed external load (or strain).

Many theoretical schemes have been proposed for the calculation of both the macroscopic overall stiffness of an aggregate of anisotropic grains, and for the calculation of *hkl*-specific stiffness. By far, the most realistic coverage of these aspects is through the so-called Kröner model [51]. For cubic systems, the calculation of overall stiffness is through the solution of a third-order polynomial in the aggregate shear modulus, see Kröner [51] or Kneer [52], and the aggregate bulk modulus is found to be identical to the single crystal bulk modulus. These two quantities relate directly to the Young's modulus and Poisson's ratio of the aggregate.

The calculation of *hkl*-specific stiffness is based on these values of the overall bulk and shear modulus; a more detailed treatment including a chronological presentation of the expressions to be evaluated is given in Ref. [49] based on which a simple computer code is easily programmed. The overall findings discussed in more detail in Ref. [49] are, as indicated by intuition previously, that all variations in *hkl* stiffnesses are less extreme than found in a single crystal; stiff directions behave in a less stiff manner and compliant directions behave less compliantly. These *hkl*-specific stiffnesses are often denoted as diffraction elastic constants (DECs) and are typically found to be within a few percent of the effective stiffness observed in the *hkl*-specific response measured by diffraction in texture-free aggregates.

An illustration of the orientation dependency of the Kröner stiffness is given in Figure 7.2 showing both the Young's modulus and Poisson's ratio as a function of the orientation parameter, *hkl*, presented in equation (5). The numerical results for ferritic iron are normalized to the Kröner predictions of aggregate properties discussed above.

The numerical values of the *hkl*-dependent Kröner stiffness are presented in Table 7.3 for selected fcc materials. Similar data for selected bcc materials are found in Ref. [49], and a comprehensive coverage for other crystal structures is found in Ref. [53].

7.4   Modelling of polycrystal deformation into the plastic regime

Modelling of the plastic deformation of polycrystalline aggregates is a well-established scientific field beginning with the first attempts made by Sachs in 1928 [54] and Taylor in 1938 [55]. These simple schemes, which were born in a time lacking today's computing power, have conceptual flaws which in some respects make them unrealistic. Nonetheless they succeeded in identifying many of the features which could be observed by experimental techniques; primarily macroscopic flow curves and texture evolution. These models are among the so-called "one-site" models, where the term "one site" refers to the principle that each grain is considered as if it does not specifically know each of its neighbors, but rather in some idealized way senses the effect of the surroundings. One of the most fundamental

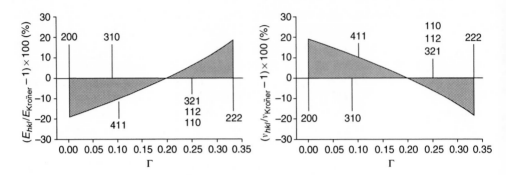

Figure 7.2 The Kröner predictions of the effective elastic response of ferritic iron along the loading direction in a polycrystalline aggregate. The effective elastic properties, E_{hkl} and v_{hkl}, are here given as a function of the orientation parameter, Γ_{hkl}, and normalized to the Kröner predictions of the overall bulk elastic aggregate properties.

Table 7.3 Tabulated values of the *hkl*-specific effective elastic properties for selected fcc materials following the Kröner modelling scheme

	(200)	(311)	(420)	(531)	(220)	(422)	(331)	(111)
Young's modulus								
Al	67.6	70.2	70.3	71.2	71.9	71.9	72.3	73.4
Cu	101.1	122.0	122.5	131.5	139.1	139.1	144.3	159.0
Ni	160.0	185.0	185.6	195.6	203.9	203.9	209.5	224.6
Fe$^\gamma$	149.1	183.5	184.4	199.5	212.7	212.7	221.8	247.9
Poisson's ratio								
Al	0.35	0.35	0.35	0.35	0.34	0.34	0.34	0.34
Cu	0.38	0.35	0.35	0.34	0.33	0.33	0.32	0.31
Ni	0.36	0.33	0.33	0.33	0.33	0.32	0.31	0.30
Fe$^\gamma$	0.34	0.31	0.31	0.29	0.28	0.28	0.27	0.24

issues in modelling the mechanics of polycrystal deformation is how to take proper account of the interaction between the grain and the surroundings. Many proposals have been put forward to take account of this interaction, and many textbooks on the topic give historical reviews of the process towards more and more realistic modelling attempts.

Currently, the most usable and operational treatment of the interaction is dealt with by the so-called elasto-plastic self-consistent (EPSC) schemes, which take full account of the changes in material properties of the surroundings brought about as a consequence of progressive plastic distortion of the aggregate. The principles behind these EPSC schemes rest on the equivalent inclusion method of Eshelby [56, 57] formulating how an elastically anisotropic inclusion in a homogeneous effective medium, accommodates deformation brought about by some external loading of the medium. In the EPSC models, each grain is considered as such an inclusion of regular shape, that is, ellipsoidal or spherical, embedded in a homogeneous effective medium (HEM) having the instantaneous deformation properties of the polycrystalline aggregate. A priori neither stresses nor strains are prescribed in the individual grains; rather we prescribe the boundary conditions on the aggregate as well as the external loading in the form of stress or strain increments. In a stepwise manner each grain is now considered from the assumption that the weighted average of the stress and strain rates of all the grains correspond to the macroscopic stress and strain rates imposed on the aggregate. The EPSC models are essentially all based on the same formalism given by Hill in 1965–1967 [58–61] which was addressed comprehensively by Hutchingson in 1970 [62]. The discussion will here be based on an implementation of this modelling scheme by Clausen in 1997 [15] and a detailed description of the implementation was given in a published Risø-R-report by Clausen and Lorentzen, also in 1997 [16].

As for many earlier modelling approaches, the original focus was on predictions of macroscopic stress/strain curves, texture and yield surfaces [63–66], and not until around 1988 were the modelling schemes addressed with specific emphasis on lattice strains and stresses in polycrystalline aggregates [67, 68]. Here we will specifically focus on the valuable insight in regard to anisotropic lattice strain response which can be modelled by using this kind of algorithm.

The EPSC-model approach appears to be a rather realistic conception of the mechanics of polycrystal deformation in many respects, but in judging its capability we must pay attention to the level of detail at which it operates. Being a "one-site" model, we may naturally not expect it to capture details at the level of the individual grain; hence it is not expected to capture direct grain-to-grain interactions. The EPSC scheme operates at a level considering whole families of identical grain orientations, and considering how such families of grains deform when being restricted by the overall properties of the surroundings. In this respect, an EPSC scheme is especially well matched in level of detail to the neutron diffraction technique for residual stress/strain characterization. The neutron diffraction technique specifically monitors the average lattice strain response of such families of grains, and hence great prospects are seen for a fruitful interaction between the two approaches: the theoretical and the experimental.

As for any modelling framework, the EPSC model requires some materials input. The most immediate materials property needed is some way of quantifying a threshold level, τ^i, of a specific slip system (i) above which crystallographic slip occurs and hence plasticity sets in. The numerical level of this threshold stress, the critical resolved shear stress, is typically selected to be around half the difference between the two largest principal stresses at first yield.

It appears natural to relate the rate of evolution of τ^i, ($\dot{\tau}^i$), to the deformation history, and it has been proposed to express this in terms of summing the contribution of slip on all slip systems in the constituent (the grain):

$$\dot{\tau}^i = \sum_j h^{ij} \dot{\gamma}^j \tag{7}$$

The hardening matrix, h_{ij}, can be defined in any way the programmer likes, but a specific definition was proposed by Hill in 1966 [60], allowing for some degree of freedom in specifying whether hardening on one slip system affects hardening on the other slip systems (so-called *latent-hardening*), or whether slip on each slip system was considered to be completely independent of the history of slip on the other slip systems (so-called *self-hardening*). This was accomplished by introducing a "q"-parameter, and expressing the hardening matrix as:

$$h^{ij} = h_\gamma (q + (1 - q)\delta^{ij}) \tag{8}$$

Little is known about whether latent- or self-hardening is the dominating effect, or to what extent one or the other should be implemented into such modelling schemes.

The specific way of introducing the deformation history into the equation is through the definition of h_γ, and again the programmer is left with a choice to make. It is natural to relate h_γ in some way to the crystallographic slip, and in some way taking account of the progressive accumulation of crystallographic slip. The simplest choice is to make h_γ depend linearly on the accumulated crystallographic slip on all slip systems of the constituents. This, however, has proven to be very inaccurate in reproducing both the elastic/plastic transition and the plastic regime as discussed by Clausen and Lorentzen [17], who then introduced additional free parameters following a somewhat more complicated expression of the form:

$$h_\gamma = h_{\text{final}} \left(1 + (h_{\text{ratio}} - 1)e^{h_{\text{exp}}\gamma^{\text{acc}}} \right) \tag{9}$$

It should be emphasized that though the calculation procedure takes its starting point based on a very physically meaningful definition of crystallographic slip using well-defined slip systems, the definition of the hardening behaviour continues to be of an empirical nature. The expressions given above relating the increment in critical resolved shear stress, τ^i, to the accumulated crystallographic slip, γ^{acc}, are empirical expressions of the fact that the critical resolved shear stress ought to increase in some way with plastic deformation much like the description of work hardening in a continuum mechanics sense. The description given above, however, has no conception of the physical mechanisms of hardening by crystallographic slip and the associated pile-up of dislocations; it is nothing more than expressing an empirical rule of evolution of the critical resolved shear stress. Nor have the various "h"-parameters given in equation (9) any physical meaning and in the calculation procedure, they are considered free fitting parameters.

Having selected these material parameters, one can determine the macroscopic deformation of the aggregate through calculations addressing each grain in the aggregate; typically a few thousand grains are considered to adequately represent the true distribution of grain orientations in a texture-free aggregate. Subsequently, we can now ask for details about stresses and strains in specific grain orientations in the aggregate and in specific specimen relevant directions. Hence, with a focus on studies where stresses and strains are characterized by diffraction, we specifically address those grain orientations which represent measurable reflections by diffraction, and we specifically address macroscopic directions of the specimen relevant to the specific loading situation. We will restrict the discussion here to the uni-axial case, where the natural macroscopic directions are along and perpendicular to the loading axis, and we specifically address a selected range of hkl reflections in those two macroscopic directions of the sample. It should be noted that such calculations are based on the original modelling schemes presented 30–40 years ago and we are merely asking questions, which were not asked then, and which have not been asked until recently. It appears fair to say that this renewed interest and use of the well-established modelling schemes, have been driven by the neutron diffraction community through its growing interests in the inherent effect of elastic/plastic anisotropy on measurement of residual stresses, whether this be for scientific or engineering purposes.

A detailed theoretical study of polycrystal deformation has been presented by Clausen et al. [28] including a detailed treatment of numerical results of a more crystallographic nature. The lattice strain evolution is here exemplified by numerical results from a simulation of uni-axial loading of stainless steel (Figure 7.3).

The response at low stress shows the well-known variations in the elastic modulus, which in this model is identical to the Kröner modelling approach. The lattice strain response is hence an effective response deviating from the single crystal response as discussed previously.

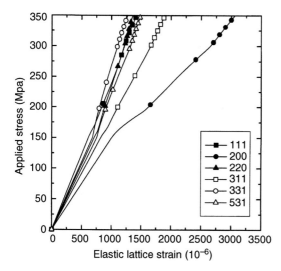

Figure 7.3 Calculated stress/strain response parallel to the tensile axis of a uni-axially loaded stainless steel sample, here given for six reflections; from [28].

A noticeable feature is the onset of nonlinearity most clearly observed for the compliant (200) reflection, which abruptly shows a dramatic increase in the rate of elastic strain accumulation. The (311) reflection shows essentially linear behaviour, while other reflections, like the (331), show a decrease in the rate of elastic strain accumulation. Upon further loading, most reflections reach another kink in the lattice strain response: again this effect is most noticeable for the (200) reflection. Notice also that the regime where the well established Kröner model renders a realistic prediction of the effective stiffness is limited to the elastic regime below 150 MPa.

To illustrate the potential detrimental impact this nonlinear effect may have on our interpretation of measured lattice strains and on our strain/stress conversion, one may imagine measuring lattice strains on a uni-axially loaded component; say, for example, diffraction data gives a measured lattice strain of 2000 $\mu\varepsilon$ measured on the (200) reflection of stainless steel. With a Kröner model based estimate of the effective elastic modulus of 149 GPa this renders a stress level of 298 MPa. However, following the highly nonlinear behaviour, exemplified by model predictions in Figure 7.3, the stress level associated with a strain level of 2000 $\mu\varepsilon$ on the (200) reflection is merely some 239 MPa. The simplified analysis using the Kröner model based effective stiffness hence gives a stress level which is 25% higher than the real stress level. Considering that this nonlinearity typically sets in long before reaching the engineering definition of yield, at the so called $\sigma_{0.2}$ limit, this complex nonlinear behaviour reduces the window for simple and safe conversion of lattice strains to stresses.

Having realized this very noticeable nonlinearity during forward loading, a reversal to zero external load makes the different families of grain orientations (here represented by the *hkl* reflections) return to different levels of residual strains. For the current loading condition and level of deformation the return curves are linear and no reverse plasticity of any measurable amount comes into play. Hence, deviation from linearity during the forward loading, corresponds exactly to the level of residual intergranular lattice strains realized once loading is removed. This is exemplified in Figure 7.4, which represents unloading conditions from different levels of forward loading as pictured in Figure 7.3.

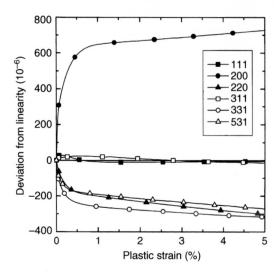

Figure 7.4 Calculated residual intergranular strains after unloading as a function of the plastic strain experienced during forward loading; from [28].

The most noticeable feature of these model predictions is the clear partitioning of residual intergranular strains into tensile or compression strains and a group showing essentially no residual intergranular strains. Similar model predictions have been made for aluminum and copper, see Ref. [28], and it shows that this partitioning into the specific groups shown in Figure 7.4, is not a general feature of fcc materials. Copper appears to behave much like stainless steel, whereas the grouping is different for aluminum. Here, the (111) reflection develops the highest tensile residual intergranular strains while the (200) reflection develops the lowest and the (331) reflection appears to develop essentially none.

The impact that this development of residual intergranular strains has on our interpretation of measured lattice strains and on our strain/stress conversion is, that for components having experienced inhomogeneous plastic deformation, any resulting macroscopic residual strain field is super imposed on an intergranular residual strain field, which varies as a function of the spatial variation in accumulated plastic strain throughout the component. As these residual intergranular strains may be relatively large as exemplified in Figure 7.4, they impinge on the correspondence between engineering estimates of the macroscopic stress/strain field, for instance based on finite element calculations, which do not include the potential presence of these residual intergranular strains. Nonetheless, they are a natural part of the total stress/strain state in the material, and they should also be taken into account.

There appears to be a tendency towards scientists searching for means of separating out these intergranular components of the stress/strain field, merely in order to reach a situation of better correspondence between finite element method (FEM) predictions and diffraction-based measurements. An example is given in Ref. [69], where the calculated residual intergranular strains as a function of plastic deformation were subtracted from the measured lattice strain profiles prior to strain/stress conversion. However, while such an approach appears to work quite well, it is merely to satisfy the engineer by showing measured strain profiles resembling calculated ones by FEM. From a scientific point of view, however, we should not try and neglect the importance of this intergranular strain contribution, but rather pursue to what extent it may affect properties and deformation mechanisms. Although it appears that the presence of intergranular stresses is simply annoying for straightforward application of diffraction techniques, at least from an engineering and industrial application point of view, actually it highlights the inherent gap in the concept of deformation between micromechanics and continuum mechanics.

7.5   Summary and discussion

Numerous recent publications exemplify the increased interest in elastic and plastic anisotropy in polcrystalline aggregates, and the effects they may have on the interpretation of diffraction data. Many theoretical and experimental investigations have dealt with these issues for materials like aluminium, copper, stainless steel, Zircaloy, Inconel and high-strength steel. This has brought about valuable insight into the nonlinear lattice strain behaviour and insight into the numerical levels and *hkl* dependency of residual intergranular strains and stresses. Early experiments were done with very little knowledge of these matters, and were overshadowed by the enthusiasm over the fact that it was actually possible to resolve the lattice strain changes to a degree where it was relevant for stress/strain determination. Most importantly this increased focus on the topic has raised the awareness of the inherent complexity of polycrystal deformation. There continues to be a strong emphasis on these matters, and it is today more the rule than the exception that experimenters address these issues in the analysis and reporting of experimental results.

The key issue appears to be to define which *hkl* reflections show linear lattice strain behaviour during loading and which show no evolution of residual intergranular strains and stresses. The research is driven by "a wish" to find *hkl* reflections, which appear to behave "decently" in all situations such that lattice strains monitored using those reflections can safely be taken as measures of macroscopic internal strains, which are readily converted into stresses by the common continuum mechanics formalism. It would naturally be to great benefit, if it was indeed possible to verify that *hkl* reflections could be found that are unaffected by elastic/plastic anisotropy in the general sense. Several experiments and numerical calculations have already shown that even for the simplest loading situations, it is not possible in the general sense to identify specific *hkl* reflections of this nature. Although numerical results of Clausen *et al.* [28] indicated that, for instance, the (113) reflection showed essentially no development of residual intergranular strains in the loading and the transverse direction, experimental results [34], however, indicated that the (111) reflection appeared to be a better choice based on minimal residual intergranular strains in both sample directions. From the early days of X-ray diffraction for residual stress measurements the neutron diffraction community has inherited the qualitative recommendation to use higher index reflections, where these complicated aspects related to elastic/plastic anisotropy and grain interaction effects appeared to be smeared out. This recommendation likely originates from experience and not from detailed experimental and theoretical investigations. Several more recent publications originating from the neutron diffraction community show that this historic recommendation does not appear to be one which can be safely adopted for all cases. Experimental results on stainless steel, for instance, show the (531) reflection to be among those which develop the highest residual intergranular strains, at least higher than many of the lower indices reflections.

The numerous recent results are definitely governing a better understanding of the elastic/plastic anisotropy, but the mechanics of polycrystal deformations is complex and we have experienced that it is nontrivial to derive general conclusions. More so, most investigations have been done on texture-free materials, and essentially all *in situ* loading experiments have only dealt with the uni-axial case.

Although it may appear that we have introduced more confusion than explanations and recommendations by the investigations into anisotropy, this does move the field forward. Much has been gained even within the past 3–5 years; not only with the aim of promoting better and more valid data analysis for engineering purposes, but also proving that neutron diffraction is a very valuable tool in more scientific oriented investigations into the general mechanisms of polycrystal deformation.

In comparison with the opportunities that X-ray diffraction brings, neutron diffraction is in particular an attractive tool when focusing on models working at a level of detail, where we do not directly consider the grain-to-grain interaction, but rather the effective volume average response of particular families of grains in the bulk of a polycrystal. Next generation polycrystal models, with direct account of grain-to-grain interaction, will benefit from the increased spatial resolution of high-energy X-rays from synchrotron sources. But as to the self-consistent type of modelling approach, much work is still awaiting use of neutron diffraction.

Acknowledgements

The present work has been performed with support from the "Engineering Science Centre for Structural Characterization and Modeling of Materials" at Risø. The centre is sponsored by the Danish Technical Research Council (STVF).

References

[1] Allen A., Andreani C., Hutchings M. T. and Windsor C. G., Measurements of internal stresses within bulk materials using neutron diffraction. *NDT-Int.*, **15**, 249–254 (1981).
[2] Pintschovius L., Jung V., Macherauch E., Schäfer R. and Vöhringer O., Determination of residual stress distributions in the interior of technical parts by means of neutron diffraction. In E. Kula and V. Weiss, editors, Residual stress and stress relaxation, *Proceedings of the 28th Army Materials Research Conference*, Lake Placid, NY, July 1981, Plenum, pp. 467–482.
[3] Krawitz A. D., Brune J. E. and Schmank M. J., Measurements of stress in the interior of solids with neutrons. In E. Kula and V. Weiss, editors, Residual stress and stress relaxation, *Proceedings of the 28th Army Materials Research Conference*, Lake Placid, NY, July 1981, Plenum, pp. 139–155.
[4] Schmank M. J. and Krawitz A. D., Measurement of stress gradient through the bulk of an aluminium alloy using neutrons. *Metal. Trans. A*, **13**, 1069–1075 (1982).
[5] Pintschovius L., Jung V., Macherauch E. and Vöhringer O., Residual stress measurements by means of neutron diffraction. *Mater. Sci. Eng.*, **61**, 43–50 (1983).
[6] Pintschovius L., Scholtes B. and Schröder R., Determination of residual stresses in quenched steel cylinders by means of neutron diffraction. In N. Hessel Andersen, M. Eldrup, N. Hansen, D. Juul Jensen, T. Leffers, H. Lilholt, O. B. Pedersen and B. N. Singh, editors, *Proceedings of the 5th Risø International Symposium on Materials Science, Microstructural Characterization of Materials by Non-destructive Techniques*, Risø National Laboratory, DK-4000 Roskilde, Denmark, September 1984, Risø National Laboratory, pp. 419–424.
[7] MacEwen S. R., Faber Jr, J. and Turner P. L., The use of time-of-flight neutron diffraction to study grain interaction stresses. *Acta. Metall.*, **31**, 657–676 (1983).
[8] Allen A. J., Andreani C., Hutchings M. T., Sayers C. M. and Windsor C. G., Neutron diffraction studies of texture and residual stress in weldments. In N. Hessel Andersen, M. Eldrup, N. Hansen, D. Juul Jensen, T. Leffers, H. Lilholt, O. B. Pedersen and B. N. Singh, editors, *Proceedings*

of the 5th Risø International Symposium on Materials Science, Microstructural Characterization of Materials by Non-destructive Techniques, Risø National Laboratory, DK-4000 Roskilde, Denmark, September 1984, Risø National Laboratory, pp. 169–174.

[9]   Allen A., Hutchings M. T. and Windsor C. G., Neutron diffraction methods for the study of residual stress fields. *Adv. Phys.*, **34**, 445–473 (1985).

[10] Holden T. M., Powell B. M., Dolling G. and MacEwen S. R., Internal strain measurements in over rolled cold-worked Zr 2.5wt% Nb pressure tubes by neutron diffraction. In N. Hessel Andersen, M. Eldrup, N. Hansen, D. Juul Jensen, T. Leffers, H. Lilholt, O. B. Pedersen and B. N. Singh, editors, *Proceedings of the 5th Risø International Symposium on Materials Science, Microstructural Characterization of Materials by Non-destructive Techniques*, Risø National Laboratory, DK-4000 Roskilde, Denmark, September 1984, Risø National Laboratory, pp. 291–294.

[11] Holt R. A., Dolling G., Powell B. M., Holden T. M. and Winegar J. E., Assessment of residual stresses in Incoly-800 tubing using X-ray and neutron diffraction. In N. Hessel Andersen, M. Eldrup, N. Hansen, D. Juul Jensen, T. Leffers, H. Lilholt, O. B. Pedersen and B. N. Singh, editors, *Proceedings of the 5th Risø International Symposium on Materials Science, Microstructural Characterization of Materials by Non-destructive Techniques*, Risø National Laboratory, DK-4000 Roskilde, Denmark, September 1984, Risø National Laboratory, pp. 295–300.

[12] Lorentzen T. and Christoffersen J., Limitations on the strain tensor determination by neutron diffraction using a position sensitive detector. *NDT-Int.*, **23**, 107–109 (1990).

[13] MacEwen S. R., Faber J. and Turner A. P. L., The use of time-of-flight neutron diffraction to study grain interaction stresses. *Acta Mater.*, **31**, 657–676 (1983).

[14] Turner P. A. and Tomé C. N., A study of residual stresses in Zircaloy-2 with rod texture. *Acta Metall.*, **42**, 4143–4153 (1994).

[15] Clausen B., Characterisation of polycrystal deformation by numerical modelling and neutron diffraction measurements. Technical Report Risø-R-985(EN), Risø National Laboratory, DK-4000 Roskilde, Denmark, 1997, 80pp.

[16] Clausen B. and Lorentzen T., A self-consistent model for polycrystal deformation; description and implementation. Technical Report Risø-R-970(EN), Risø National Laboratory, DK-4000 Roskilde, Denmark, 1997, 42pp.

[17] Clausen B. and Lorentzen T., Polycrystal deformation; experimental evaluation by neutron diffraction. *Metall. Mater. Trans. A*, **28A**, 2537–2541 (1997).

[18] Daymond M. R., Bourke M. A. M., Von Dreele R., Clausen B. and Lorentzen T., Use of Rietveld refinement for elastic macrostrain determination and for the evaluation of plastic strain history from diffraction spectra. *J. Appl. Phys.*, **82**, 1554–1562 (1997).

[19] Tomé C. N., Holden T. M., Turner P. A. and Lebensohn R. A., Interpretation of intergranular stress measurements in Monel-400 using polycrystal models. In T. Eriksson, M. Odén and A. Anderson, editors, *Proceedings of the 5th International Conference on Residual Stresses, ICRS-5*, Vol. 1, Linköping, Sweden, June 1997, Institute of Technology, Linköping University, pp. 40–45.

[20] Lorentzen T. and Clausen B., Self-consistent modelling of lattice strain response and development of intergranular residual strains. In T. Eriksson, M. Odén and A. Anderson, editors, *Proceedings of the 5th International Conference on Residual Stresses, ICRS-5*, Vol. 1, Linköping, Sweden, June 1997, Institute of Technology, Linköping University, pp. 454–459.

[21] Lebensohn R. A., Turner P. A., Signorelli J. W. and Tomé C., Selfconsistent calculation of intergranular stresses based on a large strain viscoplastic model. In T. Eriksson, M. Odén and A. Anderson, editors, *Proceedings of the 5th International Conference on Residual Stresses, ICRS-5*, Vol. 1, Linköping, Sweden, June 1997, Institute of Technology, Linköping University, pp. 460–465.

[22] Daymond M. R., Bourke M. A. M., Clausen B. and Tomé C., Effect of texture on *hkl* dependent intergranular strains measured in-situ at a pulsed source. In T. Eriksson, M. Odén and A. Anderson, editors, *Proceedings of the 5th International Conference on Residual Stresses, ICRS-5*, Vol. 1, Linköping, Sweden, June 1997, Institute of Technology, Linköping University, pp. 577–585.

[23] Pang J. W. L., Holden T. M. and Mason T., In situ generation of intergranular strains in zircaloy under uniaxial loading. In T. Eriksson, M. Odén and A. Anderson, editors, *Proceedings of the 5th International Conference on Residual Stresses, ICRS-5*, Vol. 2, Linköping, Sweden, June 1997, Institute of Technology, Linköping University, pp. 610–615.

[24] Lebensohn R. A., Turner P. and Pochettino A., Development of intergranular residual stresses in zirconium alloys. In T. Eriksson, M. Odén and A. Anderson, editors, *Proceedings of the 5th International Conference on Residual Stresses, ICRS-5*, Vol. 2, Linköping, Sweden, June 1997, Institute of Technology, Linköping University, pp. 781–786.

[25] Holden T. M., Clarke A. P. and Holt R. A., The impact of intergranular strains on interpreting stress fields in Inconel-600 steam generator tubes. In T. Eriksson, M. Odén and A. Anderson, editors, *Proceedings of the 5th International Conference on Residual Stresses, ICRS-5*, Vol. 2, Linköping, Sweden, June 1997, Institute of Technology, Linköping University, pp. 1066–1070.

[26] Holden T. M., Holt R. A. and Clarke A. P., Intergranular stresses in Incoloy-800. *J. Neutron Res.*, **5**, 241–264 (1997).

[27] Lorentzen T., Leffers T. and Clausen B., Polycrystal models and intergranular stresses. In J. V. Carstensen, Leffers T., Lorentzen T., O. B. Pedersen, B. F. Sorensen and G. Winther, editors, *Proceedings of the 19th Risø International Symposium on Materials Science, Modelling of Structure and Mechanics of Materials from Microscale to Product*, Risø National Laboratory, DK-4000 Roskilde, Denmark, September 1998, Risø National Laboratory, pp. 345–354.

[28] Clausen B., Lorentzen T. and Leffers T., Self-consistent modelling of the plastic deformation of f.c.c. polycrystals and its implications for diffraction measurements of internal stresses. *Acta Mater.*, **46**, 3087–3098 (1998).

[29] Pang J. W. L., Holden T. M. and Mason T. E., In situ generation of intergranular strains in an a17050 alloy. *Acta Mater.*, **46**, 1503–1518 (1998).

[30] Pang J. W. L., Holden T. M. and Mason T. E., The development of intergranular strains in a high-strength steel. *Int. J. Strain Anal. Eng. Des.*, **33**, 373–383 (1998).

[31] Holden T. M., Tomé C. N. and Holt R. A., Experimental and theoretical studies of the superposition of intergranular and macroscopic strains in Ni-based industrial alloys. *Metall. Mater. Trans. A*, **29A**, 2967–2973 (1998).

[32] Holden T. M., Holt R. A. and Clarke A. P., Intergranular stresses in Incoloy-600 and the impact on interpreting stress fields in bent steam-generator tubing. *Mater. Sci. Eng.*, **A246**, 180–198 (1998).

[33] Pang J. W. L., Holden T. M., Turner P. A. and Mason T. E., Intergranular stresses in Zircaloy-2 with rod texture. *Acta Mater.*, **47**, 373–383 (1999).

[34] Clausen B., Lorentzen T., Bourke M. A. M. and Daymond M. R., Lattice strain evolution during uniaxial tensile loading of stainless steel. *Mater. Sci. Eng.*, **A259**, 1724 (1999).

[35] Sayers C. M., The strain distribution in anisotropic polycrystalline aggregates subjected to an external stress field. *Philos. Mag. A*, **49**, 243–262 (1984).

[36] Allen A. J., Bourke M., David W. I. F., Dawes S., Hutchings M. T., Krawitz A. D. and Windsor C. G., Effect of elastic anisotropy on the lattice strains in polycrystalline metals and composites measured by neutron diffraction. In G. Beck, S. Denis and A. Simon, editors, *Proceedings of the 2nd International Conference on Residual Stresses, ICRS-2*, London, 1989, Elsevier Applied Science, pp. 78–83.

[37] Glocker R., Einfluß einer elastischen Anisotropie auf die röntgenographische Messung von Spannungen. *Z. Techn. Phys.*, **19**, 289–293 (1938).

[38] Bollenrath F. and Schiedt E., Röntgenographische Spannungsmessungen bei Überschreiten der Fließgrenzen an Biegestäben aus Flußstahl. *VDI Z.*, **82**, 1094–1098 (1938).

[39] Bollenrath F., Hauk V. and Osswald E., Röntgenographische Spannungsmessungen bei Überschreiten der Fließgrenzen an Zugstäben aus unlegiertem Stahl. *VDI Z.*, **83**, 129–132 (1939).

[40] Bollenrath F. and Osswald E., Über den Beitrag einzelner Kristallite eines vielkristallinen Körpers zur Spannungsmessung mit Röntgenstrahlen. *Z. Metallkde.*, **31**, 151–159 (1939).

[41] Smith S. L. and Wood W. A., A stress–strain curve for the atomic lattice of iron. *Proc. Roy. Soc.*, **A178**, 93–106 (1941).

[42] Smith S. L. and Wood W. A., Internal stresses created by plastic flow in mild steel, and stress–strain curves for the atomic lattice of higher carbon steels. *Proc. Roy. Soc.*, **A182**, 404–414 (1944).

[43] Greenough G. B., Residual lattice strains in plastically deformed metals. *Nature*, **160**, 258–260 (1947).

[44] Greenough G. B., Residual lattice strains in plastically deformed polycrystalline metal aggregates. *Proc. Roy. Soc.*, **A197**, 556–567 (1949).

[45] Greenough G. B., Internal stresses in worked metals, lattice strains measured by X-ray techniques in plastically deformed metals. *Met. Treat. A: Drop. Forg.*, **16**, 58–64 (1949).

[46] Greenough G. B., Quantitative X-ray diffraction observations on strained metal aggregates. *Proc. Metal Phys.*, **3**, 176–219 (1952).

[47] Heyn E., Eine Theorie der Verfestigung von metallischen Stoffen infolge Kaltreckens. In *Festschrift der Kaiser Wilhelm Gesellschaft zur Forderung der Wissenschaften, zu ihrem zehnjährigen Jubiläum*, Berlin, 1921. Verlag von Julius Springer, pp. 121–131.

[48] Noyan I. C. and Cohen J. B., *Residual Stress, Measurement by Diffraction and Interpretation*. MRE, Materials Research and Engineering, 1987, Springer-Verlag, New York, 1st edition.

[49] Lorentzen T., Measurement and interpretation. In *Introduction to Characterisation of Residual Stress by Neutron Diffraction*, Taylor and Francis, to be published.

[50] Nye J. F., *Physical Properties of Crystals, their Representation by Tensors and Matrices*, 1967, Clarendon Press, Oxford, 4th edition.

[51] Kröner E., Berechnung der elastischen Konstanten des Vielkristalls aus den Konstanten des Einkristalls. *Z. Phys.*, **151**, 504–518 (1958).

[52] Kneer G., Die elastischen Konstante quasiisotroper Vielkristallaggregate. *Phys. stat. sol.*, III, k331–k335 (1963).

[53] Behnken H. and Hauk V., Berechnung der röntgenographischen elastitätskonstanten (rek) des Vielkristalls aus den Einkristalldaten für beliebige Kristallsymmetrie. *Z. Metallkde*, **77**, 620–625 (1985).

[54] Sachs Z., Zur Ableitung einer Fliessbedingung. *Z. Vereines Deutsch. Ing.*, **72**, 734–736 (1928).

[55] Taylor G. I., Plastic strain in metals. *J. Inst. Metals*, **62**, 307–324 (1938).

[56] Eshelby J. D., The determination of the elastic field of an ellipsoidal inclusion, and related problems. *Proc. Roy. Soc.*, **A241**, 376–396 (1957).

[57] Eshelby J. D., Elastic inclusion and inhomogeneities. In I. N. Sneddon and R. Hill, editors, *Progress in Solid Mechanics*, 1961, North Holland, Amsterdam, pp. 89–140.

[58] Hill R., Continuum micromechanics of elastoplastic polycrystals. *J. Mech. Phys. Solids*, **13**, 89–101 (1965).

[59] Hill R., A self-consistent mechanics of composite materials. *J. Mech. Phys. Solids*, **13**, 213–222 (1965).

[60] Hill R., Generalized constitutive relations for incremental deformation of metal crystals by multislip. *J. Mech. Phys. Solids*, **14**, 95–102 (1966).

[61] Hill R., Essential structure of constitutive laws for metal composites and polycrystals. *J. Mech. Phys. Solids*, **15**, 79–95 (1967).

[62] Hutchingson J. W., Elastic–plastic behaviour of polycrystalline metals and composites. *Proc. Roy. Soc.*, **A319**, 247–272 (1970).

[63] Nemet-Nasser S. and Obata M., Rate-dependent finite elasto-plastic deformation of polycrystals. *Proc. Roy. Soc.*, **A407**, 343–375 (1986).

[64] Harren S. V., The finite deformation of rate-dependent polycrystals – i: A selfconsistent framework. *J. Mech. Phys. Solids*, **39**, 345–360 (1991).

[65] Harren S. V., The finite deformation of rate-dependent polycrystals – ii: A comparison of the self-consistent and Taylor methods. *J. Mech. Phys. Solids*, **39**, 361–383 (1991).

[66] Lebensohn R. A. and Tomé C. N., A self-consistent anisotropic approach for the simulation of plastic deformation and texture development of polycrystals: application to zirconium alloys. *Acta Metall.*, **41**, 2611–2624 (1993).

[67] Corvasce F., Lipinski P. and Berveiller M., Second order residual and internal stresses in microin-homogeneous granular materials under thermomechanical loading. In G. Beck, S. Denis and A. Simon, editors, *Proceedings of the 2nd International Conference on Residual Stresses, ICRS-2*, London, 1989, Elsevier Applied Science, pp. 399–404.

[68] Corvasce F., Lipinski P. and Berveiller M., Intergranular residual stresses in plastically deformed polycrystals. In G. Beck, S. Denis and A. Simon, editors, *Proceedings of the 2nd International Conference on Residual Stresses, ICRS-2*, London, 1989, Elsevier Applied Science, pp. 535–541.

[69] Lorentzen T., Faurholdt T., Clausen B. and Danckert J., Characterisation of residual stresses generated during inhomogeneous plastic deformation. *Int. J. Strain Anal. Eng. Des.*, **33**, 243–252 (1997).

Part 3

Measurement techniques

8 Neutron diffraction using a constant wavelength

L. Pintschovius

8.1 Introduction

Neutron stress analysis was started in the early 1980s [1–3] using instruments which were designed for very different purposes. Gradually, more and more groups joined the field, again using existing instruments which were adapted, to some degree, to meet the special requirements of neutron stress analysis. Eventually, some instruments were transformed into dedicated instruments for neutron stress analysis which generally implied substantial modifications of major components. Still, it was difficult, if not impossible, to modify the instruments to an extent as needed to exploit the potential of the method in full. It is only very recently that dedicated instruments are built from scratch which allows one to translate the ideas of an ideal strain scanner into reality. These ideas are the fruit of instrumental developments pursued by several groups working at many research reactors all over the world. In the following chapter, I will outline these ideas and will comment on the price which has to be paid if constraints do not allow one to utilize all of these ideas. In this chapter, only instruments using a constant wavelength will be considered, whereas time-of-flight (TOF) instruments will be dealt with in the following chapter. The question whether the constant wavelength technique or the TOF technique is more advantageous depends largely on the type of the neutron source, that is, whether it is a continuous or a pulsed source. So, the vast majority of diffractometers operated at research reactors use a constant wavelength.

When talking of an *ideal strain scanner*, this conception presupposes a certain field of application. As such, I consider the investigation of stress fields with a spatial resolution of typically a few cubic millimeters. Neutron stress analysis has been used as well for the determination of phase specific stresses or grain interaction stresses which are spatially homogeneous over a length scale of several centimeters. I will not consider the specific requirements of measurements of such a kind, not least because such measurements are better done using pulsed neutrons. The remaining field of neutron stress analysis is still so broad that no instrument can be conceived as an ideal strain scanner unless it is very versatile. For instance, high resolution is, in general, imperative for the determination of strains in ceramic materials because these strains are usually quite small. On the other hand, high resolution is of little help for the investigation of macrostresses in martensitic steels or other materials showing large intrinsic linewidths. I will come back to these issues later on.

For those who are not familiar with neutron diffractometers, the main components of an instrument for neutron stress analysis will be explained in the next section. Then, each of the major components will be dealt with in more detail in the following sections with the aim to point out the critical features which are necessary for a high performance of the instrument.

8.2   Overall view of a diffractometer for strain measurements

The outline of a neutron diffractometer for strain measurements is shown in Figure 8.1. A beam of polychromatic neutrons emanating from the neutron source impinges on the monochromator which reflects a monochromatic beam according to Bragg's law

$$2d_{hkl} \sin \theta_m = n\lambda \tag{1}$$

where d_{hkl} denotes the lattice spacing of the monochromator, θ_m the scattering angle at the monochromator, λ the neutron wavelength and n is a small integer number. The factor n means that the beam is not fully monochromatic but contains a certain amount of higher harmonics ($n > 1$) which, however, is rarely relevant for stress measurements. A small fraction of the beam hitting the sample is diffracted. As the incident beam is monochromatic, the diffracted beam is confined to a well-defined direction, again given by Bragg's law. Finally, the diffracted beam is recorded by a neutron detector.

Collimators placed before and after the monochromator and after the sample define the angular spreads of the neutron beam α_i and β_i in the horizontal and in the vertical plane, respectively. The relatively wide (≈ 4 cm) beam generated by a typical neutron monochromator is narrowed down by a slit, a few millimeters wide, placed just before the sample.

Similarly, the diffracted beam is confined by another slit placed just after the sample. These two slits define the gauge volume which is normally much smaller than the sample itself (for details, see Section 8.5.1). Stress mapping over the entire sample is done by moving the sample stepwise with respect to the gauge volume.

A realistic outline of a modern strain scanner [4] is shown in Figure 8.2. Development of neutron diffractometers for strain measurements has not yet resulted in a standard

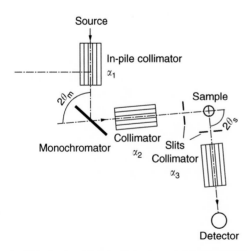

Figure 8.1 Schematic diagram of a neutron diffractometer.

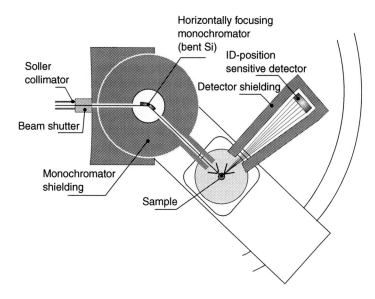

Figure 8.2 Outline of the neutron strain scanner at the LVR-15 reactor at Rez. Note that the relatively small monochromator take-off angle $2\theta_m$ of this particular instrument is offset by the use of a focusing monochromator to achieve high resolution (reprinted from [4]).

design, so that Figure 8.2 depicts just one example of the many different instruments around the world.

8.3   Production of monochromatic neutrons

8.3.1   *The beam line*

A neutron diffractometer for stress analysis should be located at a thermal beam tube to allow one to extract neutrons of small as well as of large wavelengths. Unfortunately, such positions are not always available, whereas it is usually rather easy to get access to a position at a neutron guide. In the case that the guide is fed directly by the reflector and not by a cold source, a diffractometer for stress analysis located at such a guide can compete fairly well with an instrument located at a beam tube: the relatively small horizontal divergence of a neutron beam emanating from a thermal neutron guide is in many cases not detrimental because it is needed to achieve high resolution anyway; the small vertical divergence can be overcome by using vertically focusing monochromators. An example of such an instrument used very successfully for neutron stress analysis is the diffractometer D1A at the high flux reactor of the ILL, Grenoble. If, however, the neutron guide is fed by a cold source (as is the case, e.g. for the strain scanner G5.2 of the LLB, Saclay), this entails the serious drawback that the flux of neutrons with wavelengths shorter than 2 Å is prohibitively low. As a consequence, there will be often just one set of lattice planes {hkl} of the material under investigation suitable for strain measurements (for instance, in the case of ferritic steels, just the {110}-reflection line). As a consequence, it is impossible to investigate the effects of elastic and plastic anisotropies.

8.3.2   The monochromator housing

The monochromator housing should allow one to continuously vary the neutron wavelength in order to achieve scattering angles $2\theta_s \approx 90°$ for all the materials to be investigated. A scattering angle $2\theta_s \approx 90°$ is usually chosen to achieve the best spatial resolution for a given width of the slits before and after the sample. Hence, it follows that the condition $2\theta_s \approx 90°$ need not be met strictly, but only approximately so. Therefore, it is no serious drawback if the instrument allows one to vary the monochromator angle $2\theta_m$ only in steps of $\sim 10°$, provided these changes can be performed with little loss of beam time. If for some reason $2\theta_m$ is fixed, the monochromator housing should accommodate at least two, or better still three to five monochromators with different d-spacings with the possibility of a rapid interchange of these monochromators to select a suitable neutron wavelength. A rather elegant solution to this problem is the use of a properly cut monochromator crystal, of which the reflecting planes $\{hkl\}$ are chosen by a proper rotation (as is done, e.g. on the diffractometer D1A of the ILL, Grenoble).

A rapid interchange between different monochromators is desirable as well for instruments with continuously variable take-off angles $2\theta_m$, because resolution considerations impose constraints on the choice of $2\theta_m$ (see the next section). The range of $2\theta_m$ should extend to at least $\sim 90°$, better $\sim 110°$, again for reasons of resolution. Unfortunately, most available locations impose serious constraints on the maximum 2θ achievable due to space problems. In the case where $2\theta_m$ is limited to about $60°$ or even less, the use of bent perfect crystals as a monochromator is particularly attractive because high resolution and good luminosity can be achieved with this technique for relatively small $2\theta_m$ values by a proper choice of the bending radius (see Section 8.3.3.2).

A set of two or three easily interchangable Soller collimators before the monochromator is of considerable utility in achieving a high resolution or a high flux in keeping with the requirements of a particular experiment. As is explained in the next section, this comment applies only to the case of mosaic crystals as monochromators, whereas no Soller collimators should be placed in front of bent perfect crystals.

8.3.3   The monochromator

8.3.3.1   Mosaic crystals

Mosaic crystals, that is, crystals which are made up of small blocks which show a spread of lattice misorientations relative to one another, are the standard choice for the monochromator of a powder diffractometer. For this reason, they have been used throughout the development of neutron stress analysis. Mosaic crystals are now gradually replaced by bent perfect crystals for many applications, but they will not come completely out of use in the foreseeable future. The most widely used monochromator materials are pyrolitic graphite (PG), copper, and hot-pressed germanium, because they were found to give the highest peak reflectivities. In principle, other materials like Be are attractive as well, but they are not readily available in the form of large mosaic crystals.

It has long been known [5, 6] that a powder diffractometer offers the highest luminosity for a certain resolution if the scattering angles at the monochromator $2\theta_m$ and at the sample $2\theta_s$ are about the same:

$$2\theta_m \approx 2\theta_s \tag{2}$$

When condition (2) is fulfilled, the horizontal divergence α_2 between the monochromator and the sample can be opened up without affecting the resolution (in this sense, it is the focusing condition for flat mosaic crystals). In neutron stress analysis, however, α_2 has often to be kept fairly low ($<1°$) for the sake of a precise definition of the gauge volume shaped by slits in the incident beam. With decreasing α_2, the resolution depends less sensitively on matching $2\theta_m$ with $2\theta_s$. Furthermore, the optimum is shifted towards larger $2\theta_m$ (for instance, for $2\theta_s = 90°$ and a collimation $0.2°-0.5°-0.2°$, the optimum resolution is found for $2\theta_m \approx 100°$). From these considerations it follows that the monochromator d-spacing should be chosen such as to bring $2\theta_m$ into the range 80–120° for the wavelength needed to obtain $2\theta_s = 90°$. If the monochromator housing does not permit take-off angles in this range, all that can be done is to maximize $2\theta_m$ within the given constraints to come at least relatively close to the optimum working conditions. Further, two or three in-pile collimators defining α_1 should be available to achieve high or medium resolution and high intensity, as needed. I note that the need for a tight in-pile collimation cannot be replaced by tight collimations before and after the sample to arrive at a high resolution. So an in-pile collimator of the order of $0.2°$ is highly desirable for a neutron diffractometer for stress analysis (if the instrument is fed by a neutron guide, the natural collimation will be of the right order for thermal neutrons, so no in-pile collimator is needed).

In general, a considerable gain in intensity can be achieved by vertical focusing without loss in resolution. There is, however, a price to be paid for vertical focusing which is the loss in definition of the gauge volume in the vertical direction due to penumbra effects (see, e.g., the discussion in Ref. [7]). Since there are many applications where the gauge volume is made very long in the vertical direction anyway so that penumbra effects are not critical, monochromator crystals for neutron stress analysis should all be made vertically focusing.

8.3.3.2 Bent perfect crystals

As has been pointed out by Vrána *et al.* [8], bent perfect crystals are an attractive alternative to mosaic crystals as monochromators for neutron strain scanners. An outline of such a monochromator is shown in Figure 8.3. The curvature is achieved by a four-point bending scheme, while keeping the material in the elastic regime. A neutron traveling through the crystal crosses the reflecting planes at slightly different angles when going from the front to the rear side leading to a finite effective mosaic width η despite the fact that the crystal is perfect. However, the strength of the material severely limits the achievable curvature and thereby the achievable η. So, typically, the effective mosaic spread of a bent perfect crystal is only a few minutes of arc, that is, much smaller than that of mosaic crystals used as monochromators (which is typically $25'$!). On the other hand, the small effective mosaic width, which would lead to an unacceptably low neutron flux at the sample position for a mosaic crystal, is compensated by a high peak reflectivity (close to unity for most cases of practical interest) and, in particular, by the possibility to achieve perfect focusing in real space as well as in

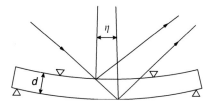

Figure 8.3 Schematic of bent perfect crystal monochromator. η denotes the effective mosaic spread.

reciprocal space by choosing a proper bending radius. As is explained in Refs. [4, 8], perfect focusing can be achieved without fulfilling condition (2), and so this technique is particularly attractive for instruments where $2\theta_m$ is limited to moderate values. For instance, Mikula *et al.* [4] report having achieved a resolution $\Delta d/d < 2 \times 10^{-3}$ for $2\theta_m = 57°$. Even when using bent perfect crystals, however, violating condition (2) does have a certain drawback: for $2\theta_m \neq 2\theta_s$, a focusing bent perfect crystal selects a non-parallel input beam. Therefore, the maximum useful divergence α_2 between the monochromator and the sample will be limited by the available in-pile divergence (which should not be reduced by an in-pile Soller collimator; I note that the in-pile Soller collimator shown in Figure 8.2 is in place only for technical reasons, but is detrimental to the performance of the instrument when a bent Si crystal is used as a monochromator [9]). In practice, however, this is not a serious handicap, as α_2 will rarely be chosen larger than $1°$ for the sake of a good definition of the internal gauge volume.

At present, Si is the standard choice for neutron monochromators, because it is readily available in large size as perfect crystals and sustains rather large stresses elastically, so that the bending radius can be changed reversibly. Ge appears, in principle, to be even better suited than Si because of the larger neutron scattering power of Ge, but is much more expensive and also more brittle. Vertical focusing can be achieved along with horizontal focusing by elastically bending a curved stack of Si lamellae [10–12]. Recently, as many as 26 vertically stacked lamellae have been assembled in a bending unit to produce a vertically and horizontally focusing monochromator [13].

The concept of bent perfect crystals sounds so convincing that one may ask whether bent perfect crystals will be the best choice in all cases. The major limitation comes from the fact that the effective mosaic spread cannot be made larger than a few minutes of arc without breaking the crystals – at least with the present technology. Therefore, if the experiment cannot benefit from the high resolution offered by this technique because of an inherently strong line-broadening, mosaic crystals will provide a considerably higher intensity for about the same total linewidth.

8.3.3.3   Microfocusing monochromators

The term "microfocusing" has been used by Popovici and Yelon [14, 15] for a set-up where a wide neutron beam is focused onto a narrow line at the sample position. In principle, microfocusing could be achieved by just one bent crystal, but due to the large divergence of a typical neutron beam hitting the monochromator this would require a very short monochromator-to-sample distance which is impractical. Therefore, Popovici and Yelon [14, 15] proposed using a double monochromator in antiparallel $(+, +)$ reflection as outlined in Figure 8.4. Crystal 1 is placed at the usual position of a monochromator surrounded by heavy shielding, whereas crystal 2 is placed roughly 50 cm away from the sample [16]. With pneumatically bent perfect Si crystals, the width of the neutron beam at focus could be brought down to as little as 2 mm. The use of spherically bent Cu mosaic crystal led to a considerably larger beamwidth ($\approx$5 mm), on the one hand because of the finite mosaic width before bending, and on the other hand because Cu crystals can be bent only plastically which leads to an anisotropic broadening of the mosaic width. The beam height at focus reported in Ref. [16] is rather large (several centimeters), but might be reduced by making the vertical curvature much stronger than the horizontal one. Popovici and Yelon note that crystal 1 of the double monochromator might be flat, which makes it easy to convert a standard monochromator to a microfocusing set-up.

Focusing in real space is, of course, only of help for strain measurements if focusing in reciprocal space can be achieved at the same time. As has been outlined in Refs. [14, 15],

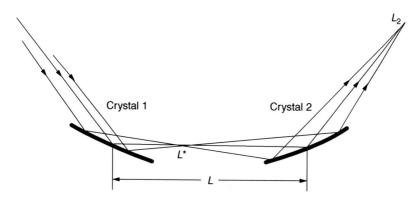

Figure 8.4 Geometry of double antiparallel reflection by bent crystals for neutron microfocusing (reprinted from [15]).

focusing in reciprocal space can be achieved with a microfocusing monochromator as well as by a proper choice of the distances L, L^* and L_2 (see Figure 8.4) and of the bending radii of the crystals for a particular scattering angle $2\theta_s$. Test measurements, largely in line with computations, yielded a minimum linewidth of about $0.6°$ for a scattering angle $2\theta_s \approx 80°$ [16]. This means that microfocusing does not allow one (at least at present) to achieve high resolution, but the available resolution is sufficient for many applications. Likewise, the spatial resolution achievable with microfocusing is not good enough for many applications, but for others it is (I note that the beam width at focus cannot be efficiently reduced by slits because of the strong convergence of the beam). Finally, one has to say that a microfocusing monochromator has to be adjusted very carefully for a particular wavelength and scattering angle $2\theta_s$, because the minimum of the resolution versus $2\theta_s$ curve is very narrow (see, e.g., figure 10 of Ref. [15]). As a consequence, it is not easy to switch from one d-spacing to another, so that in practice only a limited number of materials will be accessible. In summary, microfocusing will not become the standard technique for neutron stress analysis, but for a certain field of applications it will speed up the measurements by an order of magnitude or even more.

8.4 Detectors

8.4.1 *Standard He^3-detectors*

In the beginning, neutron stress measurements were performed with standard He^3-detectors in combination with Soller collimators or slit systems. In recent years, they have been largely replaced by position sensitive detectors (PSDs) although available PSDs are not always a better choice than optimized single detectors. To begin with, the efficiency of a standard He^3 detector (2.5 cm$\emptyset$) is close to unity, whereas the efficiency of some PSDs is as low as 50% [17]. Second, a single He^3-detector can be made very long, so that the vertical acceptance is very large ($\pm 13°$). As has been pointed out in Ref. [18], the large vertical acceptance does not affect the linewidths in $2\theta_s$ as long as $2\theta_s \approx 90°$, which is the usual case in neutron stress analysis. Furthermore, the loss of definition in terms of $\sin^2 \psi$ due to the large vertical acceptance is rather small, as long as measurements are made along principal strain directions – again the standard case in neutron stress analysis (in this case, the interval

$\pm 13°$ translates into an interval $-0.05 < \sin^2 \psi < 0.05$). For these reasons, the large vertical acceptance of a single He^3-tube may make it competitive with an available PSD for a number of applications – at least at present, where PSDs rarely meet all the requirements of neutron stress analysis (see the next section).

8.4.2   Position sensitive detectors

Position sensitive detectors are now in widespread use for strain measurements. Such a PSD has to be only one-dimensional (1D) and has to cover an angular range of a few degrees only. On the other hand, it should have a good angular resolution ($0.1°$), a large vertical acceptance ($\geq 10°$), a high efficiency ($\approx 90\%$) and a low electronic background. Although it seems technically feasible to meet all these requirements to a large extent, PSDs actually used for strain measurements often fall considerably short of what is desirable. In particular, most available PSDs have a rather small vertical acceptance. For instance, the PSD of the instrument shown in Figure 8.2 has a vertical acceptance of $\approx \pm 1.5°$ only [9], and that of the strain scanner of the LLB, Saclay, is $\approx \pm 3°$ [19].

The major reason for the gap between the performance of actual PSDs and what seems technically feasible is the fact that the development of a special PSD for neutron strain scanners appeared to be too costly. To my knowledge, the PSD which comes up closest to the requirements of stress measurements is the one described by Berliner *et al.* [20]. It is a linear position sensitive proportional counter detector array, whose height can be made to any desired vertical acceptance by assembling the necessary number of elements. There are some losses associated with the many detector shells in the sensitive area, but an area-averaged efficiency as high as 73% at a neutron wavelength of $1.0\,\text{Å}$ has been achieved nevertheless. Probably, costs will be the major road block for the widespread use of such a PSD, but it would certainly boost the performance of many instruments.

8.4.3   Shielding

In neutron stress measurements, counting rates are often quite low as a consequence of very small gauge volumes and/or strong absorption associated with long neutron flight paths within the component investigated. Therefore, detectors must be more thoroughly shielded than for standard powder diffraction measurements. The aim of preventing stray neutrons entering into the detector is greatly facilitated by surrounding the entire flight path of the neutrons from the monochromator shielding to the entrance slits and from the exit slit to the detector by some shielding material as indicated in Figure 8.2. Shielding materials are primarily chosen for their stopping power of neutrons. However, it should be checked whether there is some need for shielding against gamma rays as well. For instance, the high background counting rate observed with the PSD at the G5.2-diffractometer at Saclay [21] has been drastically lowered by additional lead shielding.

8.5   Sample environment

8.5.1   Slits for defining the gauge volume

Apart from the special case of a microfocusing monochromator, the gauge volume within the sample is defined by slits both in the incoming and in the scattered beam (see Figure 8.5). The shape of the gauge volume need not to be a cube (Figure 8.5, right). As indicated in Figure 8.5,

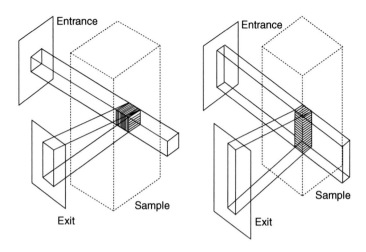

Figure 8.5 Creation of the gauge volume using slits in the incident and in the scattered beam.

it is generally advantageous to define the vertical dimensions of the gauge volume by the entrance slit only, that is, to use a large vertical divergence on the detector side to gain intensity.

In order to keep penumbra effects to a minimum, entrance as well as exit slits should be placed as close to the sample as possible. Tightly placed slits require, of course, great care to avoid collisions with the sample when changing the orientation or the position of the sample. Nevertheless, the user should not forget how stringent the requirements for the slit positions are if a spatial resolution less than 2 mm is aimed at. For instance, the image of a slit 1 mm wide will be blurred to $\approx$2 mm at a distance $l = 10$ cm when the beam has a divergence of 1°. Proper placing of the slits is facilitated by mounting them on optical guides aligned along the incoming and along the scattered beam, respectively.

In the case that the sample is too bulky to allow one to put the slits close to the gauge volume, the ultimate means is to use convergent collimators instead of slits to define the gauge volume. Radial collimators were first developed for a TOF spectrometer [22] because it is a prerequisite to exploit the very large useful range in 2θ of a TOF instrument ($\approx$15°). Based on this experience, a collimator with a focal length of 150 mm giving a spatial resolution of about 1 mm has been manufactured by Euro Collimators Ltd for the strain scanner at the ILL, Grenoble [23]. When compared to a set-up using a slit placed very close to the gauge volume giving the same spatial resolution, the radial collimator entails a loss in intensity by a factor $\approx$2.5 which is the price to be paid for having a long focal length. Using Cd masks at the exit of the radial collimator, the spatial resolution can be further improved to 0.73 mm at the expense, of course, of a further drastic loss in intensity by a factor $\approx$3.

Usually, the gauge volume will be completely immersed in the sample. However, when stresses near surfaces or interfaces are of interest, it is often desirable to shift the gauge volume so close to the surface/interphase that it is only partially immersed. In this way, the distance between the surface and the centroid of the gauge volume might be made considerably smaller than the slit width. There is, however, a serious drawback with this procedure apart from the obvious loss in intensity: in general, partial immersion of the gauge volume is associated with systematic errors, sometimes called pseudostrains (see Figure 8.6). This effect, which has been reported several times in the literature (e.g. [24–26]), has been studied in detail by

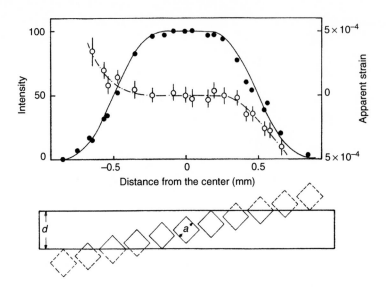

Figure 8.6 Intensity (filled circles) and pseudostrain (open circles) for a stress-free sample versus the distance from the sample center to the center of the gauge volume (reprinted from [25]).

Pluyette *et al.* [27, 28] and Piant-Garot *et al.* [29, 30] (see also Chapter 14 of this book). It is found that the effect depends not only on the parameters of the instrument (collimations, θ_m, θ_s) but also on the angle between the surface and the incident beam. That is to say, the pseudostrains will be different in reflection or transmission geometry. They will further strongly depend on the distance between the slits and the gauge volume: placing the slits as close as possible to the sample will not only keep the loss in definition due to penumbra effects to a minimum but at the same time, keep the systematic errors associated with a partial immersion of the gauge volume to a minimum. In any case, it is recommended to study the effect for the actual configuration of the measurement on a stress-free reference sample and then to correct for it in the evaluation of near-surface strains.

8.5.2   *Positioning the sample along x–y–z*

Positioning of the sample along $x–y–z$ should be computer controlled with an accuracy of $\sim$0.01 mm or better. The translation table should be able to accommodate rather bulky and heavy ($\geq$50 kg) items. Dedicated stress diffractometers are now (2002) being designed with 500 kg capability. Diffractometers of the *Tanzboden*-type have the special advantage that distances between monochromator and sample as well as between sample and detector can be chosen at will, so that these distances can be kept to a minimum for small samples but can be increased to handle very bulky samples as well. Precisely aligned laser beams (as installed, e.g., at the strain scanner of the HMI, Berlin) are of great help in locating the gauge volume relative to the sample position.

8.5.3   *Eulerian cradles*

Many neutron stress measurements are restricted to measurements along principal strain directions inferred from the sample geometry. Such measurements will not benefit from an Eulerian cradle, whereas others certainly do, in particular when measurements are to be performed in arbitrary directions to determine the principal axes of the stress tensor. Rather large Eulerian cradles (diameter up to 1 m) of different designs are now commercially available. A description of an Eulerian cradle equipped with an x–y–z positioning table can be found in Ref. [31].

8.5.4   Loading devices

As for X-ray stress analysis, tensile test rigs are a very useful complement to neutron strain scanners to study the behavior of the material as a function of the applied stress. For instance, the strain scanner at the LLB, Saclay, has been installed with a device having the capacity to apply tensile loads up to 20 kN [19]. The load may be applied cyclically with a frequency up to $\nu = 6$ Hz. Other facilities such as ISIS (UK) and LANSCE (US) have devices with capacities of 50 kN and 100 kN respectively. The relatively large dimensions of a neutron diffractometer make it rather easy to develop loading devices for other loading conditions as well. For instance, devices for applying a torsional stress have been described in Refs. [31, 32].

8.5.5   Controlling the temperature

Technical components are often used at elevated temperatures, so that it is of great interest to study the residual strains or load strains at elevated temperatures as well. An elegant way of achieving temperatures up to 1700°C during a strain measurement has been reported by Reimers [33]: the sample is heated with the help of quartz-halogen lamps with a maximum power of 2000 W, whereby the light is transferred by specially designed mirrors to the sample located at focus. A similar furnace is available at the ILL, Grenoble [23].

8.6   Computer control and data processing

An efficient use of the beam time in neutron stress measurements requires computer control of all the major parameters, that is, rotation angles as well as sample position, possibly also the load applied to the sample. For dedicated instruments this goal is now largely achieved. Data acquisition and data analysis is relatively simple for diffractometers using single detectors. With the advent of PSDs, this task is far from being trivial, and hardware and software development is being pursued for each of the systems actually employed (see, e.g., Ref. [20]). The aim is not to burden the user with corrections for non-uniform detector efficiency and nonlinearities of the position response. Further, the system should allow the user to select a particular angular range and to scan the integrated intensity over this range versus the sample position, as it is often quite time-consuming to line up the sample with respect to the location of the gauge volume.

Finally, speeding up strain measurements by using optimized monochromators and optimized PSDs generates lots of data within a short time, and so it is highly desirable to automate peak fitting not only for the most basic case of a single peak sitting on a flat background, but for other cases as well. From the rapid progress in commercially available software it can be expected that this goal will be universally achieved in the near future.

8.7   Concluding remarks

In this chapter, I have tried to summarize present-day knowledge about the optimum design of a neutron diffractometer for strain measurements. Although a shortage of resources – in terms of manpower as well as in terms of funds – has often slowed down instrumental developments considerably, substantial progress in instrument performance has been achieved at many places during the last decade: existing facilities have been upgraded with impressive gain factors, or dedicated instruments are being built from scratch. This means that the potential of neutron diffraction for stress analysis is now much better exploited than in the early days, when instruments had to be used which were designed for a completely different purpose. This should give the technique a strong boost in the near future.

References

[1]   Allen A. J., Andreani C., Hutchings M. T. and Windsor C. G., Measurement of internal stress within bulk materials using neutron diffraction, *NDT Int.*, October, 249–254 (1981).

[2]   Krawitz A. D., Brune J. E. and Schmank M. J., Measurement of stress in the interior of solids with neutrons, in *Residual Stress and Stress Relaxation*, E. Kula and V. Weiss (eds), Plenum, New York, 1982, pp. 139–155.

[3]   Pintschovius L., Jung V., Macherauch E., Schäfer R. and Vöhringer O., Determination of residual stress distribution in the interior of technical parts by means of neutron diffraction, *Residual Stress and Stress Relaxation*, in E. Kula and V. Weiss (eds), Plenum, New York, 1982, pp. 467–482.

[4]   Mikula P., Vrána M., Lukáš P., Šaroun J., Strunz P., Ullrich H. J. and Wagner V., Neutron diffractometer exploiting Bragg diffraction optics: a high resolution scanner, in *Proceedings of the ICRS-5*, June 16–18, 1997, Linköping, Sweden, 721 (1998).

[5]   Caglioti G., Paoletti A. and Ricci F. P., Choice of collimators for a crystal spectrometer for neutron diffraction, *Nucl. Instrum.* **3**, 223–228 (1958).

[6]   Caglioti G. and Ricci F. P., Resolution and luminosity of crystal spectrometers for neutron diffraction, *Nucl. Instrum. Methods* **15**, 223–228 (1962).

[7]   Webster P. J., Spatial resolution and strain scanning, in *Measurements of Residual and Applied Stress Using Neutron Diffraction*, M. T. Hutchings and A. D. Krawitz (eds), Kluwer Academic Publishers, Dordrecht, 1992, pp. 235–251

[8]   Vrána M., Lukáš P., Mikula P. and Kulda J., Bragg diffraction optics in high resolution strain measurements, *Nucl. Instrum. Methods A* **338**, 125–131 (1994).

[9]   Mikula P., private communication.

[10] Popovici M. and Yelon W. B., A high performance focusing silicon monochromator, *J. Neutron Res.* **5**, 227–239 (1997).

[11] Stoica A. D., Popovici M., Spooner S. and Hubbard C. R., in *Proceedings of the Sixth International Conference on Residual Stress – ICRS-6*, Oxford, July 10–12, 2000, pp. 1264–1271.

[12] Minakawa N., Morii Y. and Shimojo Y., A focusing monochromator of neutron diffractometer for residual stress analysis, in *Proceedings of the Sixth International Conference on Residual Stress – ICRS-6*, Oxford, July 10–12, 2000, pp. 1107–1111.

[13] Popovici M., private communication.

[14] Popovici M. and Yelon W. B., Design of microfocusing bent-crystal double monochromators, *Nucl. Instrum. Methods A* **338**, 132–135 (1994).

[15] Popovici M. and Yelon W. B., Design of microfocusing with bent crystals, *Z. Kristall.* **209**, 640–648 (1994).

[16] Popovici M. and Yelon W. B., Focusing monochromators for neutron diffraction, *J. Neutron Res.* **3**, 1–25 (1995).

[17] Pyzalla A., private communication.

[18] Pintschovius L., Optimisation d'un diffractomètre pour la caractérisation des contraintes, in *Analyse des contraintes résiduelles par diffraction des neutrons et des rayons X*, A. Lodini and M. Perrin (eds), Commissariat à l'énergie atomique, 1996, pp. 129–136.

[19] Ceretti M., The new strain-dedicated neutron diffractometer at the Laboratoire Léon Brillouin, in *MAT-TEC 97*, A. Lodini (ed.), IITT-International, 1997, pp. 47–53.

[20] Berliner R., Charlton D., Yelon W. B., Krawitz A. D., Winholtz R. A., Manback B. C., Fjellfaag M. and Steinsvoll D., Hardware and software system design for high resolution linear position sensitive proportional counter detector arrays, in *Proceedings of the International Workshop on Data Acquisition Systems for Neutron Experimental Facilities*, Dubna 2–4, 1997, p. 199.

[21] Pintschovius L., Schreieck B. and Eigenmann B., Neutron, X-ray and finite element stress analysis on brazed components of steel and cemented carbide, *Textures Microstruc.* **33**, 263–278 (1999).

[22] Johnson M. W., Edwards L. and Withers P. J., ENGIN – A new instrument for engineers, *Physica B* **234–236**, 1141–1143 (1997).

[23] Pirling T., A new high precision strain scanner at the ILL, *Mater. Sci. Forum*, **312–324**, 206–211 (2000).

[24] Webster P. J., Mikks G., Wang X. D., Kang W. K. and Holden T. M., Impediments to efficient through-surface strain scanning, *J. Neutron Res.*, **3**, 27 (1995).

[25] Pintschovius L., Neutron diffraction methods, in *Structural and Residual Stress Analysis by Non-destructive Methods*, V. Hauk (ed.), Elsevier, Amsterdam, 1997, pp. 495–519.

[26] Wang D. Q., Harris I. B., Withers P. J. and Edwards L., Near-surface strain measurement using neutron diffraction, in *Proceedings of the ECRS-4*, S. Denis *et al.* (eds), June 4–6, 1996, Cluny, France, pp. 69–77.

[27] Pluyette E., Thesis, Champagne-Ardenne University, France, 1997.

[28] Pluyette E., Sprauel J. M., Lodini A., Perrin M., Ceretti M. and Todeschini P., Residual stresses evaluation near interfaces by means of neutron diffraction: modelling a spectrometer, in *Proceedings of the ECRS-4*, S. Denis *et al.* (eds), June 4–6, 1996, Cluny, France, pp. 153–163.

[29] Piant-Garot A., Thesis, Champagne-Ardenne University, France, 1996.

[30] Piant-Garot A., Lodini A., Holden T. M., Rogge R. and Braham C., in *Proceedings of the ICRS-5*, T. Ericsson, M. Odén and A. Andersson (eds), June 16–18, 1997, Linköping, Sweden, pp. 244–249.

[31] Pyzalla A. and Reimers W., Eigenspannungsanalyse an Werkstoffen und Bauteilen, *Materialprüfung (Mater. Testing)* **40**, 303–310 (1998).

[32] Allen A. J., Hutchings M. T., Windsor C. G. and Andreani C., Neutron diffraction methods for the study of residual stress fields, *Adv. Phys.* **3**, 445–473 (1985).

[33] Reimers W., Spiegelofensysteme für Spannungsanalysen bei hoher Temperatur, *Druckschrift des Hahn-Meitner-Instituts*, Berlin (1997).

9   Neutron pulsed source instrumentation

M. W. Johnson and M. R. Daymond

9.1   Introduction

Pulsed source diffractometers differ from their continuous flux counterparts in one simple respect: the time-of-flight of each neutron is measured, and used to determine its wavelength. Using this method a single, stationary detector placed at a scattering angle of 2θ is able to record an entire spectrum $I(\lambda)$ and determine the peaks in the scattering intensity when the Bragg condition $\lambda = 2d \sin \theta$ is met. In practice, more than one detector is used to increase the count rate of the instrument, and the individual spectra added together to provide an effective spectrum of $I(\lambda/2 \sin \theta)$.

Pulsed source instruments designed to measure strains in materials, like their continuous flux counterparts, achieve this by measuring the normal distance $d(hkl)$ between the $\{hkl\}$ lattice planes within a small volume of the sample – the gauge volume. The strain at this point is then calculated by comparing this measurement with the corresponding $d(hkl)$ within an un-strained part of the material ($d_0(hkl)$). An engineering instrument on a pulsed source is therefore a specialised neutron diffractometer with three important attributes:

- an ability to measure the atomic lattice spacings to a high precision for multiple diffraction peaks,
- an ability to make such measurements on a small 'gauge volume' within the component under study, and
- an ability to position the gauge volume accurately with the component.

The neutron is a very penetrating probe, and using such a specialised diffractometer it is possible to measure the strain, and hence stress, distribution along a line, within a plane, or within the volume of the component under study. Moreover, by rotating the sample the strain may be measured along any particular direction, thus enabling the vector strain field to be mapped out in three-dimensions.

In the remainder of this chapter, we first consider the nature of time-of-flight diffraction instruments and their characteristics. We then look at a specific example of a neutron strain scanning instrument, in this case the ENGIN instrument at the ISIS facility. Since two of the key aspects of an engineering instrument are the definition of the gauge volume and the measurement of the lattice parameter, these topics are given separate sections, together with a discussion of instrument alignment. Finally a description is given of how a time-of-flight instrument design may be optimised to deliver the best performance for a given source characteristic.

9.2   Time-of-flight powder diffractometers

Time-of-flight powder diffractometers are generally sited on accelerator-based spallation sources such as the ISIS facility at the Rutherford Appleton Laboratory (Oxfordshire, UK), Figure 9.1, though there is a pulsed reactor source, the JINR at Dubna in Russia. It is also possible to carry out time-of-flight measurements at continuous flux reactors making use of choppers, as at the GKSS source at Geesthacht in Germany.

At the ISIS facility, currently the brightest of its type in the world, a 50 Hz synchrotron accelerates protons to 800 MeV and delivers them to a tantalum target where, through the spallation [1] process, approximately 15 neutrons (from 1 to 800 MeV in energy) are produced for each incident proton. To be of use in neutron scattering experiments, these high energy neutrons must be moderated to energies in the range 1–1000 meV, and this is accomplished by small ($\sim100 \times 100 \times 50$ mm) hydrogenous moderators (CH_4, H_2O or H_2) placed near to the tantalum target and illuminating the entrance to the 20 beam-lines surrounding the neutron source.

The essential components of a diffractometer on a pulsed neutron source are shown diagramatically in Figure 9.2. Neutrons originate from the moderator (M) in short pulses (of between 5 and 50 µs) and travel to the sample (S) where they may scatter into a detector (D) situated at an angle of 2θ to the straight-through beam. The neutrons originating from the moderator have a wide energy range from a few meV up to many eV, which corresponds to wavelengths in the range 0.1–10 Å. The precise shape of the intensity versus energy (or wavelength) distribution depends on the moderator used, and the source characteristics. For the three ISIS moderators (CH_4 at 90 K, H_2O at 300 K and H_2 at 20 K) the distributions are shown in Figure 9.3. The number of neutrons at different wavelengths which reach the sample will be affected by both the moderator and the characteristics of the neutron optics between moderator and sample (Figure 9.4).

The moderated neutrons in the range 1–1000 meV have velocities comparable to the speed of sound in air, and hence the neutrons take milliseconds to travel from the moderator to the detector, a distance usually between 10 and 100 m.

A polycrystalline sample, such as most engineering components, will diffract only those neutrons that satisfy Bragg's law;

$$\lambda = 2d_{hkl} \sin \theta \tag{1}$$

and since the wavelength λ of a neutron is related to its velocity by the expression:

$$\lambda = h/m_n v$$

$$\lambda \, [\text{Å}] \approx 3956/v \, [\text{m/s}] \tag{2}$$

where h is Planck's constant and m_n the mass of the neutron. The time-of-flight (t) of those neutrons which are diffracted is given by:

$$t \, [\mu s] \approx 505.5 d_{hkl} \, [\text{Å}] \, L \, [\text{m}] \sin \theta \tag{3}$$

Thus if the detected neutron count is plotted as a function of time (Figure 9.5) it will exhibit a series of peaks corresponding to the different d_{hkl} lattice planes in the material. The shape of the peaks in this diffraction pattern, and hence the resolution of the diffractometer, are determined by the time distribution of the neutrons leaving the moderator (Figure 9.6) and any variation in the path lengths taken by the neutrons reaching the detector. This latter contribution is known as the 'geometrical' contribution to the resolution function.

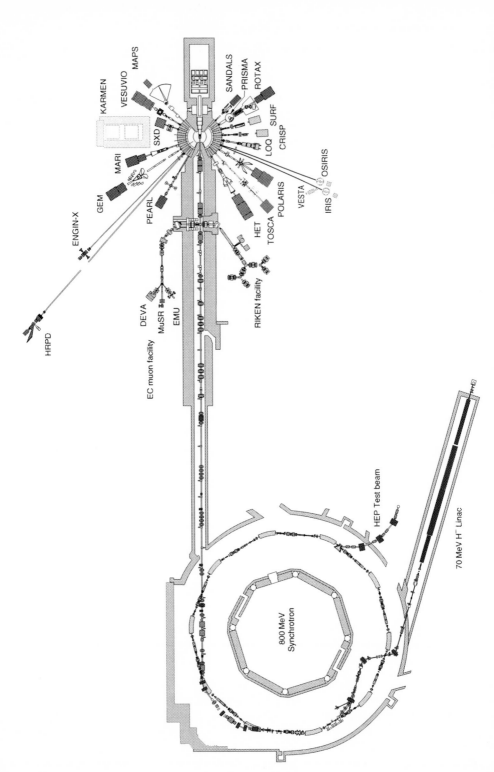

Figure 9.1 The ISIS pulsed neutron source. The diameter of the synchrotron is 50 m. The ENGIN instrument is one of the stations on the PEARL beam-line.

Figure 9.2 The basic components of a time-of-flight diffractometer.

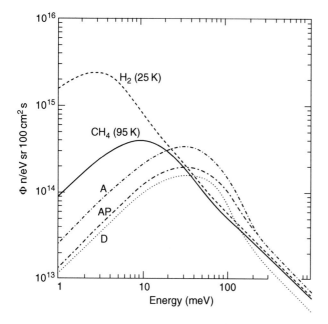

Figure 9.3 Calculations of the intensity distribution against neutron energy for different ISIS moderators [1]. The H_2 moderator is optimised for intensity and has a lower resolution than the methane moderator. 'A', 'AP' and 'D' represent various options for an ambient water moderator. The ENGIN instrument at ISIS views the methane moderator.

By measuring the time-of-flight (t) from the moderator to the detector, and knowing the diffraction angle (2θ) and path length L the lattice spacing d_{hkl} for the particular set of $\{hkl\}$ lattice planes within the sample may be determined. In time-of-flight instruments, a wide wavelength range (0.5–5 Å) is generally used so that a large number of diffraction peaks are recorded simultaneously (see Figure 9.5). The information from all these diffraction peaks may be analysed separately as in continuous flux sources, or used to establish an average unit cell parameters a, b, and c using the Rietveld refinement technique [2] (see below).

Having determined a it is then, in principle, straightforward to calculate the strain (ε) at that point in the sample from the relationship:

$$\varepsilon = (a - a_0)/a_0 \tag{4}$$

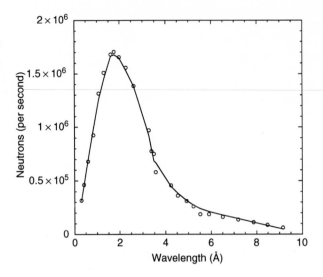

Figure 9.4 The flux distribution as a function of wavelength measured at the end of the guide on the HRPD diffractometer at ISIS (100 m flight path, nickel guide, methane moderator).

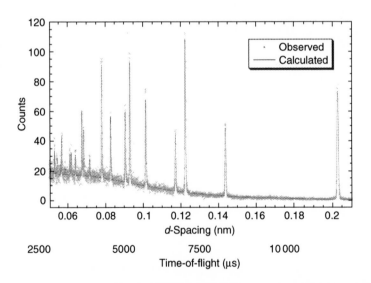

Figure 9.5 A diffraction spectrum collected from a steel sample at a pulsed neutron source. A Rietveld calculation fit to the experimental data is also shown.

where a_0 is the reference lattice parameter. Typically this reference lattice parameter would be for an unstressed material. It should be noted that the strain (ε) thus measured is actually a vector quantity and is measured along the direction $\mathbf{q}$ defined by the expression:

$$\mathbf{q} = \mathbf{k}_2 - \mathbf{k}_1 \tag{5}$$

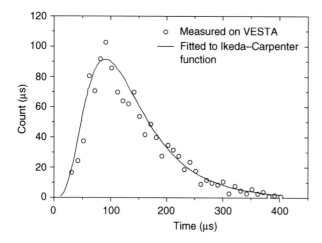

Figure 9.6 The time distribution of 6 Å neutrons emerging from the H_2 moderator, as measured on the VESTA instrument at ISIS using a monochromator crystal. The distribution has been fitted to the Ikeda–Carpenter function (see Section 9.5.2.1).

The directions of k_1 and k_2 being along the incident and scattered directions of the neutron path, with their magnitudes being defined by $k = 2\pi/\lambda$.

9.2.1 Contributions to resolution

One of the most important characteristics of a pulsed source moderator is the time distribution of the neutron pulse, that is, the spectrum of neutron wavelengths produced by the interaction of the fast incident neutrons with the moderating material. This plays a key role in determining the resolution of the instrument. The time distribution is typically modelled as a truncated exponential broadened by Gaussian and Lorentzian components; its shape is influenced by the shape of the moderator, the moderating material and its temperature, its 'coupling' to its environment, and whether it is 'poisoned'. Poisoned moderators contain a thin membrane of neutron absorbing material which has the effects of reducing the effective size of the moderator for neutrons in the thermal energy range. The shape of the distribution for $\lambda = 6$ Å from the hydrogen moderator at ISIS is given in Figure 9.6.

The shape of this pulse from the moderator controls one fundamental contribution to the resolution with which a measurement can be made. A second contribution comes from the so called 'geometrical resolution' term.

Since the illuminated sample volume is small compared to the moderator, the variation in path lengths for neutrons travelling from moderator to sample to detector are largely due to the variation in angle. That is, the angular contribution to resolution is defined by the moderator size as viewed from the sample, or in the case of a guide, the size of the guide aperture as viewed from the sample. It thus defines the uncertainty in angle of incidence of the neutron on the sample. There is a second angular contribution to the resolution function, which is the size of the detection element as viewed from the sample, however this is typically matched in size to the moderator contribution to the resolution.

The intrinsic resolution of the instrument will be a combination of these factors, the moderator to sample distance, the guide properties and the size of the detector elements (or pixellation). These points are discussed in more depth in Ref. [3].

9.3 Instrument description

9.3.1 Overview

In order to understand the functioning of a time-of-flight instrument for measuring strain, we will describe the ENGIN instrument here in some detail.

In designing an instrument to meet the three objectives listed in Section 9.1, a number of principal design parameters must be decided upon. In some cases (e.g. count rate and gauge volume) a technical compromise must be decided upon, in other cases the main constraint is cost. The principal design parameters of ENGIN were chosen such that:

- strain within samples could be measured to $\pm 50 \times 10^{-6} (\pm 50 \mu \varepsilon)$;
- gauge volumes within samples could be resolved down to ~ 2 mm cube;
- samples could be scanned ± 13 cm (x, y, z);
- samples could be positioned to $< \pm 0.1$ mm;
- samples up to 250 kg could be handled;
- samples' temperatures could be controlled up to 1400 K;
- samples' loads could be controlled up to 50 kN.

The ENGIN instrument is situated on the S9 beam-line at the ISIS pulsed neutron source, situated at the Rutherford Appleton Laboratory, UK (Figure 9.1). The ENGIN instrument consists of six major components. A collimated incident pulsed neutron beam, a large $xyz\omega$ positioner (to position the gauge volume within the sample), secondary flight-path radial collimators (which define the gauge volume), and neutron detectors. In addition two telescopes and a laser/CCTV system have been installed to enable the sample to be accurately aligned, and its motion monitored. These components are illustrated in Figures 9.7–9.9. Ancillary equipment for altering the environment of small samples, include a horizontally mounted, custom-built 50 kN Instron hydraulic loading rig, which comes with high temperature capability using a radiant furnace.

The most innovative aspect of the instrument's design was the first use of two large radial collimators to define the gauge volume within the sample. The collimators each have 40 vanes made from gadolinium oxide coated Mylar, which constrain the detectors to receive neutrons from a small volume (just under 2 mm in size, along the beam direction). For very large samples, one of the collimators can be rotated in the diffraction plane through 90° away from the sample, to give a larger sample space than the normal 300 mm region between the collimators.

The detectors have three horizontal rows, each of 45 elements, giving 135 elements per bank. Thus there are a total of 270 detectors altogether. The spatial dimension of elements in the diffraction plane is 5 mm, which with a sample to detector distance of 750 mm matches the instrumental resolution $\Delta \lambda / \lambda = 0.7\%$.

In addition, a small class 2 semiconductor laser is suspended above the centre of rotation of the positioner, which is the main fixed reference point, and a TV camera looks down at this point, its output being visible on two monitors, one in the blockhouse and one in the control cabin.

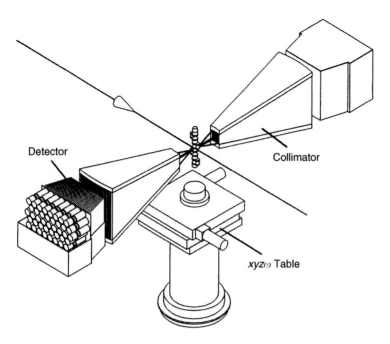

Figure 9.7 Schematic view of the ENGIN instrument.

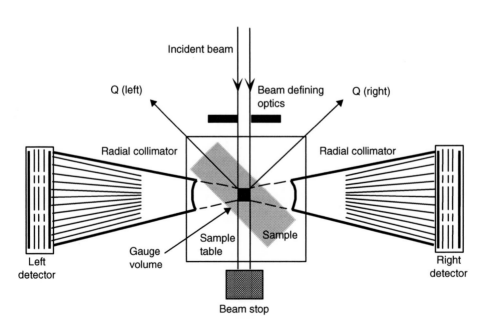

Figure 9.8 Plan view of the ENGIN instrument [4]. The arrows marked 'Q' indicate the directions in which strain is measured by the two detectors.

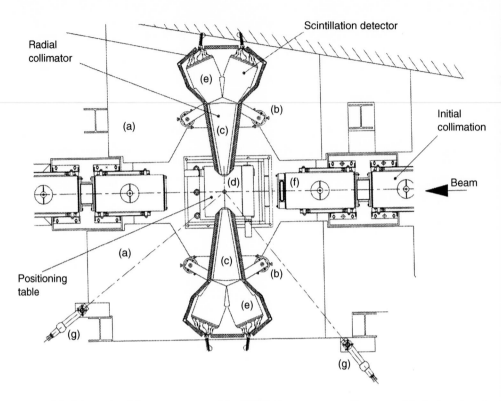

Figure 9.9 Schematic of the components of the ENGIN instrument. (a) Two sets of decked steelwork act as supports for the collimators, and provide a working platform for sample changing. (b) Two collimator mounts, with alignment blocks and the facility to rotate collimator through 90° and lock collimator in position. (c) Two radial collimators each with 40 gadolinium oxide coated Mylar foils, giving 41 line of sight paths for neutrons. Input area 46 mm wide by 60 mm high. Detector face area 230 mm by 300 mm, shielded by boron carbide castings. (d) A four-axis positioning device comprising an $x–y$ table mounted on a rotating stand-off, the complete assembly movable in the z direction by a geared drive. The $x–y$ table has 125 mm of movement about the centre point in each direction. The z drive has a total of 250 mm movement. The off-axis load carrying capacity is 250 kg. (e) Detectors. Each system consists of 135 detectors arranged in three rows of 45 detectors stacked above each other. (f) An incoming beam slit definition system consisting of a motorised translation stage carrying a frame supporting a pair of fixed slits forming a square or rectangular aperture. (g) Alignment telescopes. There is also a closed circuit TV system with two monitors, one in the control cabin and one in the blockhouse. A class 1 laser is installed shining vertically down onto the centre of rotation, thus providing a rough marking of the position of the gauge volume.

9.3.2   *Incident collimation*

Along the path of the beam from the moderator to the sample, sets of coarsely aligned B_4C apertures limit the neutrons reaching the sample area, minimising background without obscuring the view of the moderator from the sample. The final incident neutron beam colli-mation, which defines the gauge volume, is provided by a pair of orthogonal, interchangeable slits made from sintered boron carbide. A range of such slits (0.5, 1, 2, 3, 4, 5, 10 and 25 mm wide, each 30 mm long) has been made, allowing a wide range of square and rectangular beams to be provided. The whole assembly is cantilevered from a motorised, linear translation stage, so that it can be brought close to the sample. Translation of the slits is performed by a stepper motor system (so that no vibration will be transmitted to the slit assembly) and under commands similar to those of the main $xyz\omega$ positioner.

9.3.3 Positioner

The positioner has been designed to carry objects up to 250 kg and position them to within 0.01 mm in x and y, 0.001 mm in z and with an angular accuracy of 0.01° in ω. It consists of an $x-y$ table with a number of mounting holes in it to enable samples to be firmly fixed to it. The x and y axes can move 125 mm on either side of their normal datum point. The $x-y$ table is mounted on a vertical column which can move in the z direction over a total range of 300 mm. The maximum height of the table top is 310 mm below the neutron beam. The column sits on a direct drive motor which rotates the whole assembly in the ω direction through slightly less than 360°.

The system accepts movement input commands and displays output positions in millimetres or degrees, accurate to 0.01 mm or 0.01°. The motion of the positioner has also been programmed using 's-curve profiling' to reduce the acceleration forces felt by the sample. Since it is possible to drive samples into the collimator, pressure sensitive pads have been installed on the shielding of the collimator to stop the positioner if this occurs.

9.3.4 Detector collimation

In front of each detector is a radial collimator [5]. This consists of 40 vertical foils of GdO coated mylar film arranged radially about a fixed focal point. The foils are opaque to thermal neutrons, so that only neutrons which originate from near this focal point are detected. The dimensions of the radial collimator are given in Figure 9.10. The foil thickness, including paint, is $\sim$100 μm.

9.3.5 Detectors

As can be seen in Figure 9.6, ENGIN has two banks of detectors, each centred on a Bragg angle (2θ) of ±90°, enabling the strain in the specimen to be measured simultaneously in two

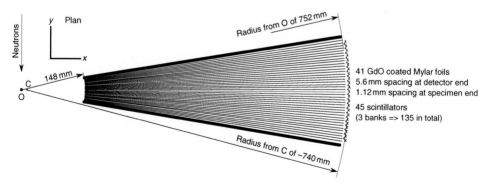

Figure 9.10 A scale diagram (plan view) of various components of the right hand collimator and a detector bank.

perpendicular directions. Each detector bank consists of three detector modules containing 45 individual elements 5 mm wide and 100 mm high. The 45 elements consist of 'V'-shaped scintillators fibre-coupled through a ^2C_n code to 10 photo-multiplier (PM) tubes [2]. A simultaneous signal from any pair of PM tubes then uniquely specifies a neutron detection event at one of the 45 detector elements. Then three modules are stacked vertically within each detector bank to produce 45 detector elements, each 5×300 mm. The detector therefore covers the angular range 81.4–98.6° in 2θ on each side of the incident neutron beam. The relative positions of the elements are determined using a standard powder sample, allowing the combination of spectra from multiple detectors (see Section 9.5.1). The overall efficiency of the detector modules is about 50% at 1 Å.

The neutron detection signals from each element are recorded individually so that the precise scattering angle ($\pm 0.19°$) and time of flight (within 32 ns) is recorded for each detected neutron.

9.4   The gauge volume

When measuring the strain within an engineering component, it is essential to know *where* the measurement has been taken. For this reason the *position* of the gauge volume must be well characterised. A detailed understanding of the *shape* of the gauge volume is also necessary since changes in the shape of the gauge volume can affect the measured value of the strain.

The gauge volume of a pulsed instrument is defined by the intersection between the incoming neutron beam and the space viewed by the radial collimators in front of the neutron detectors. A common simplification is to consider the gauge volume to be described by a cuboid with perfectly sharp edges. However, even for the simplest of collimation systems, this is an oversimplification.

In order to discuss the characteristics of the gauge volume in detail, it is necessary to define it more precisely. A new standard for neutron measurements of stress is being prepared at the time of writing, and the definitions below follow the practice recommended in the draft standard. However, earlier publications may use slightly different terminology. The definitions are equally applicable to pulsed and continuous flux neutron sources.

We define the *nominal gauge volume* (NGV) to be that volume of space defined by the intersection of the incident and diffracted beams, assuming that they follow perfect, parallel transmitted paths through the defining apertures for both incident and diffracted neutrons (Figure 9.11a). It is thus a top-hat (i.e. a two-dimensional step) function. The centroid of the NGV is the geometric centre of this volume, and is coincident with the nominal measurement position of the diffractometer.

The *instrumental gauge volume* (IGV) is the volume of space defined by the actual neutron beam paths through the defining apertures, taking into account beam divergence and the incident beam intensity profile (Figure 9.11b). The response function will not be a simple top-hat. The IGV is defined by experiment or by simulation calculations, and will typically be expressed in terms of the contour of a fraction of maximum intensity. The dimensions can also be defined in terms of a full width at half maximum (FWHM). The difference between instrumental and nominal gauges may be particularly evident for small nominal gauge volumes. It should be noted that the IGV and NGV are properties of the diffractometer itself. The centroid of the IGV is the intensity-weighted centre of this volume.

Finally, the *sampled gauge volume* (SGV, sometimes *sampling* gauge volume, Figure 9.12) is that part of the instrumental gauge volume from which measurements are obtained in an

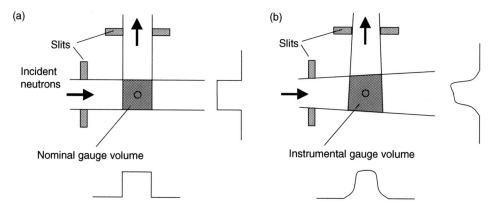

Figure 9.11 Plan views of the nominal and instrumental gauge, shown at $2\theta = 90°$ for simplicity. Below and beside each plan view is an idealised schematic indicating a possible intensity cross section. The centroid of the gauge is indicated by the 'O'.

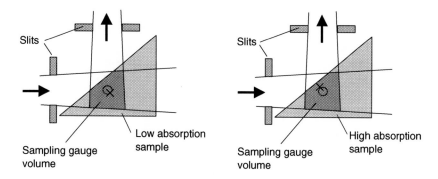

Figure 9.12 The sampling gauge for low absorption and high absorption materials. The 'O' indicates the centroid of the instrumental gauge volume, the 'X' the centroid of the sampling gauge volume, which are displaced relative to each other.

actual experiment, over which the strain measurement is averaged. It is affected by

- partial filling of the instrumental gauge volume,
- attenuation of neutrons within the sample, and
- the wavelength distribution of neutrons across the incident beam.

If the IGV is filled by a non-absorbing material the SGV and IGV are equivalent. The centroid of the SGV is the weighted centre of the gauge, taking the effects listed above into account. That is, it is the single position which on average the gauge volume is centred at during a given diffraction measurement.

The SGV and its centroid can be determined by simulation or calculation. It should be noted that it may be necessary to make corrections to the obtained lattice parameter in order to determine the 'true strain' in cases when the SGV and IGV are not coincident (see below). Traditional methods of measurement of the instrumental gauge volume involve scanning a small probe through the instrumental gauge volume, for example, a thin (0.25 mm) wire.

The SGV of a collimator can also be more strictly defined by means of the *spatial resolution function* (SRF) [5]. The SRF can be defined at a particular point **r** within the sample by the expression:

$$SRF(\mathbf{r}) = P_i(\mathbf{r}) P_s(\mathbf{r}) P_d(\mathbf{r}) \qquad (6)$$

where $P_i(\mathbf{r})\delta V$ is the probability that a neutron is incident on a small volume δV surrounding the point **r**, $P_s(\mathbf{r})\delta V$ is the probability that the neutron will be scattered within δV, and $P_d(\mathbf{r})\delta V$ is the probability that, having been scattered it will be detected. Thus, equation (6) shows how the SRF will vary according to the incident beam collimation, absorption, or scattering within the sample, the absence of material within the geometrical gauge volume, and the effect of the collimation on the diffracted beam. To illustrate the principles involved a simplified model of a time-of-flight instrument is considered here, and the shape of the 'IGU' calculated in the horizontal plane that passes through the incident beam centre. We will assume that the pulsed source moderator has a 100×100 mm (uniformly intense) face that illuminates the instrument, and that this moderator is directly viewed without the use of neutron guides to increase the neutron flux. A slit of width '*s*' is placed at some distance from the geometrical centre of the instrument. The extreme rays of this arrangement are shown in Figure 9.13, and a representation of $P_i(\mathbf{r})$ in the vicinity of the instrument centre from this geometry is shown in Figure 9.14.

The radial collimator is assumed to be of length 600 mm starting 150 mm from the instrument centre, with an inter-vane spacing of 5.6 mm at the back of the collimator (i.e. 750 mm from the instrument centre). A plan view of this collimator is shown schematically in Figure 9.10, and the corresponding function $P_d(\mathbf{r})$, the predicted collimator response function in the vicinity of the instrument centre from this geometry, is shown in Figure 9.15. It can be seen that it is approximately prismatic in shape with the axis of the prism normal to the incident beam direction. This is a useful feature since it means that the *x* direction of the sampling volume is determined effectively by the width of the incident beam slits (allowing for dispersion) and is not significantly affected by the collimator. The cross-section of the resolution function in the *y* direction (across the face of the collimator) is triangular at the centre, becoming approximately Gaussian a few millimetres away from the focal point.

The actual measurement volume of the instrument (i.e. the IGV) is defined by the multiplication of these two probability density functions, and is shown in Figure 9.16. This can be compared with an experimental determination of the function in Figure 9.17.

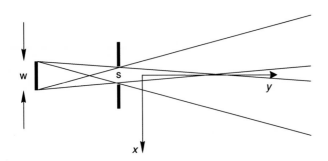

Figure 9.13 Ray diagram showing divergence associated with a slit mechanism.

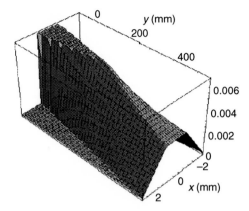

Figure 9.14 Calculation of $P_i(\mathbf{r})$, the probability density function for neutrons passing through a 2 mm slit. More than 200 mm from the slit (y direction), the function broadens noticeably.

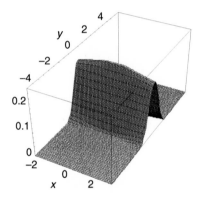

Figure 9.15 The collimator response function of ENGIN $P_d(\mathbf{r})$. The x direction (perpendicular to beam path) has a range -2 to 2, the y direction has a range -4 to 4, that is, the line of maximum probability lies perpendicular to that in Figure 9.14.

9.4.1 Corrections to lattice spacing measurements from partial filling of the gauge

It should be emphasised again that the sampling gauge volume is not an instrumental constant but will vary according to the *specimen being measured*. For example, if the specimen is a strong absorber of neutrons, the sampled gauge centroid will be biased towards those areas of the gauge volume for which the total path length of neutrons in the specimen is shortest. An extreme case is where the gauge volume is incompletely filled (as happens when making near surface measurements). Here the diffraction contribution is of course zero everywhere the specimen is absent. This variation of the sampling gauge volume with specimen position, shape and material means that the region of the specimen represented by the measured lattice spacing will also vary.

The consequence of this is that the measured lattice spacing may vary as a function of the specimen geometry and material, independent of any residual stresses, if the gauge volume is

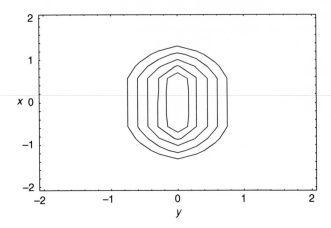

Figure 9.16 Contour plot showing the probability distribution function for detection of neutrons, that is, $P_i(\mathbf{r}) \times P_d(\mathbf{r})$.

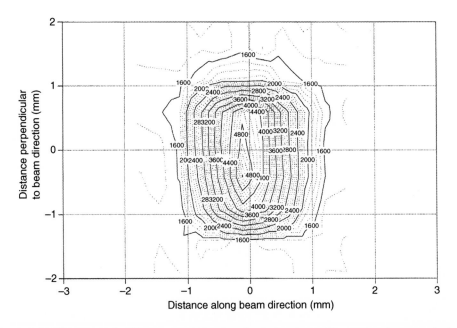

Figure 9.17 The measured collimator resolution function in the plane of the neutron beam, equivalent to Figure 9.16. The result is shown as a contour map provided by scanning a 0.4 mm diameter vertical nylon thread through the beam. The horizontal width of the distribution is defined by the radial collimators, the vertical width of the distribution is defined by 2 mm slits positioned 50 mm in front of the nominal measurement position. The background is ~1400 counts, the peak height ~5000 counts. The half-height dimensions of the measured area are 2 mm (slits) and 1.4 mm (collimator).

incompletely filled or filled with a strongly absorbing sample. These effects are collectively referred to as *geometrical pseudo-strains*.

Fortunately, these effects are not large in the most common engineering materials, such as steel and aluminium, except where the geometrical gauge volume is only partially filled.

The need to predict and account for this variation of measured lattice spacings means that a detailed model of the behaviour of the detection and collimation system must be made where these effects may be present.

Since the time-of-flight is given by:

$$\tau = c d_{hkl} l \sin \theta \tag{7}$$

and

$$\frac{\partial \tau}{\tau} = \frac{\partial d}{d} + \frac{\partial (l \sin \theta)}{l \sin \theta} \tag{8}$$

we may write:

$$\text{apparent strain} = \text{lattice strain} + \text{pseudo-strain} \tag{9}$$

where d_{hkl} is the lattice spacing, l is the total neutron path length (moderator to detector), θ is the Bragg diffraction angle and c is a constant (see equation (3)).

From the above equation we can see that the measured peak shift has two contributions: the lattice strain, which we are aiming to measure; and the variation of $l \sin \theta$. Were the moderator, gauge volume and detectors to be point-like objects, then $l \sin \theta$ would be a constant. However, in reality the instrumental components, including the gauge volume, have a finite size and the value of $l \sin \theta$ will have a distribution of values. The $l \sin \theta$ term in the above equation therefore represents a weighted average of $l \sin \theta$ over the whole gauge volume. The weighting function will be the detected intensity at each point in the gauge volume, that is, the SRF.

We have already seen that changes of absorption coefficient or incompletely filled gauge volumes affect the sampling gauge volume; they will therefore cause a change in the average $l \sin \theta$ and therefore the time-of-flight of the detected neutrons. If this peak shift is not to be mistaken for a change in the lattice parameter, we must be able to predict these geometrical pseudo-strains and thus correct for them. It is possible to predict the geometrical pseudo-strains due both to absorption coefficient and incompletely filled gauge volumes using computer models of the scattering process.

9.5   Determining lattice spacings

The final output we require from a strain scanning measurement is a map of a lattice spacing d_{hkl}, or an average lattice spacing a, as a function of position within the component under study.

The starting point for just one point on this map is the recorded data from one of the ENGIN detector modules consisting of 135 spectra from each of the 135 individual detector elements making up the module. The process of calculating the lattice spacing map from this large amount of raw data therefore takes two steps:

(i)   adding the individual spectra together to form a single diffraction pattern (*focusing*),
(ii) calculating the lattice cell size from this diffraction pattern using a least-squares method.

These steps may be accomplished with the crystallographic data analysis software available at the ISIS facility, which has been modified to automate the processing of large numbers of measurements.

9.5.1   Focusing

The purpose of the *focusing* routine is to add together the data from each of the individual detectors, taking account of the difference in $l \sin \theta$ between them.

Before this can be done, the values of $l \sin \theta$ for each detector element must be measured. This is done by means of a calibration procedure which is performed from time to time on the instrument. In this procedure a diffraction pattern of a known standard (such as silicon or ceria) is taken with sufficient counts that the data from each detector element can be used to measure the values of $t_{hkl} (= 505.5 L d_{hkl} \sin \theta)$ for the $\{hkl\}$ diffraction peaks of the powder diffraction pattern, from which the average value of $l \sin \theta$ for that element can be determined. In practice the individual t_{hkl} are not determined, but a Rietveld refinement (see Section 9.5.2) is made of the individual diffraction patterns to establish the average value of $l \sin \theta$ for that element.

Having established these $l \sin \theta$ values for each detector element, this quantity is used to re-scale the diffraction pattern from each detector element onto a '*d*' rather than '*t*' scale. With all patterns on a common scale, the individual patterns may be added together. By carrying out this summation, implicitly the diffraction pattern obtained is an average over a range of 2θ values, in the case of ENGIN, $17.2°$. That is, the strain obtained is not strictly the one in a direction corresponding to $\theta = 45°$, but an average from $\theta = 40.7°$ to $49.3°$. The effect however of this averaging on the strains obtained can be shown to be negligible, provided the detector array is not too large [6].

9.5.2   Refinement of the lattice dimensions

Given a powder pattern such as that shown in Figure 9.5, there are essentially two broad approaches that may be used to determine an average value for the unit cell size a measured in the direction **q** (defined by equation (5)): either to determine a from individual peaks, or from the entire pattern. The first method has the advantage of simplicity, but has the major drawback that it ignores the large quantity of information relating to a from the rest of the powder pattern. However, by analysing multiple peaks using one lattice parameter we run into complications if different lattice reflections behave differently, as will be discussed.

9.5.2.1   Refinement of single peak positions

An example of the time distribution for neutrons with a particular wavelength leaving the moderator has been given in Figure 9.6. It will be seen that the shape is roughly Maxwellian, being asymmetric with a sharp leading edge and an exponential tail. The precise form of the shape arises from the moderation process that the neutrons undergo in the moderator. The neutrons that enter the moderator from the spallation target have energies in the region of 1 MeV. The neutron cross-section of the hydrogenous materials used in moderators means that these high-energy neutrons travel only a few millimetres between collisions, losing approximately half of their energy at each collision. The pulse of fast neutrons that reach the moderator may therefore be viewed as a dilute, cooling gas. The time at which neutrons reach the surface of the moderator that faces the beamline therefore depends on the geometry of the moderator, together with density and cross section of the moderator material contained within it.

A number of theoretical studies have been made of this process, with the conclusion that a satisfactory description of the peak shape is given by a convolution of a Gaussian, a Lorentzian and a complex switching exponential term representing the decaying contributions from 'fast' and 'slow' neutrons [7]. This shape thus depends upon several parameters, some of which change with wavelength, that is, across the time-of-flight spectrum.

The peak shape recorded from a powder sample will be closely related to that described above, but not identical. This is because we have initially considered only the time distribution of the neutrons leaving the surface of the moderator. The diffraction peak will be further broadened by the differing path lengths that the neutrons have taken through the instrument, and the distribution of d-spacings in the sample contributing to the particular peak. The latter contributions may be Gaussian (e.g. from strain gradients) or Lorentzian (e.g. from particle size effects). In practice the predominant contribution is usually Gaussian, and simply adds to the Gaussian component present from the moderator.

It is clearly vital that the parameters describing the instrumental contribution to peak shape are known so that they may be separated from the effects of the measured samples. This is achieved by calibration measurements on an 'ideal powder' such as silicon (Si) or ceria (CeO_2), and may be augmented by single crystal measurements.

Figure 9.18 shows a typical diffraction peak shape from a pulsed source; the asymmetry of the peak is clear. It is sometimes asked 'where is the real centre of these peaks?', since the mathematical description of the centre (marked by the arrow) is different from the maxima in intensity. As well as least squares methods, it is possible to determine the centre of gravity of a peak from its first moment, since broadening by even functions such as Gaussians or

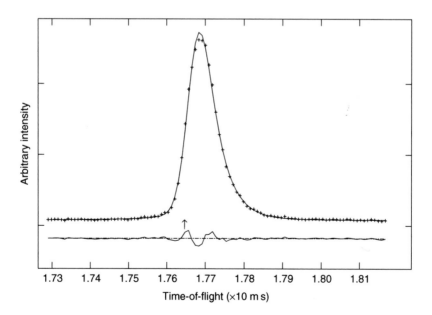

Figure 9.18 A diffraction peak from a measurement on a silicon powder at the ENGIN instrument. A fit to the data shows the calculated position of the peak (marked by the arrow), and the residual difference between fit and observed data (bottom line).

Lorentzians does not affect this position. What is important is that a quantitative description of the time distribution of the peak shape is given for $L = 0$ (i.e. the moderator surface) and the time difference between this distribution and that arriving at the detector measured.

Measuring for shorter or longer periods obviously affects the resulting uncertainty on the d value obtained. It is found that at short times the uncertainty on the value of d (σ_d) approximately follows an expected $t^{-0.5}$ behaviour, but at longer times the uncertainty returned by the least squares procedure appears to 'saturate', with increasing counting time not resulting in an increased precision of measurement. This is due both to the presence of a background, and also the accuracy of the peak shape description, which is sufficient to refine individual peak positions to an accuracy of 20 parts in 10^6, equivalent to around 1/5000 of the peak width. For pulsed sources with appropriately chosen moderation, the sharp rising edge of the diffraction peak aids accurate peak position determination.

9.5.2.2   Rietveld refinement of the entire powder pattern

While there are instances where the strain in a material may be determined from an individual d_{hkl}, the availability of the entire diffraction pattern on a pulsed source instrument means that strain is generally obtained from analysis of the entire pattern [8]. This reduces the uncertainty in the determination of the lattice parameter for a given measurement time, since more information (i.e. multiple diffraction peaks) is used in the fitting process.

The technique of obtaining the unit cell and crystallographic structural parameters by performing a least-squares fitting procedure to the entire diffraction pattern was first published by Rietveld [2]. This technique initially assumed the pattern to be derived from an untextured powder sample of a single phase material. Thus the formal definition of the Rietveld process is the minimisation of the function:

$$S_y = \sum_i w_i (y_i(\text{obs}) - y_i(\text{calc}))^2 \qquad (10)$$

where $y_i(\text{obs})$ is the observed (measured) intensity at data point i in the spectrum, $w_i = 1/y_i(\text{obs})$, and $y_i(\text{calc})$ the intensity predicted from the assumed crystallographic structure. The uncertainties on the fitted parameters are obtained from the co-variance matrix [9]. While the original Rietveld method may be applied straightforwardly to powder samples, engineering samples often present particular difficulties. In these, the diffraction pattern may be complicated by texture, the anisotropic effects of stresses or by multi-phase samples.

9.5.2.2.1   TEXTURE

The effect of texture is to change the relative intensities of the individual Bragg peaks recorded in a diffraction pattern. To incorporate these effects, which are often present in engineering samples, a modification of the Rietveld method, known as a Pawley refinement [10] can be used for analysing full diffraction patterns. A Pawley refinement accommodates the variation in peak intensities by allowing the *intensity* of individual reflections to vary freely, while the peak positions are determined in the usual manner from the unit cell dimensions.

Since it is the unit cell parameters that are of prime interest in strain measurements, the Pawley refinement is perfectly suited to this purpose. If information on the texture is required, this may be qualitatively obtained by comparing the intensities of the recorded Bragg peaks with those calculated for the ideal powder sample. However, it is also possible to make a quantitative evaluation of the texture in the sample. Since each diffraction pattern provides information about the intensity of many diffraction peaks, for the case of higher symmetry samples only a relatively small number of sample orientations may be required to allow an approximate evaluation of texture, using the March–Dollase technique [11], or more advanced techniques such as spherical harmonics [12] or WIMV [13].

9.5.2.2.2 MULTI-PHASE MATERIALS

Many interesting materials (both from a scientific and engineering viewpoint) consist of more than one crystalline phase. This presents difficulties when analysing the diffraction profiles since there will be a distinct set of Bragg peaks for each phase present. In such situations it is not uncommon for peaks from different phases to overlap. It is also possible that extra diffraction peaks will be present in the pattern recorded due to scattering from sample environment equipment. There are three approaches to the profile refinement of multi-phase diffraction spectra: ignore the presence of the second phase, eliminate the second phase peaks from the refinement, or simultaneously refine all the phases. The latter is by far the preferred option if the crystallographic structure of all phases is known. In cases where peak overlap is not significant, or occurs only for a few peaks of the phase of interest, the difference in strain obtained for the three approaches is generally negligible.

9.5.2.2.3 SIMULTANEOUS REFINEMENT OF ALL PHASES

Figure 9.19 shows the results of a multi-phase refinement on a ceramic layer composite, consisting of alternate layers of Al_2O_3 and ZrO [4]. It can be seen that the fitted line is very close to the experimental data, even in regions with significant overlapping of peaks. It should be emphasised that the data from this ceramic composite is impossible to analyse using single

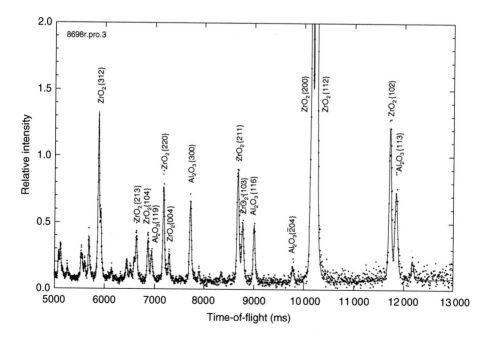

Figure 9.19 Multi-phase Rietveld refinement of an alumina–zirconia layered composite.

phase profile fitting. Without multi-phase Rietveld software it would, in fact, also have been difficult to satisfactorily analyse this data using single peak fitting methods.

The rationale behind carrying out fits on multiple diffraction peaks at a pulsed source is that increasing the number of peaks in the refinement increases the accuracy of the measured lattice parameter, simply because more data points are available in the fit. Empirically, on a material such as steel, the count times required to obtain a given strain accuracy for a Pawley or Rietveld fit are from 30% to 50% of the time which might be required to obtain the same accuracy on a fit to one of the single peaks, dependent on the particular peak intensity. This number is worsened when counting statistics are poor, in which case the peak shape may be particularly poorly defined when only one peak is observed, and the Rietveld fit can out-perform single peak observations by a factor of 6 (i.e. 15% of the time is required).

However, the use of this technique raises a complication, since the effect of stresses on individual lattice peak spacings is *not* uniform. Differences in response of the variously oriented diffraction planes within individual grains are caused firstly by the anisotropy of the elastic stiffness of individual grains (elastic anisotropy), and secondly by the anisotropy of the relaxation mechanisms on a granular level, as slip occurs preferentially on certain slip systems (plastic anisotropy). That is, the elastic strain response of individual grains within the polycrystal is dependent on their orientation and, since each diffraction peak is produced by a different family of suitably oriented grains within the polycrystal, this is reflected in the strains determined using individual diffraction peaks. In practice, the anisotropy strains are small enough that they do not interfere with the use of the entire diffraction pattern in a Pawley or Rietveld refinement, which typically assumes that no such anisotropy is present. Indeed such a fit generally avoids the problems that may be encountered when individual single peaks are used [8, 14]. Further, analysis of the different lattice strains caused by anisotropy provides a tool to probe the effects of plasticity on materials, both at a microstructural [15], and at a macroscopic [8] level. It is also possible to include the effects of elastic anisotropy at least directly into the Rietveld refinement process [8].

9.6   Optimisation of time-of-flight engineering instruments

While ENGIN is an excellent instrument, developments in neutron optics and the theory of instrumentation mean that a fully optimised instrument could now be built with an order of magnitude increase in performance. This is achieved by:

(i)   the use of a figure of merit (FOM) in the design process itself, and
(ii) the use of an optimally configured super-mirror guide and detector arrays to maximise the FOM.

Such an instrument has been designed (ENGIN-X), and at the time of writing is being commissioned at the ISIS facility, Rutherford Appleton Laboratory.

9.6.1   FOM for a single peak

Neutron diffractometers are generally built as 'all-purpose' instruments, and their designs are compromises which balance the competing requirements to measure the intensities, positions and widths of diffraction peaks simultaneously. In the case of an optimally designed engineering strain scanner such compromises are not necessary, since the overriding requirement of the instrument is the accurate measurement of a lattice parameter at a known location within the material under study.

To enable different instruments to be compared it is reasonable to define an FOM such that an increase of a factor of two in the source illuminating an instrument results in a factor of two increase in the FOM. It is also necessary to take into account the uncertainty of the result obtained. Hence the most useful high-level definition of a FOM for a strain measuring instrument will be *'the inverse of the time taken to measure a d-spacing to a given uncertainty'*.

d-Spacings are obtained from the observed diffraction patterns by a 'least-squares' fitting procedure, and it has been shown by Sivia [16] that in the situation of an isolated Gaussian peak, the time (t) taken to measure (with an accuracy of σ) the position of a peak is:

$$t \propto w^2/I\sigma^2 \tag{11}$$

where w is the width of the peak, and I the intensity recorded in unit time. Hence the FOM required for an instrument concerned solely with measuring peak position may be simplified to:

$$\text{FOM} = I\sigma^2/w^2 \tag{12}$$

if the peaks were Gaussian in shape and well separated.

A similar, analytic, result for the FOM required when an *arbitrary* peak shape is fitted by least squares has now been derived [17], and so equation (12) may be used quite generally, to establish the performance of an instrument designed to measure a lattice parameter.

The veracity of this result has also been demonstrated empirically [18]. This has been done by deriving the position of a large number of experimentally measured peaks by least-squares fitting. These were measured at different facilities (reactor and pulsed neutron sources, and an X-ray synchrotron source) on different materials.

If $t = k(w^2/I\sigma^2)$ (i.e. equation (11)), then $\ln(\sigma/w) = k' - (1/2)\ln(It)$. Thus, plotting the uncertainty in position over the width against the integrated counts in the peak, as shown in Figure 9.20, it will be seen from the form of this figure that equation (11) is at least reasonably well confirmed (the slope is in fact −0.46, not −0.5).

From this, it is clear that the FOM (equation (12)) must be maximised in the design of an optimised strain scanning instrument, and that the ratio of the FOM to other instruments quantifies its increased speed of measurement.

9.6.2 Maximising the flux of neutrons – solid angle

One way in which the FOM may be maximised is by utilising the fact that the primary measuring position in a strain scanning instrument is at a scattering angle of 90°. At this scattering angle the widths (w) of the peaks in the diffraction pattern (and hence FOM) are insensitive to changes in the vertical angle of incidence of the incident beam [19]. Thus, while Liouville's theorem dictates that we cannot increase the flux of neutrons per unit solid angle incident on the sample (over that emanating from the moderator), we can increase the *total flux* of neutrons usefully incident on the sample by increasing the vertical divergence of the beam incident on the sample. In an optimised strain scanning instrument the vertical divergence of the neutron beam would therefore be increased to the maximum that can be achieved by the use of super-mirror guides above and below the incident beam.

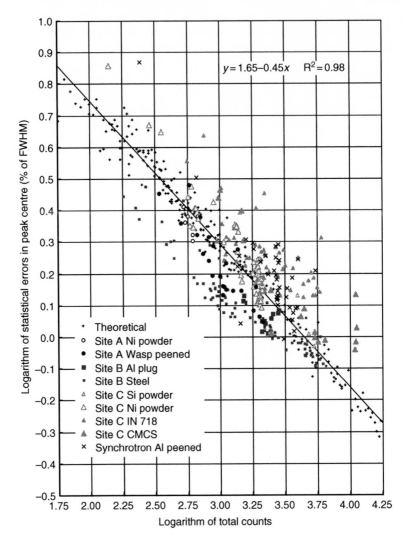

Figure 9.20 $\ln(It)$ versus $\ln(\sigma/w)$ for a series of diffraction strain measurements [18]. (See Colour Plate IV.)

However, the length of the primary flight path, and the angular divergence of the neutron beam in the horizontal plane play dominant roles in defining the widths of the diffraction peaks (w). A detailed optimisation of strain scanning instrument design [17] has shown that the FOM is maximised when the primary flight path of the instrument is approximately 50 m, and the horizontal angular divergence set to 100/50 000. In this case the appropriate divergence has been achieved by foreshortening the side walls of the neutron guide at a distance of 38 m. The combined effects of these design features have produced FOM increases over the existing ENGIN instrument of an order of magnitude.

It should finally be noted that it is possible to define FOMs for the determination of position, width and amplitude of a peak even in the presence of a significant background [17].

References

[1] Taylor A. D., *SNS Moderator Performance Predictions*, Report No. RAL-84-120 (1984).

[2] Rietveld H. M., *J. Appl. Crystallogr.* **2**, 65–71 (1969).

[3] Johnson M. W., Daymond M. R., *J. Appl. Crystallogr.* **35**, 49–57 (2002).

[4] Johnson M. W., Withers P. J., Edwards L., Priesmeyer H. and Rustichelli F., *The Precise Measurement of Internal Stress within Materials using Pulsed Neutrons*, Report No. RAL-TR-96-068 (1996).

[5] Withers P. J., Johnson M. W. and Wright J. S., *Physica B* **292**, 273–285 (2000).

[6] Daymond M. R., *Physica B* (2001).

[7] Ikeda S. and Carpenter J. M., *Nucl. Instrum. Methods* **A239**, 536–544 (1985).

[8] Daymond M. R., Bourke M. A. M., Von Dreele R. B., Clausen B. and Lorentzen T., *J. Appl. Phys.* **82**, 1554–1562 (1997).

[9] Young R. A. (ed.), *The Rietveld Method*, International Union of Crystallography, Oxford University Press, Oxford, 1993.

[10] Pawley G. S., *J. Appl. Crystallogr.* **14**, 357–361 (1981).

[11] Dollase W. A., *J. Appl. Crystallogr.* **19**, 267–272 (1986).

[12] Von Dreele R. B., *J. Appl. Crystallogr.* **30**, 517–525 (1997).

[13] Matthies S., *Phys. Status Solidi* B **92**, 135–138 (1979).

[14] Daymond M. R., Bourke M. A. M. and Von Dreele R. B., *J. Appl. Phys.* **85**, 739–747 (1999).

[15] Clausen B., Lorentzen T., Bourke M. A. M. and Daymond M. R., *Mater. Sci. Eng.* **259**, 17–24 (1999).

[16] Sivia D. S., *Data Analysis – A Bayesian Tutorial*, Oxford University Press, Oxford, 1996.

[17] Daymond M. R. and Johnson M. W., Optimisation of the design of a neutron diffractometer for strain measurement, in *Proc. 15th Intl. Collaboration on Adv. Neutron Sources*, **1**, 499–503, Tsukuba, Japan, 2000.

[18] Edwards L., Fitzpatrick M. E., Daymond M. R., Johnson M. W., Webster G. A., O'Dowd N. P., Webster P. J. and Withers P. J., ENGIN-X: A neutron stress diffractometer for the 21st century, in *Proc. 6th Intl. Conf. on Residual Stresses*, **2**, 1116–1123, Oxford, 2000.

[19] Johnson M. W., Edwards L. and Withers P. J., *Physica B: Condens. Matter* **234–236**, 1141–1143 (1997).

10 Use of synchrotron X-ray radiation for stress measurement

P. J. Withers

10.1 Introduction

Conventional X-ray and neutron diffraction techniques are well established and can provide the engineer, as well as the materials scientist, with valuable information about the state of residual stress in new materials and in engineering components. Both have their limitations; X-rays sample only a very shallow (a few microns) surface layer, while neutron strain measurements are characterised by relatively low intensities giving rise to slow rates of data acquisition and a spatial resolution of around 1 mm. Modern X-ray synchrotron sources on the other hand can provide very intense narrow beams of highly collimated and highly penetrating energetic X-ray photons (Figure 10.1). The low level of angular divergence and narrow energy bandwidth leads to peak widths that are symmetric and inherently very narrow ($\sim$0.01° full-width half-maximum compared with a degree or so for neutrons), and wavelengths can be selected down to less than 0.1 Å. As discussed in Chapter 2, attenuation falls off sharply with a shortening of wavelength such that at X-ray energies between 40 and 80 keV

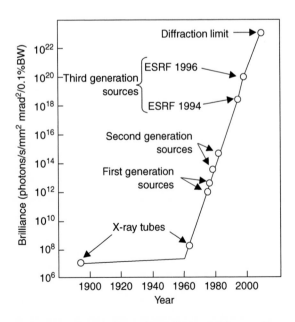

Figure 10.1 Historical overview of X-ray brightness (from www.esrf.fr).

Table 10.1 Table of attenuation lengths (mm) to 66% attenuation for neutrons, and synchrotron and laboratory X-rays

Attenuation lengths (mm)	Al	Ti	Fe	Ni	Cu
Thermal neutrons	100	17	8	5	11
ID15 (ESRF) 150 keV	27	13	6.5	5	5
ID11 (ESRF) 60 keV	11.5	2.9	1.0	0.7	0.7
BM16 (ESRF)/16.3(SRS) 40 keV	6.5	1.0	0.36	0.25	0.23
Cu radiation	0.07	0.01	0.004	0.02	0.02

($\sim$0.4–0.15 Å) X-ray path lengths in aluminium of many centimetres are possible and significant distances in relatively heavy elements are also practicable ($\sim$5 mm in Ni) (Table 10.1). Furthermore, beams as small as 1 μm can be used to form a high spatial resolution probe [1].

At the present time, very few studies have been undertaken using synchrotron radiation, so that the engineering potential of the technique still remains largely untapped. It is anticipated that the development of synchrotron strain scanning will fill the important near-surface gap between what is possible with neutrons and what is accessible with traditional X-ray techniques. The aim of this chapter is to introduce the basic techniques that have been used to date and to review a number of exemplar experiments that illustrate some of the opportunities that synchrotron X-ray diffraction will open up as the technique matures.

10.2 The basic method

Like all the diffraction-based methods discussed in this book, synchrotron techniques exploit Bragg's equation. Unlike conventional X-ray diffraction, the wavelength can be selected over a very wide range. The relationship between X-ray energy (E) and wavelength (λ) is $E = hc/\lambda$ where h is Planck's constant and c the speed of light. This gives

$$\lambda \text{ (in Å)} \approx 12.4/E \text{ (in keV)}.$$

At the very high energy end (typically $>$80 keV), wavelengths as short as 0.15 Å can be used giving rise to excellent penetration, but very low scattering angles ($\theta \sim 2°$), while at lower energies (typically around 35 keV) scattering angles more reminiscent of conventional X-ray diffraction are employed, but with very much increased intensity.

As with any new technique, the best methods will take time to be fully defined and optimised. To date, at least three more or less different methods have been tried successfully namely (Figure 10.2) (i) $\theta/2\theta$ scanning in either the reflection or transmission geometry, with or without an analyser crystal; (ii) using high-energy monochromatic photons in transmission with a two-dimensional (2D) detector; and (iii) white beam methods using an energy sensitive detector. In reality, the distinctions between these are somewhat arbitrary and various aspects of the methods, as described below, can be blended to form hybrids. In all cases, just as for neutron strain scanning, the gauge volume is defined by apertures and any spatial variation in strain mapped by translating the sample across the beam.

10.3 Traditional $\theta/2\theta$ scanning

Of the three methods, this is the closest to that of the conventional monochromatic X-ray and neutron diffraction methods. A small sampling volume is defined by apertures placed on the incident and the diffracted beams (Figure 10.3). Because the diffracting angles are

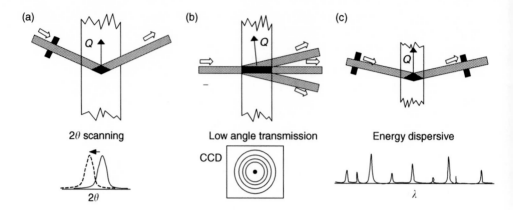

(a)   Q

20 scanning

20

(b)   Q

Low angle transmission

CCD

(c)   Q

Energy dispersive

λ

Figure 10.2 Schematic showing three different strain measurement geometries, the strain measurement direction (parallel to the *Q* vector) and the form of data they record.

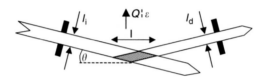

Figure 10.3 Definition of the sampling gauge.

usually considerably lower than for X-ray or neutron diffraction, the gauge volume is typically diamond shaped, being much longer than normal to the scattering vector *Q* (the strain measurement direction) than parallel to it. The length of the gauge normal to the scattering vector is given by:

$$l_g = (l_i/2 \sin \theta + l_d/2 \sin \theta) \approx (l_i + l_d)/2\theta \quad \text{(at small angles)} \quad (1)$$

where θ is in radians, l_i is the incident aperture width and l_d the width of the diffracted beam aperture (Figure 10.3). Taking the Al(111) peak as an example (plane spacing $d_{(111)} = 2.34$ Å) using an incident energy of 15 keV gives a Bragg angle (θ) $\approx 10°$ and l_g is about three times the sum of the two slit widths, while for an incident energy of 80 keV ($\theta \approx 2°$) l_g is about 14 times the sum of the aperture widths. This means that for apertures of 50 μm the gauge is around 300 μm long at $2\theta = 20°$, but 1400 μm at 4°. Of course the distinction between moderate and high energy is somewhat arbitrary, but the important point is that large diffracting angles do allow for better spatial resolution along the incident beam. From the viewpoint of penetration, reflections should be chosen to give large diffraction angles since these give better penetration in reflection (at a given energy) due to the shorter path lengths compared to glancing angle paths.

The introduction of strain independent shifts in the angle at which the peak centre is recorded when the gauge is only partially filled has been well documented for neutron diffraction (Chapter 14). These effects are also observed for synchrotron diffraction. Recent work has shown that the placement of a crystal analyser before the detector in a conventional triple axis arrangement is beneficial (Figure 10.4). It increases the angular discrimination of

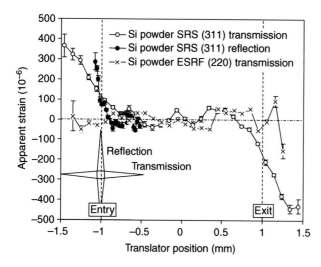

Figure 10.4 Shifts in the *d* spacing recorded for an Al powder as it is scanned in transmission and reflection on 16.3(SRS) without an analyser and in transmission at BM16 (ESRF) with an analyser. The surfaces of the powder are indicated by the dashed lines and the gauge shape/orientation for transmission and reflection is also indicated. The shifts are of geometric origin and occur up until the point at which the gauge is totally filled. The magnitude of the surface effect is markedly reduced by incorporating an analyser crystal on the diffracted beam (carried out in collaboration with Webster P. J., Owen R. A., Browne P. A. and D. Hughes).

the detector to the diffracted beam and therefore markedly decreases the extent of the anomalous peak shifts when the gauge volume passes through a surface. Though these shifts can be corrected for if the diffracting geometry and gauge position are known [2, 3], it is clearly preferable to undertake surface measurements using an analyser. As illustrated in Figure 10.4, the surface effect is essentially anti-symmetrical upon entry and exit of the gauge volume from the sample. As a result, in cases where it is not appropriate to use an analyser the surface effect can be removed by taking the mean of the entry and exit curves for symmetrical stress fields (Figure 10.5). More generally, surface effects can be accounted for by rotating the sample by 180° so that the entry side becomes the exit and vice versa and taking the mean of the two scans.

Unlike the traditional laboratory X-ray method, increased penetration means that the technique can be used in either reflection or transmission geometry. In fact because of the relatively small scattering angles, much greater depths can normally be probed in transmission than reflection. This is because for reflection at grazing angles the path length becomes large ($= 2 \times$ depth$/\sin \theta$) even at small depths. The surface effect is less pronounced for reflection measurements, because the entry of the gauge volume occurs over a much shorter distance in reflection than in transmission (for which the gauge is elongated normal to the surface) (Figure 10.4).

The following examples have been chosen to illustrate four important aspects of synchrotron strain scanning; the importance of accurate sample location, the high spatial resolution achievable, the ability to map large areas quickly and efficiently and the intergranular/macrostress effects evident at small gauge volumes.

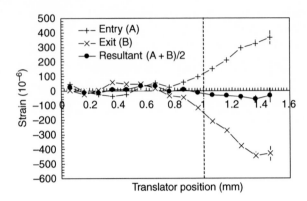

Figure 10.5 When a symmetric strain field is expected for transmission measurements made from edge to edge, the surface effect can be removed by taking the mean of the entry and exit curves, as carried out here for the data in Figure 10.4. More generally, the mean of two scans can be used, where the entry and exit surfaces have been switched over.

10.3.1   *Peened plate*

Near-surface stresses can be very important for the prolongation or shortening of component life. One way to prolong life is to put the surface into compression by peening. Steep in-plane strain profiles have been measured by the $\theta/2\theta$ method for a heavily shot-peened hiduminium Al alloy [4]. The alloy is used extensively in the aerospace industry and, by altering shot-peening parameters, material, size, speed and duration, the magnitude and depth of the peening effect can be varied. Measurements were made using 35 keV (0.35 Å) X-rays. The incident beam cross-section was 0.1×5 mm, with 0.1 mm slits in front of the detector, to define the small 'gauge volume' that is used to scan the sample. The measurements in Figure 10.6 show excellent near-surface capability compared with neutrons and tremendous depth penetration compared with conventional X-rays. Data collection rates were reported to be between 10 and 100 times faster than for neutrons with up to five times the spatial resolution. Note that the data in Figure 10.6(a) have been inferred from strain measurements made normal to the peened surface on the basis that the out-of-plane stress is zero. Near to the surface an up-turn is clearly resolved; this is indicative that the surface has been over-peened [5]. Because the synchrotron beam has such a narrow inherent peak width it is an ideal tool for the study of changes in peak shape and width. Peak widths provide a measure of microstrain and plastic deformation and give an indication of the characteristically low level of instrumental broadening. Figure 10.6(c) shows the peak width variation, from 0.06° near the surface to 0.02° internally, for the Al samples as a function of depth from the peened surface. These results provide supporting quantitative evidence that the more intense the peening intensity the greater is the depth of the plastically deformed layer.

In cases such as this where there are very steep gradients of strain (equivalent to 2000 MPa/mm), an uncertainty in the position of the surface of only 250 μm can lead to errors of ±500 MPa even with an analyser in place to minimise surface strain shifts. Consequently, precise location of the surface is essential. One means of achieving this is to undertake the near-surface strain measurements in reflection in the normal manner but then to back-calculate the sample location by fitting the peak intensity to that anticipated theoretically.

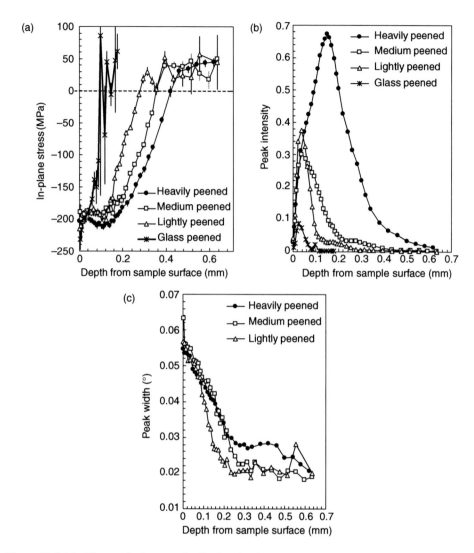

Figure 10.6 (a): The residual stress distribution for four Al sheets shot peened to different Almen intensities with metal and glass shot. (b) The decrease in peak intensity with depth of penetration. (c) The changes in peak width. In (a) the error bars are smaller for the peened region than in the parent, because the finer crystal mosaic size gives better grain sampling statistics [4].

Because of the sharp increase in attenuation as the gauge penetrates deeper below the surface (Figure 10.6b), it is possible to locate the surface to within $\pm 10\,\mu$m [6].

10.3.2 *TIG welding*

Until recently, aerospace aluminium alloys have been regarded as unweldable. TIG welding is currently being re-examined as a means of welding these materials. While the weldability issues are being resolved by microstructural control of the weld zone [7], large residual

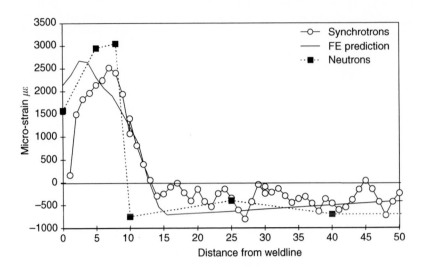

Figure 10.7 Linescans lateral to an autogeneous bead-on-plate TIG weld on a 3.2 mm Al alloy plate at a position of 50 mm from the start of the weld made using 100 μm slits (i.e. giving an approximately 100 × 1000 μm diamond shaped gauge). Neutron (using a square cross-section 1 × 1 × 20 mm gauge volume) and finite element predictions are also shown for comparison.

stresses and weld distortion remain significant obstacles to the practical adoption of the technique. Test TIG welds in 3.2 mm thick Al plate have been examined at both high and intermediate resolution by synchrotron X-ray diffraction. The strain field caused by the TIG weld has been examined at high spatial resolution using 100 μm × 5 mm apertures on BM16 at the ESRF [8]. In order to ensure that the results were not affected by stress-free variations in lattice spacing caused by the thermal treatment of the alloy in the heat affected zone of the weld, a series of stress-free matchstick samples were cut from an identical weld. These provided a stress-free reference as a function of lateral position (y). The synchrotron measurements in Figure 10.7 show a point to point scatter which is much larger than the statistical error quoted by the fitting routine and which is larger than that recorded by neutron diffraction. These variations are due to the smaller gauge volume used for the synchrotron measurements. This results in a statistical sampling of a relatively small number of grains within the sample. When three linescans are taken 100 μm apart and the results averaged then a smoother variation is achieved in line with the neutron measurements and the finite element predictions (Figure 10.8a); that this scatter does not arise from a peak-fitting error is illustrated by the similarity of three repeat linescans measured at the same locations (Figure 10.8b), which except in the weld itself (where the grains are much bigger) shows agreement to within the error bars. This illustrates an important issue when taking strain measurements using such a high resolution probe; one needs to select a gauge volume according to whether one wishes to sample the grain to grain variations in stress, or one wishes to acquire a statistically representative average of an ensemble.

Since synchrotron strain measurements can be acquired in seconds, the technique is very well suited to the production of area maps. This makes it an ideal tool for the validation of finite element model predictions. This is exemplified by the area map in Figure 10.9 of the elastic strain field around a TIG weld [9] alongside which is a corresponding finite element

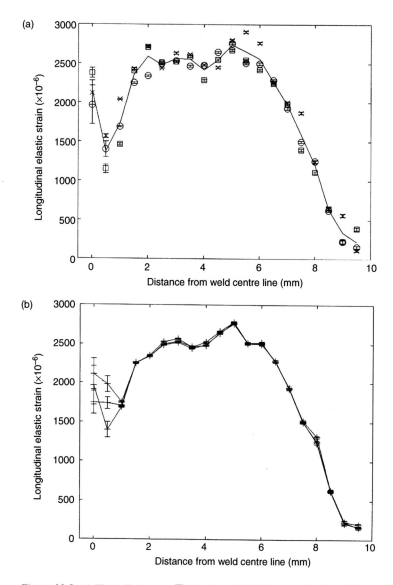

Figure 10.8 (a) Three linescans (□, ○, ×) made 100 μm apart using 100 μm slits in the vicinity of the linescan shown in Figure 10.7 along with the average (solid line). (b) Three repeat linescans made at the position of one of the scans in (a) [8].

model strain map. In Figure 10.10, the capability to produce detailed strain maps is illustrated for the end point of the TIG weld.

10.4   The transmission method using a two-dimensional detector

In many ways, strain scanning at high energies using a 2D detector has much in common with diffraction in a transmission electron microscope (low scattering angles, transmission

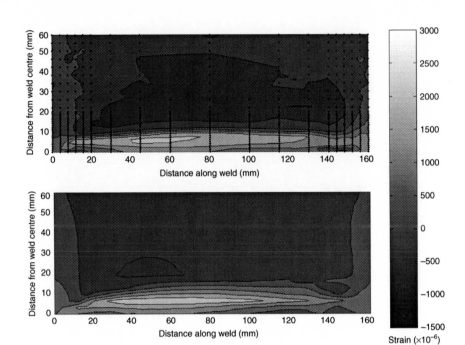

Figure 10.9 Direct comparison of finite element predictions (bottom) and synchrotron strain scanning
measurements (top) made on BM16 at the ESRF for the longitudinal strains in an 8 mm
TIG weld in 3.2 mm thick butt-welded plate (courtesy of BAe systems) [9]. The locations
of the (600) measurement (1 × 1 mm area) are marked by black spots.

geometry, etc). As a natural consequence of the high energies and the low scattering angles, a
transmission geometry is employed (Figure 10.2b). Data collection rates can be accelerated
by using CCD detectors to collect large 2D diffraction patterns (Figure 10.11). Because the
diffraction angles are small, the Q diffraction vector is almost perpendicular to the incident
beam, that is, the strain is sampled in the plane normal to the beam. This means that changes in
the horizontal and vertical diameters in the diffraction rings can be used to provide information
about the strains in the horizontal and vertical in-plane directions simultaneously. Through
the use of apertures on the incident beam it is possible to achieve very high lateral resolution
($\sim$20 × 20 μm), but at present it is very difficult to restrict the gauge along the direction
of the beam. This is a problem not just in terms of spatial resolution, but also in terms of
strain resolution. Without any defining apertures on the diffracted beam, the peak position is
sensitive both to the diffracting angle and the distance from the CCD from which it originates.
This means that if the diffracting volume moves towards the CCD it is accompanied by a
stress-free decrease in the measured ring radius, while if the volume moves away from the
detector the ring radius increases. This effect gives rise to apparent strains if not accounted
for. This effect is clearly evident in the ring radii data measured for the elastic strain in a
distorted TIG welded plate (Figure 10.12a). It can be eliminated by relocating the sample
for each measurement in a linescan to ensure that the distance of the gauge volume to the
CCD detector is essentially constant, but for distorted plates this can take many minutes
while the strain measurement itself takes only a few seconds. One solution suggested by

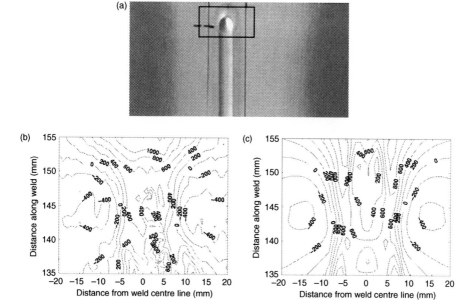

Figure 10.10 (a) Photograph showing location of high spatial resolution map (b) of strain around the end of an autogeneous bead-on-plate TIG welding pass on 3.2 mm thick Al plate, compared with finite element predictions (c) [8]. The measurements show the concentration of residual strain around the end (at (0,145)) of an autogeneous 8 mm wide bead on plate TIG weld.

R. A. Owen is to stick a copper sheet to the surface of the sheet and to use this as a reference. The variation in the reference ring radius can then be used to adjust for any change in detector to sample distance (Figure 10.12b). As the figure illustrates, greater accuracy is achieved at large sample-detector distances, both because corrections for sample location are relatively smaller and because the detector is more sensitive to angular shifts at large distances.

There are at least two methods [11] of obtaining some information about the location through-thickness of the diffracting event but retaining the acquisition of complete diffraction rings:

- Conical slit apertures [12]; a conical slit is simply an extension of the normal gauge defining slits to 2D (Figure 10.13a). For any given crystal system it is possible to set up concentric conical apertures spaced so as to obtain a number of peaks simultaneously. The conical apertures only allow diffraction from a small gauge volume to reach the detector. The best slits currently are around 20 μm wide. This is best suited to samples which give true powder diffraction rings (Figure 10.14).
- X-ray tracing; here the principle is to allow diffraction from all through-thickness depths, but to translate the detector along the direction of the incident beam (Figure 10.13b). In this way, it is possible to trace a straight through the centre of mass of corresponding spots recorded at different distances to determine the centre of mass of the diffracting region. This is best suited to spotty few-grain situations.

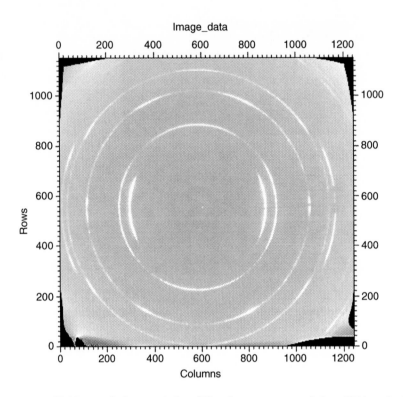

Figure 10.11 A typical transmission diffraction pattern recorded on ID11 at the ESRF at high energy using a CCD [10].

For thick relatively large-grained samples, uncertainty as to the effective position along the direction of the beam of the diffracted signal can introduce significant scatter into the strain measurements. That the sample is large grained relative to the gauge volume is evidenced by spottiness in the CCD diffraction pattern. Except in severe cases this is not a serious problem as the CCD data can be collapsed ('caked') into a series of n line profiles by summing azimuthally over a given angular range ($360/n$). This has the effect of averaging out the spots to form smooth line profiles. If the sheet is thin then the uncertainty in position is necessarily small and all the contributing spots will lie at a similar radius. However, if the transmitted path length is large, then there is considerable longitudinal uncertainty as to the origin of each spot and this will mean that each will lie at different radii. Because for large grained materials, these profiles are dominated by a relatively small number of bright spots, their peak position can be shifted according to whether the dominant diffracting spots originate from the front or the back of the sample. This introduces an uncertainty into the strain calculation of not more than the thickness/detector distance. At a CCD distance of 500 mm this might be $5/500 = \pm10\,000\mu\varepsilon$ for a 5 mm thick sample – much greater than the strain accuracy required. Except in the extreme cases, it is very unlikely that all the diffracting grains would come from either the front or the back of the sample, meaning that this is probably pessimistic by a least factor of 10; nevertheless it does seriously inhibit the strain accuracy achievable for thick samples (>2 mm). There are two means of obviating this problem: (i) to increase the sample to detector distance (Figure 10.12b); and (ii) to introduce

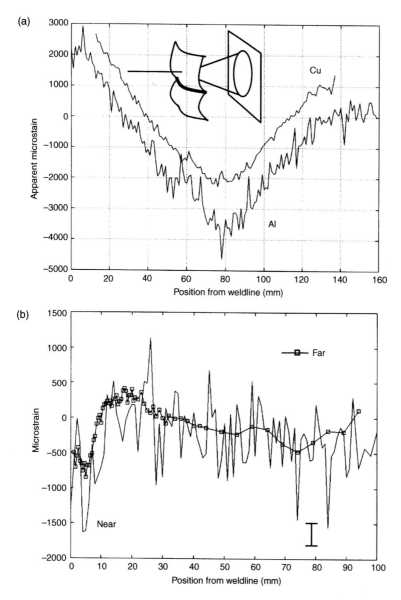

Figure 10.12 (a) Diffraction ring radii as a function of lateral distance from a distorted Al TIG welded plate (inset). Much of the variation is due to the change in gauge volume-detector distance caused by distortion of the plate. This has been corrected for by the placement of a reference thin sheet of Cu on the back of the welded plate. By subtracting the variation in the ring radii for the Cu and Al rings it is possible to correct the variation in gauge volume-detector distance to derive (b) the corrected longitudinal strain variation (here shown with the detector near (~500 mm) to and far (>3000 mm) from the sample.

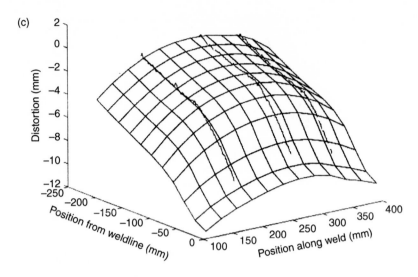

Figure 10.12 (c) The distortion of the plate for comparison with direct distortion measurements. Representative peak fitting error bars are indicated.

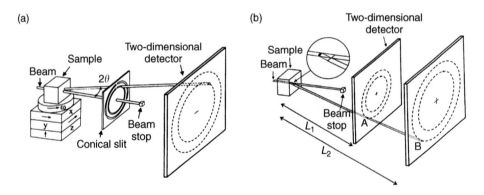

Figure 10.13 Defining the through-thickness gauge dimension using: (a) a conical aperture; and (b) X-ray tracing [11].

a defining (preferably conical) aperture into the diffracted beam. In this case, the uncertainty drops to approximately the aperture width/ring radius (typically $\sim 25/100\,000 = \pm 250\mu\varepsilon$ for a 25 µm slit and 100 mm ring diameter).

Of course, under stress the Debye–Scherrer cones are distorted from producing circular rings to ellipses on the detector. Strain can be extracted in a number of ways from the data. Firstly, the strain in a given direction can be assessed simply by measuring the change in radius (or diameter). However, it is often more useful to analyse the complete rings. By fitting the diffraction rings to ellipses it is possible to deduce the principal in-plane strain directions (the major and minor axes) and their angles to the laboratory frame (Figure 10.15) [13]. Another approach is to use a $\sin^2\psi$ plot to extract the axial and transverse responses (Figure 10.16) [14].

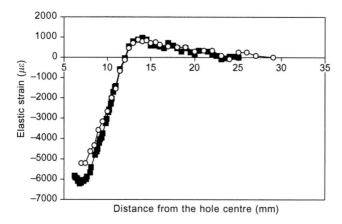

Figure 10.14 Strain measurements made in the vicinity of a cold expanded 6.2 mm hole in a 10 mm Al plate made at the mid-thickness of the plate using neutrons (open circles) and with a cone aperture on ID11 at the ESRF (■). Zero strain has been taken as the far field strain in each case. Presumably the different strains recorded near the surface of the hole are due to the finer lateral spatial resolution of the synchrotron method.

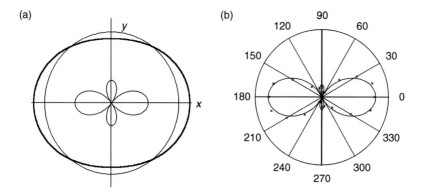

Figure 10.15 (a) An illustration of bi-axial strain state. The unit circle (thin line) is distorted into the ellipse (thick line) by the application of principal strains ε_x and ε_y. The strain polar plot (inner contour) is obtained from the difference between the ellipse and the circle. (b) A strain polar plot representing the strain in the Al matrix of an Al/SiC composite in an axially loaded condition. The continuous curve shows a fit to the experimental data points (markers) [13].

10.4.1   Quenched and bent bar

An example of using the eccentricity of the Debye–Scherrer diffraction cone to measure the principal in-plane strain directions is given by the measurement of the thermal macrostress field caused by quenching and bending an Al/SiC bar sample using a $200 \times 200\,\mu m$ beam [13]. The diffraction cone was segmented ('caked') into 24 sectors and the diffracted intensity azimuthally binned to produce 24 line profiles at intervals of 15. The changes in the diameter of the diffraction ring were then recorded at these angles to calculate the strain (Figure 10.17a).

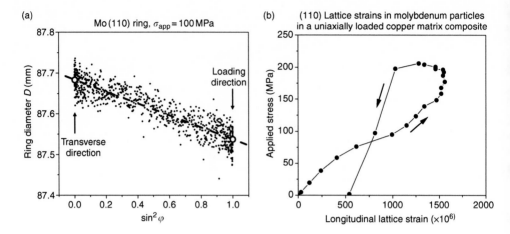

Figure 10.16 (a) Example of a sine-square plot for the (100)Mo phase in a Cu–Mo composite loaded to 100 MPa. (b) A plot showing data extracted using the analysis of (a) for the longitudinal loading response of the Mo phase [14].

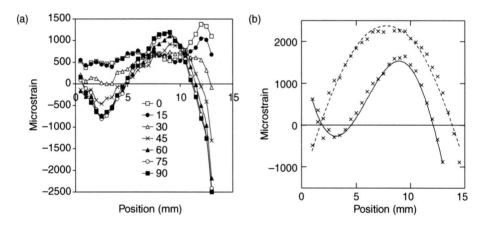

Figure 10.17 (a) The microstrain variation in the Al matrix of a Al/SiC particulate composite across a 14 mm wide quenched and bent plate as a function of the measurement angle; the 90 curve corresponds to the in-plane strain, the 0 curve to the out-of-plane strain. (b) The in-plane strain field after bending plastically (solid line) in three-point bending. The characteristic bending residual stress pattern has been biased by the initial quench residual stress field (dashed curve) which encourages compressive plastic flow [13].

The variation in the in-plane through-thickness caused by bending can be seen by comparing the distribution with that after quenching in Figure 10.17b.

10.4.2   Ti/SiC 'imaging' high spatial resolution

In principle, the transmission technique is capable of very high spatial resolution, provided that the grain size is sufficient to give reasonable quality diffraction rings. This is exemplified by the measurement of strain in an individual SiC 140 μm diameter monofilament of a Ti/SiC

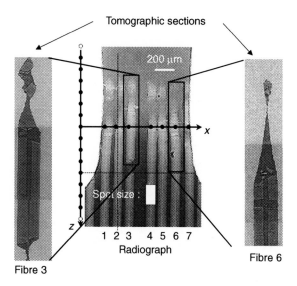

Figure 10.18 Synchrotron X-ray radiograph of the Ti/SiC composite test-piece showing the fibre arrangement and the footprint of the $100 \times 200\,\mu\text{m}$ incident beam spot [14]. The spot was translated in $140\,\mu\text{m}$ steps along lines including each of the seven fibres as indicated by the spacing of the black dots. Non-destructive X-ray tomographic sections show the damage around fibres 3 and 6.

composite material [15]. The composite is shown in (Figure 10.18). A beam of 200 (along fibre axis) by $100\,\mu\text{m}$ (lateral to fibre axis) was used and the sample scanned across the beam to locate the position of each fibre. Despite the very small gauge volume, the ultra-fine grain size of the CVD monofilaments meant that excellent powder ring profiles were obtained. The resulting strain measurements along two individual fibres are shown in the composite as it was loaded. The effect of the prior fibre break in fibre 3 is clearly seen in Figure 10.19. Another advantage of obtaining diffraction ring data is that a number of phases can be studied simultaneously [14, 15].

10.5 Energy dispersive white beam methods

As for the $\theta/2\theta$ method, apertures are placed up and downstream of the sample to restrict the gauge volume, but here a white beam is used (energies range from around 40 to 300 keV). After the exit slits the energies of the diffracted photons are analysed using a germanium detector having an energy resolution of around $\Delta E/E \sim 10^{-4}$. In this case, the whole diffraction profile (Figure 10.20) is obtained as a function of energy rather than angle, which remains fixed at a low value. With slits of $80\,\mu\text{m}$ and a scattering angle of $2\theta = 7.5°$ the measurement volume is around $1.65 \times 0.08\,\text{mm}$. This technique is particularly well suited to the study of complex crystal or multiphase systems, such as for thermal barrier coatings.

10.5.1 Thermal barrier coatings

Wear, oxidation and thermal barrier coatings and interlayers are becoming increasingly important in the race to design high-performance materials at economic cost. Internal stresses within the coating are important as they can give rise to spallation or cracking, seriously limiting

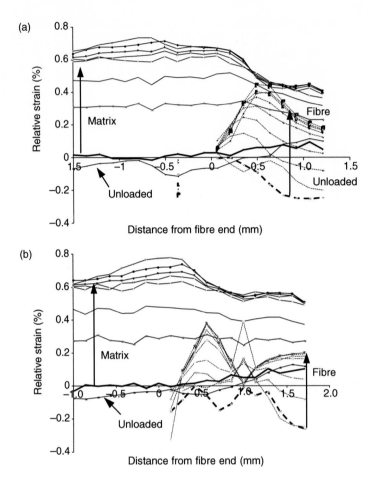

Figure 10.19 Axial strains measured along the lines of fibres: (a) 1 and (b) 3 of the composite
shown in Figure 10.15 made as the composite is progressively loaded using a fine
synchrotron beam on ID11 at the ESRF. At each loading increment 20 measurements
were made along each line giving 140 measurements per load. Strain measurements
for the variation of axial matrix (continuous) and fibre (dashed) strain along the lines
of fibres 1 and 3. Each line represents an increase in the applied load. The initial
residual strains are indicated by the bold solid lines [15].

component life. The stresses within such coatings or interlayers are very difficult to measure with X-rays because of environmental problems, or because the important layers are
sub-surface, while neutron diffraction is often simply unable to detect the layers due to poor
scattered intensity for thin layers or of insufficient spatial resolution to measure the strain
gradients.

In Figure 10.20, the white beam energy dispersive method has been used to investigate a
1.2 mm CoNiCrAlY bond coat/2 mm Fe substrate system on ID15 at the ESRF. The change
from CoNiCrAlY diffraction peaks to those characteristic of Fe is clearly discerned. By
fitting the peaks and comparing with stress-free materials, the internal stresses in a zirconia/
CoNiCrAlY bond coat/steel substrate system are compared with theroetical predictions in

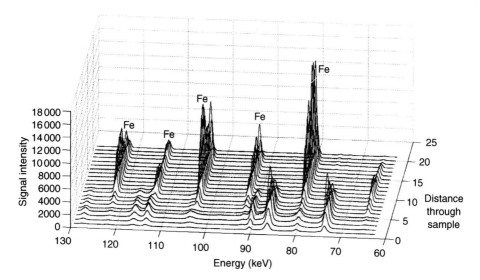

Figure 10.20 A series of energy dispersive profiles made on ID15 at the ESRF at 200 micron steps through a 1.2 mm CoNiCrAlY bond coat/2 mm Fe substrate system for a thermal barrier coating.

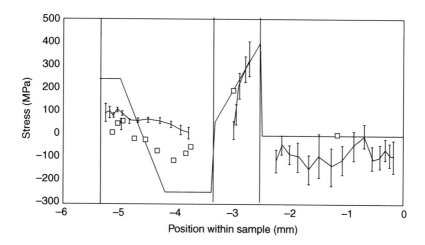

Figure 10.21 In-plane theoretical stresses (straight lines), transmission measurements (with error bars) in layered system of a Zirconia topcoat (0 to −2.5 mm), a CoNiCrAlY bond coat (−2.5 to −3.35 mm), and a steel substrate (−3.35 to −5.35 mm). Superimposed are neutron measurements (□) for a similar system but with the topcoat surface at −0.4 mm [18].

Figure 10.21. The energy dispersive synchrotron method has been pioneered at the ESRF by a group at HMI, Berlin [16, 17].

10.6 Postscript

Many scientific and technological questions have to be resolved before synchrotron strain scanning can take its place as a routine and reliable engineering tool alongside X-ray and neutron diffraction. For example, the very narrow beams (~50 μm) offer the possibility of very fine resolution spatial scanning, but give rise to experimental difficulties. These include problems of interpretation relating to the smaller number of grains illuminated within the sampled volume. Furthermore, at such high photon energies the diffracting angles are very small. This means that, while the sampling volume can be very well defined laterally, it is often less well defined along the length of the beam (usually no better than 1 mm resolution). Furthermore, with the increased spatial resolution come much tighter demands on the absolute positioning of the test-piece with respect to the beam (±5 μm, c.f. ±100 μm for neutron beams). New alignment and positioning procedures need to be developed to cope with these demands, as do faster and more automatic analysis procedures. Opportunities exist to measure the in-plane principal stresses and their orientation with respect to the component axes with a single measurement, but the reduction of the data will be non-trivial. Developments at the ESRF and elsewhere will lead to focused 50 μm microbeams of ultra-high intensity. If the experimental and analytical challenges can be surmounted, there can be no doubt that synchrotron strain scanning could revolutionise the analysis of residual stress engineering problems.

Acknowledgements

Much of the work reviewed here was developed at Manchester by the author with Alex Owen and Mark Pinkerton as part of an EPSRC/MOD-funded grant in collaboration with Prof. Peter Webster, Peter Browne, Gordon Mills and Darren Hughes at Salford and through a long-term research proposal at the ESRF with Dr Alex Korsunsky and Karen Wells of Oxford University, and their contributions are gratefully acknowledged. In addition the tireless contributions are acknowledged of Gavin Vaughan, Ann Terry, Åke Kwick, Bridget Murphy, Steve Collins, who as instrument scientists at the ESRF and SRS helped to undertake many of the experiments. The collaboration of the Risø (Henning Poulsen and Ulrich Lienert) and the Hahn Meitner Institute groups (especially Dr Anke Pyzalla) is also appreciated. All of the ESRF reports listed in this chapter can be found on the ESRF website at www.esrf.fr.

References

[1] Lienert U. *et al.*, Focusing optics for high-energy X-ray diffraction. *J. Synch. Radiat.*, **5**, 226–231 (1998).

[2] Wang D., Harris I. B., Withers P. J. and Edwards L., Near-surface strain measurement using neutron diffraction, 1996, in *4th European Conference on Residual Stress*, Cluny: Soc. Française de Mét. et de Mat.

[3] Webster P. J., Mills G., Wang X. D., Kang W. P. and Holden T. M., Impediments to efficient through-surface strain scanning. *J. Neutron Res.*, **3**, 223–240 (1996).

[4] Webster P. J., Browne P. A. and Mills G., Residual stresses in peened Hiduminium, 1999, in *5th European Conference on Residual Stress*, Holland.

[5] Webster P. J., Mills G. and Kang W. P., Residual stresses in aluminium alloy, *European Synchrotron Research Facility Annual Report*, 1996.

[6] Webster G. A., Neutron diffraction measurements of residual stress in ring and plug, Versailles Project on Advanced Materials and Structures TWA20, *Tech. Report* No. 38 ISSN 1016-2186, 2000.

[7] Norman A. F., Drazhner V. and Prangnell P. B., Effect of welding parameters on the solidification microstructure of autogenous TIG welds in an Al–Cu–Mg–Mn alloys, *Mater. Sci. Eng.*, **A259**, 53–64 (2000).

[8] Owen R. A., Preston R. V., Withers P. J., Shercliff H. R. and Webster P. J., Neutron and synchrotron measurements of residual strain in TIG-welded aluminium alloy 2024, *Mater. Sci. Eng.*, in press (2002).

[9] Preston R. V., Shercliff H. R., Withers P. J., Hughes D., Smith S. and Webster P. J., Finite element analysis and synchrotron X-ray measurement of residual strain in TIG welded Al alloy 2024, submitted to *Metall. Mater. Trans.* (2002).

[10] Daymond M. R. and Withers P. J., A synchrotron radiation study of internal strain changes during the early stages of thermal cycling in an Al/SiC$_W$ MMC. *Scripta Mater.*, **35**, 1229–1234 (1996).

[11] Jensen D. J. and Poulsen H. F., Recrystallisation in 3D, in *Recrystallization – Fundamental Aspects and Relations to Deformation, 21st Risø International Symoposium*, 2000, Risø, Denmark.

[12] Nielsen S. F., Wolf A., Poulsen H. F., Ohler M., Lienert U. and Owen R. A., A conical slit for three-dimensional XRD mapping. *J. Synch. Radiat.*, **7**, 103–109 (2000).

[13] Korsunsky A. M., Wells K. E. and Withers P. J., Mapping two-dimensional state of strain using synchrotron X-ray diffraction *Scripta Mater.*, **39**, 1705–1712 (1998).

[14] Wanner A. and Dunand D. C., Synchrotron X-ray study of bulk lattice strains in externally-loaded Cu–Mo composites, *Metal. Mater. Trans.*, **31A**, 2949–2962 (2000).

[15] Maire E., Owen R. A., Buffiere J.-Y, and Withers P. J., A synchrotron X-ray study of a Ti/SiCf composite during in situ straining. *Acta Mater.* **49**, 153–163 (2001).

[16] Pyzalla A. and Reimers W., Study of stress gradients using synchrotron X-ray radiation, in *Analysis of Residual Stress by Diffraction using Neutron and Synchrotron Radiation*, this volume.

[17] Reimers W. *et al.*, Evaluation of residual stresses in the bulk of materials by high energy synchrotron diffraction. *J. Nondest. Eval.*, **17**, 129–140 (1998).

[18] Thompson J. A., Matejicek J. and Clyne T. W., Modelling and neutron diffraction measurement of stresses in sprayed TBCs, in *Superalloys 2000 (TMS)*, 2000.

11 The use of neutron transmission for materials analysis

H.-G. Priesmeyer

11.1 Introduction

A white thermal neutron spectrum, when transmitted through polycrystalline material, is considerably reshaped by so-called Bragg edges. Bragg edges were first investigated by Fermi and collaborators [1] and found an application in neutron filters. Cassels [2] gives a theoretical summary on total neutron scattering cross-sections including Bragg edges.

In transmission, every scattered neutron contributes to the signal, because it leaves the direct beam into the full solid angle (cf. Figure 11.1). Thus, Bragg-edges can be seen faster in the time-of-flight spectrum than conventional diffraction peaks, for which a specimen is seen under the solid angle of the detector only. Combined with the availability of high intensity pulsed sources (LANSCE, ISIS), this opens the possibility of doing both transient and stroboscopic measurements. The method is relevant to different research areas, like the dynamics of phase transformations, dynamic stress influences at impact, or materials under extreme external conditions. Residual stress states in materials develop as a very complicated, while highly coupled, thermal/metallurgical/mechanical process during welding, casting or forging or as a result of other manufacturing processes: Bragg-edge analysis can contribute to the understanding of these processes and their interrelation.

An advantage of neutron transmission as compared to ordinary diffraction is the easier setup of the experiment, due to the fact that the sample can be placed anywhere between the source and the detector, while spatial information is preserved which allows two-dimensional imaging. The Bragg-edge method may become a unique tool for the investigation of strain mapping, structural phase transitions and crystal structure tomography. Recent developments

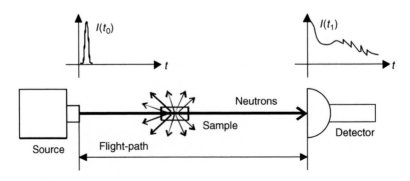

Figure 11.1 Schematic of the transmission setup [3].

have an impact on the determination of the stress-free reference lattice plane distances and strain field imaging.

11.2   The physics of neutron transmission and the generation of Bragg edges

Transmission spectra from polycrystalline materials exhibit Bragg edges instead of Bragg peaks (cf. Figure 11.2). These are steep increases in the transmitted neutron intensity, due to the fact that the angular regime of scattering from a certain lattice plane is limited by $\sin \theta = 1$ ($2\theta = 180°$): once backscattering is reached, the particular set of lattice planes will no longer contribute to coherent scattering. (It must be noted that in diffraction experiments the meaning of 'transmission' differs from the transmission considered here: if the scattered beam leaves the specimen from where the primary beam enters, it is called 'reflected', if it leaves to the opposite side, it is called 'transmitted'.) Full potentiality of Bragg-edge transmission is reached with pulsed sources where a whole spectrum can be attained using the time-of-flight method. Bragg-edge transmission on a steady-state neutron source was used by Strunz *et al.* [4] to determine strain in prestressed steel specimens.

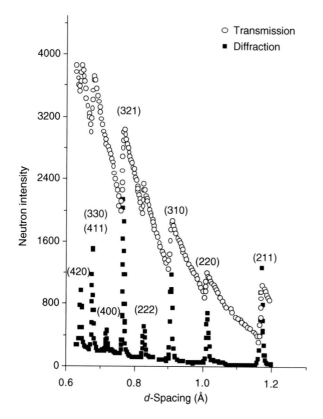

Figure 11.2 Indexed time-of-flight spectra for steel: upper curve = transmission; lower curve = diffraction. (Adapted from [5].)

Bragg edges represent the sudden discontinuity in the scattering cross-section, whenever certain lattice planes are excluded from scattering since the Bragg equation is no longer fulfilled. Johnson and Bowman have pointed out the high resolution and statistical accuracy advantages of Bragg-edge transmission [6]. For Fe samples, they were able to distinguish scattering from lattice planes (hkl) up to $n = h^2 + k^2 + l^2 = 196$, with all planes being distinguishable below $n = 90$. This high resolution in neutron time-of-flight transmission geometry results from its correspondence to backscattering with zero reflection power for all $\lambda > \lambda_{hkl} = 2 d_{hkl}$, with the difference that the detectors are placed in the forward direction.

At a powerful pulsed neutron source like LANSCE or ISIS, information about the crystal structural properties can be attained from every neutron pulse. Since each pulse contains the full thermalized neutron spectrum, a large number of Bragg reflections can be observed simultaneously.

The d-spacings are calculated from time-of-flight channels using the relation:

$$d_{\mathrm{tr}}(nm) = 0.1978 \cdot \frac{t(\mu s)}{L(mm)} \tag{1}$$

where t is the measured flight time needed for a neutron of a certain speed to traverse a flight-path of length L.

Transmission is defined as the ratio of neutron intensity $I(\lambda)$ reaching the detector to the neutron intensity $I_0(\lambda)$ of the incident beam:

$$T(\lambda) = I(\lambda)/I_0(\lambda) = \exp\left\{-\sum(\lambda) \cdot x\right\} \tag{2}$$

where $\sum(\lambda)$ is the macroscopic total cross-section and x is the specimen thickness.

The macroscopic total cross-section is the sum of the coherent-elastic scattering contribution, which is responsible for the Bragg edges, and a number of other cross-sections like incoherent and inelastic scattering and absorption.

The macroscopic coherent elastic cross-section is calculated from the microscopic coherent elastic cross-section:

$$\sum_{\mathrm{coh\,el}} = 6.023 \times 10^{23} \times \sigma_{\mathrm{coh\,el}} \times \rho/A \; [cm^{-1}]$$

where ρ is the density in g/cm^3 and A the atomic weight in g/mol.

The microscopic coherent-elastic cross-section per atom is given by:

$$\sigma_{\mathrm{coh\,el}} = \frac{\lambda^2 N_c}{2n} \sum \left(m_{hkl} |F_{hkl}|^2 d_{hkl}\right) \tag{3}$$

determined by the neutron wavelength λ, the number of unit cells per unit volume N_c, the number of atoms per unit volume n and the multiplicity factor m_{hkl}.

The summation is taken over all lattice planes hkl satisfying $d_{hkl} \geq \lambda/2$.

The structure factor is given by

$$F_{hkl} = \sum \overline{b_j} \exp[2\pi i (hx_j + ky_j + lz_j) \exp(-W)] \tag{4}$$

where $\overline{b_j}$ is the coherent-bound scattering length (averaged over isotopes and nuclear spins), $\exp(-W)$ the Debye–Waller factor and x_j, y_j, z_j are the coordinates of the positions of atom j within the unit cell.

The transmission spectra from specimens which are either very thin or very thick cannot carry much useful physical information. In both cases, the contrast which a Bragg edge can produce in the transmitted spectrum is low, so that the edge may be swamped by the statistical fluctuations of the neutron counts per channel. An optimum specimen thickness with respect to edge contrast can be calculated from the macroscopic cross sections for coherent scattering, incoherent scattering and absorption. They are tabulated for all elements or can be calculated for alloys individually from the tabulated microscopic cross-sections σ.

$$x_{opt} \simeq \frac{2}{\Sigma_{coh(d>2\lambda)} + \Sigma_{abs} + \Sigma_{incoh}} \quad \Sigma \text{ in cm}^{-1}$$

One gets $x_{opt} = 1.7$ cm for iron and $x_{opt} = 19$ cm for aluminium.

If the actual height of a certain Bragg edge is taken into account the following expression can be derived for x_{opt}:

$$x_{opt} = \frac{1}{\Sigma_{hkl}} \ln\left(1 + \frac{\Sigma_{hkl}}{\Sigma_{coh(d>2\lambda)} + \Sigma_{incoh} + \Sigma_{abs}}\right)$$

where Σ_{hkl} is the total cross-section for the hkl plane [7].

11.3   The precision limits of Bragg-edge parameters

In the field of residual stress assessment using neutron diffraction, lattice spacings must be determined with a relative error of less than 10^{-4}. The fractional random error in d-spacing, needed to find the error in strain, is given by

$$\frac{\Delta d}{d} = \sqrt{\left(\frac{\Delta t}{t}\right)^2 + \left(\frac{\Delta L}{L}\right)^2 + (\cot(\theta)\,\Delta\theta)^2} \tag{5}$$

where Δt is the width of the time channel (order of magnitude: $1-5\,\mu s$), ΔL is a combination of the moderator and detector thicknesses (order of magnitude: $2-3$ cm) and $\Delta\theta$ is the spread

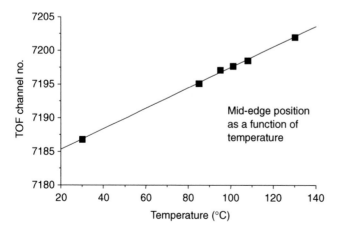

Figure 11.3 Thermal expansion shift of the position of the (311)-edge in aluminium yielding $\Delta d/d = 5 \times 10^{-5}$ for a flight-path length of 11 m.

of the collimation of the incoming beam, which can be considerably reduced by the use of a neutron guide to minutes of arc, so that the angular contribution to the resolution may be neglected. At the 60 m flight-path at LANSCE, the fractional random error is 5×10^{-5}, resulting in excellent precision of the edge position determination. The precision can be validated by measuring the thermal shift of the edge position in a temperature range where a linear relationship with a slope of the order of $10^{-5}/°C$ is expected (Figure 11.3).

11.4   Single-shot capability

With the advent of intense pulsed neutron sources driven by powerful accelerators, the possibility of doing transient as well as stroboscopic measurements with neutrons has become feasible. Such experiments require data to be taken with a single neutron pulse, either because the time of the relevant experimental conditions is so short or one would like to compare diffraction spectra within successive time intervals. Real-time changes of the material properties which influence the position, width and height of the Bragg edges (like phase changes, anisotropic strain development or thermal expansion) can be observed with a time resolution between $<100\,\mu s$ and several milliseconds. First experiences with this new method have been reported [8–10].

Figure 11.4 shows a section of a typical current mode single-shot transmission spectrum for α-iron (bcc) of 2.5 cm thickness, collected at the 10 m detector position within 7 ms [11, 12]. Improvements concerning the signal-to-background ratio can be expected from the 'pseudo current-mode' detection, described by Knudson *et al.* [13], while the best results can be expected from fast direct counting [14]. In order to further improve the statistical accuracy of the data, multiple single-shots can be accumulated.

Transient deformation of a crystal structure has been seen for the first time in an experiment (cf. Figure 11.5), where a lead zirconate titanate piezocrystal (PZT-5A) was subjected to a positive voltage pulse of $+300\,V$ and $1\,\mu s$ duration, which covered one of the more prominent Bragg edges in this material. This edge was chosen because it is close to the

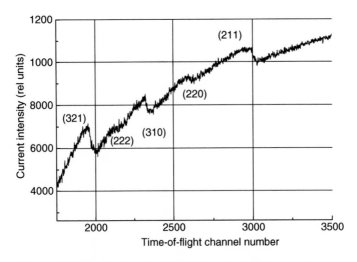

Figure 11.4 Partial single-shot spectrum at 10 m flight-path [11].

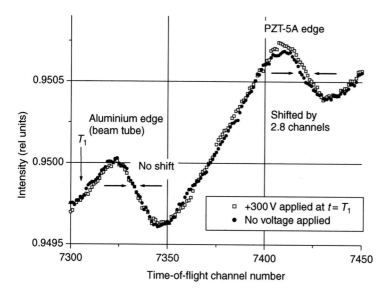

Figure 11.5 Demonstration of high timing resolution in Bragg-edge transmission in piezoceramic
material [11], as explained in the text.

nearby (220)-edge in ordinary aluminium, caused by the neutron beam tube front- and rear-
end covers. Out of all Bragg edges found in the transmission spectrum of PZT-5A, only
the one falling into the time range of the applied voltage pulse is shifted by 2.8 channels
corresponding to a positive (tensile) strain of 4×10^{-4} or $400\,\mu\varepsilon$ ($1\mu\varepsilon = 10^{-6}$). The time
resolution in this case is determined by the time interval between adjacent edges, which is of
the order of $10-100\,\mu$s.

11.4.1 Data analysis

Information relevant to edge position, height and slope can be extracted from the transmission
spectra. The shape of Bragg edges is generally a convolution of the contributions of the natural
width and the mosaic spread of lattice spacings, the resolution function and strain gradient
and texture influences. Evaluation procedures similar to conventional Rietveld refinement
have been developed [14, 15] (Figure 11.6).

11.5 Two-dimensional strain measurements by Bragg-edge transmission

Meggers [16] has for the first time used Bragg-edge transmission to determine the strains
developing in a rotating gyroscope by centrifugal force. The lattice plane information provided in transmission is an average over the transmitted path, so that the technique is not
well suited to producing three-dimensional maps of stress. The advantage of the technique
lies in the analysis of two-dimensional stress fields. In such a case a great deal of information
can be deduced by taking a series of images. An 'image' taken in a tensile stress experiment
applying the tensile force normal to the neutron beam measures the lattice distortion caused
by the Poisson contraction of the specimen, as can be seen in Figure 11.7.

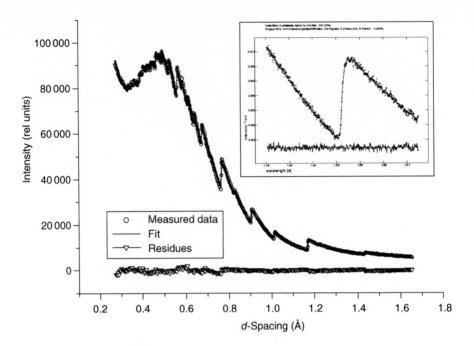

Figure 11.6 Analysis of LANSCE data by Rietveld refinement [14] ('residues' are the differences between measured data and fit).

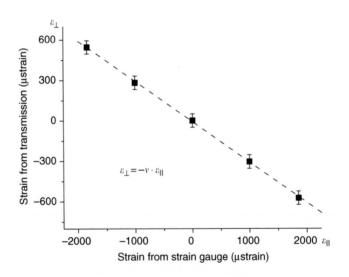

Figure 11.7 Comparison of strains measured from the transmission spectrometer and a strain gauge [17].

The strains from the transmission spectra are equal to $-\nu$ (Poisson's ratio) times the strains from the strain gauges.

Since strain measurements rely on precise determination of the edge positions, special care must be taken to consider the influence of structural materials present in the neutron beam line. Some Bragg edges from the aluminium covers of the evacuated beam tubes, for example, coincide accidentally with edges from bcc iron.

11.6 Time-resolved phase transformations

Recent research on the development of residual stresses includes their calculation by finite element methods, for which certain material properties need to be known. These include the time development of the buildup of incompatible phases as well as the interrelation between phase transformation kinetics and internal stress, so-called transformation strains.

Neutron-stroboscopy has been shown to be a powerful tool for investigating real-time solid state phase transformations, which occur during thermal treatment of materials [18, 19]. Isothermal transformation and cooling transformation diagrams which describe the kinetic microstructural behaviour are needed for model calculations of residual stress development in welding. Transformation progress is usually investigated either by microscopic examination or by measurement of physical material properties using electrical conductivity or dilatometry. The conventional experimental methods are rather complicated, since they require many samples to be subjected to different thermal treatments. With neutrons, all measurements can be done using a single sample in a tube furnace, which holds it at given temperatures for the time of phase transformation (Figure 11.8). The volume changes which accompany the phase changes are a source of residual stresses, termed transformation stresses. Strong influences of these stresses on the phase changes and vice versa can exist.

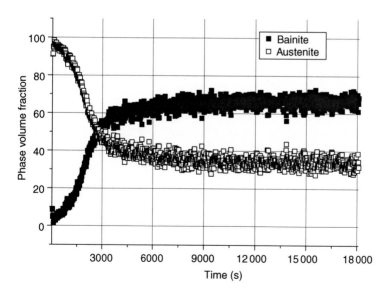

Figure 11.8 Result of a real-time phase transformation measurement from austenite to bainite (data were taken every 15 s).

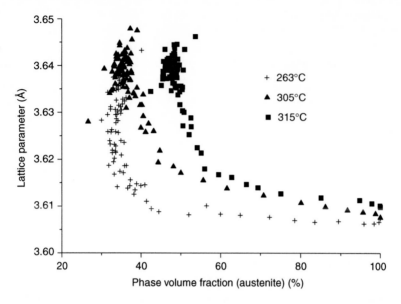

Figure 11.9 The dependence of the lattice parameter on the austenite phase volume fraction during the transformation process as a consequence of the amount of dissolved carbon [14].

The measured edge heights must be normalized with respect to either the initial height (for phases with decreasing volume fractions) or to the final height (for phases with increasing volume fractions). One may also calibrate the measurements by measuring phase volume fractions in specimens before and after heat treatment using conventional neutron powder diffraction.

Deformations of the crystal structure during the phase transformation can also be determined from Bragg-edge transmission, as is seen in Figure 11.9.

11.7   Temperature measurement by Doppler broadening of nuclear resonances

Many of the possible applications of single pulse transmission (like solid-state phase changes, development of transformation stresses, thermal stresses during cooling after welding) require knowledge of the temperature of the specimen. Fowler and Taylor [20] as well as Mayers *et al.* [21] have shown that Doppler broadening of low-lying neutron resonances can be exploited to determine temperatures. Stalder [22] has made a series of experiments using ferritic steel together with a 17 μm thick gold foil (Figure 11.10). Figure 11.11 shows a neutron time-of-flight spectrum containing both edges and a resonance.

Yuan *et al.* [24] have extended the method to determine temperatures in a wide range of dynamically changing systems, like materials through which a shock wave is passed, materials subject to shock-driven pore collapse, frictionally heated interfaces, and metal jets.

Both resonance Doppler broadening and Bragg-edge shift are functions of the temperature of a specimen. While the position of the Bragg edges is influenced by the combined effects of thermal lattice expansion and strain, a resonance is only subjected to thermal Doppler broadening. In technological applications it may not be possible to mix dopant and

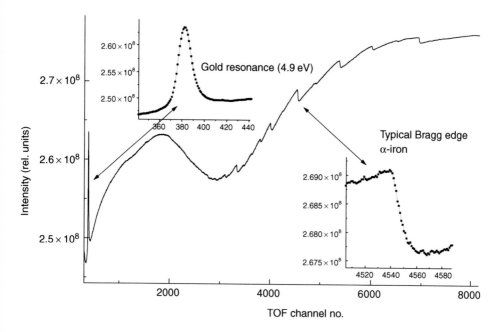

Figure 11.10 Simultaneous current mode recording of the 4.9 eV gold resonance and several edges of bcc iron [23]. Temperatures can be derived from the Doppler broadening of the gold resonance.

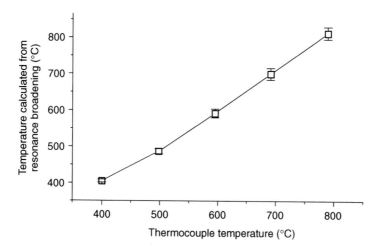

Figure 11.11 Comparison between temperatures derived from thermocouple and gold resonance broadening measurements [22].

alloy without changing the metallurgical properties of the latter. Bragg edges may vary in height for two reasons: the Debye–Waller factor will decrease the diffracted intensity with increasing temperature and diffusion-controlled phase changes may reduce the volume fraction of a certain phase. In steel up to 600°C the intensity reduction can be fully attributed to the Debye–Waller factor. The principle feasibility of the method has been demonstrated by a series of measurements from room temperature to 1000°C.

11.8   Determination of the stress-free reference lattice spacing d_0

As already stated above, the advantage of the transmission technique lies in the analysis of two-dimensional stress fields. In such a case, a great deal of information can be deduced by taking a series of images of a thin specimen. If this image is combined with at least two other images taken from the same area of the specimen at specific angles to the beam, it is possible to determine the stress-free reference lattice spacing d_0.

Using the conventional definition of the azimuth and polar angles (ϕ, ψ), for each set of lattice planes hkl the following expression describes the relation between strain and stress:

$$\frac{d_{\phi,\psi} - d_0}{d_0} = \frac{1+\nu}{E} \left\{ \sigma_{11} \cos^2(\phi) + \sigma_{12} \sin(2\phi) + \sigma_{22} \sin^2(\phi) - \sigma_{33} \right\} \sin^2(\psi)$$

$$+ \frac{1+\nu}{E} \sigma_{22} - \frac{\nu}{E} \left\{ \sigma_{11} + \sigma_{22} + \sigma_{33} \right\}$$

$$+ \frac{1+\nu}{E} \left\{ \sigma_{13} \cos(\phi) + \sigma_{23} \sin(\phi) \right\} \sin(2\psi) \tag{6}$$

For two-dimensional stress fields (no shear stresses), this reduces to:

$$\frac{d_{\phi,\psi} - d_0}{d_0} = \frac{1+\nu}{E} \sigma_\phi \sin^2(\psi) - \frac{\nu}{E} \left\{ \sigma_{11} + \sigma_{22} \right\} \tag{7}$$

where the stress in a certain direction ϕ is given by:

$$\sigma_\phi = \sigma_{11} \cos^2\phi + \sigma_{22} \sin^2\phi \tag{8}$$

The stress acting in the direction $\phi + 90°$ is given by:

$$\sigma_{\phi+90} = \sigma_{11} \sin^2\phi + \sigma_{22} \cos^2\phi \tag{9}$$

Adding the two equations leads to the identity

$$\sigma_\phi + \sigma_{\phi+90°} = \sigma_{11} + \sigma_{22} \tag{10}$$

Measurements at the angular positions $(0, 0)$, (ϕ, ψ), and $(\phi + 90°, \psi)$ result in lattice spacings

$$d_{0,0}, d_{\phi,\psi}, d_{\phi+90°,\psi}$$

Using equation (7)

$$\frac{d_{\phi,\psi} - d_0}{d_0} - \frac{d_{0,0} - d_0}{d_0} = \frac{d_{\phi,\psi} - d_{0,0}}{d_0} = \frac{1+\nu}{E} \sigma_\phi \sin^2(\psi) \tag{11}$$

the expression

$$\sigma_\phi = \frac{E}{1+\nu} \left(\frac{d_{\phi,\psi} - d_{0,0}}{d_0} \right) \frac{1}{\sin^2(\psi)} \tag{12}$$

follows.

A similar expression holds for the $\phi + 90°$ direction:

$$\sigma_{\phi+90°} = \frac{E}{1+\nu}\left(\frac{d_{\phi,+90°,\psi} - d_{0,0}}{d_0}\right)\frac{1}{\sin^2(\psi)} \tag{13}$$

If equations (12) and (13), considering equation (10), are inserted into the general expression for the normal strain,

$$\varepsilon_{33} = \frac{d_{0,0} - d_0}{d_0} = \frac{1}{E}(\sigma_{33} - \nu(\sigma_{11} + \sigma_{22})) = -\frac{\nu}{E}(\sigma_\phi + \sigma_{\phi+90°}) \tag{14}$$

the strain-free reference value d_0 can be calculated:

$$d_0 = (d_{\phi,\psi} + d_{\phi+90°,\psi} - 2d_{0,0})\frac{\nu}{1+\nu}\frac{1}{\sin^2(\psi)} + d_{0,0} \tag{15}$$

Three measurements (at $(0, 0)$, (ϕ, ψ) and $(\phi + 90°, \psi)$) are necessary to approximate d_0, independent of the magnitude and direction of an existing two-dimensional stress field within the material. The 'imaging capability' allows one to measure d_0 simultaneously at different positions, resulting in a strain radiograph. Considering the fact that a whole image can be taken in very short time, the method will have an extraordinary potential for rapid d_0 determination. This is of special importance in the case of strain gradients caused by different crystallographic or chemical compositions, as may be expected in welds, for example.

A simple experiment has been performed to compare the unstrained lattice spacing derived under stress from the above equations with the lattice spacing measured under no stress [25]. The proposed technique is especially suitable for welds, where a big change in the chemical composition of the material is expected from point to point in the sample. If a thin slice is cut from the weld, the d_0 profile can, in principle, be defined by measuring the transmission of the sample in three different orientations even when the principal stress directions are unknown. The uniaxial loading experiment performed by Steuwer et al. [25] has proved the capabilities of the proposed technique, leading to the development of a new pixellated detector for exploiting the unique advantages of the transmission geometry.

The experimental values of c^*, indicating the angles ψ^* for which the unstrained lattice parameter can be measured directly, are shown in Figure 11.12. These values are close to the theoretical values (0.217 and 0.225) predicted.

11.9 Bragg-edge transmission and texture

The presence of texture can considerably change the shape of a neutron transmission spectrum. Both the peak heights and the areas in the spectra between edges show characteristic distortions in this case. As has already been explained, a Bragg edge can be understood as a threshold of a sudden – wavelength dependent – sample transparency change for neutrons which meet grains with orientations characterized by normals of the corresponding (hkl) crystal plane parallel to the incoming neutron beam. Obviously, if, in a given textured sample (i.e. not randomly oriented grains) and for a given beam direction, there happen to be no grains at all with such orientations, no edge will be seen. The more grains are in the right position, the more pronounced the edge will be.

The corresponding probability of grains in a textured sample with (hkl) normals parallel to a sample direction $(\vec{y})$ is described in texture analysis by so-called pole figures $P_{hkl}(\vec{y})$ [26].

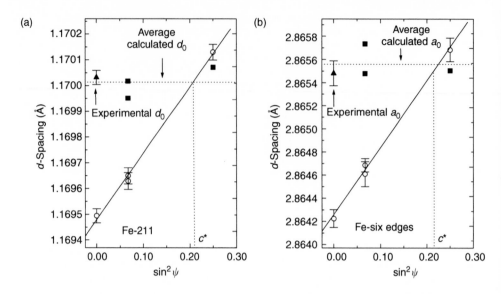

Figure 11.12 (a) Measured lattice spacing and calculated d_0 from the fitting of the Fe-211 Bragg edge. (b) As (a), but performing a multiple-edge fitting of the six lowest edges [25].

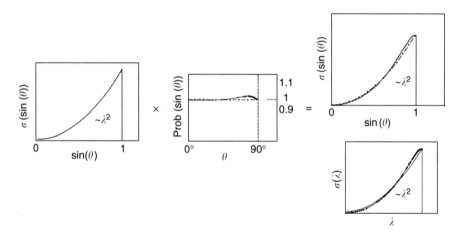

Figure 11.13 Distortion of the coherent elastic cross-section by non-random tilts of lattice planes.

Only for a random texture, with $P_{hkl}(\vec{y}) \equiv 1$, can the shape of a Bragg edge as determined by the scattering cross-section and the instrumental resolution function be expected.

In analogy to Rietveld analysis in diffraction, texture factors (proportional to the pole figure) have to be considered for textured samples, modifying the corresponding coherent-elastic scattering cross-sections, which get different weights at different wavelengths, as is demonstrated qualitatively in Figure 11.13.

Only if all planes $\{hkl\}$ are randomly distributed (prob $(\sin(\alpha)) \equiv 1$) can the ideal shape of Bragg edges be expected. Figure 11.14 shows a part of a typical measured spectrum.

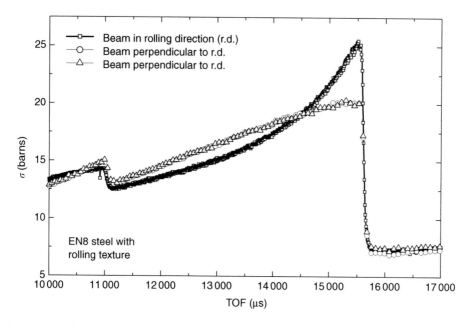

Figure 11.14 The coherent elastic scattering cross-section derived from transmission measurements on a rolled steel plate in three orthogonal directions [7].

For precision measurements, combining Rietveld and texture analysis (RITA-concept [26]), it is possible to extract both structure and texture information from diffraction spectra. Recently this combined analysis has been further developed to include texture-dependent strain/stress and phase analysis [27]. In principle the same program can be proposed for the analysis of neutron time-of-flight transmission spectra, also. However, how many spectra will be needed to get the necessary amount of data has still to be investigated by systematic experimental and theoretical studies.

11.10   Strain imaging

Neutron transmission combined with a two-dimensional pixellated detector has recently been used successfully to image strain fields [28]. The information gained from such images is comparable to photo-elastic pictures and can serve to identify critical areas in specimens, which then deserve more detailed investigation. The spatial resolution is determined by the pixel size of the detector. To measure the following example a detector array of one hundred $2 \times 2 \, mm^2$ scintillating detectors was used (Figure 11.15).

11.11   Conclusions

Bragg edges are a characteristic feature of neutron transmission spectra of polycrystalline materials. They carry useful information about lattice distortions, phase volume fractions and texture. They are characterized by their position, slope and height. These parameters provide averages along the line of the neutron beam through the material investigated.

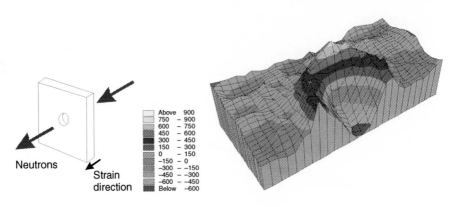

Figure 11.15 Map of the strain field around a 6 mm diameter cold-expanded hole in a ferritic steel, measured by the Bragg-edge transmission technique at ISIS, UK. (See Colour Plate V).

One of the earliest Bragg-edge transmission experiments on cold-worked brass was made in order to distinguish lattice distortion broadening from particle size broadening [29]. With the low-resolution instrumentation available in the 1950s, differences in the broadening of the (111) and (200) reflections due to material anisotropy could be seen.

With modern high-intensity pulsed neutron sources the technique can now be improved to become a unique additional tool to support residual stress characterisations. The experimental setup for Bragg-edge experiments is of convincing simplicity. Developments in the near future will make the method useful for imaging structural components over large neutron beam cross-sections, in order to identify problem areas (e.g. large stresses or stress gradients). Because of its efficient use of the available neutron intensity, the method is well suited to investigating kinematic processes in the time domain of seconds or less. If progress is made in understanding the influence of texture on the shape of the spectra, it is probable that rapid qualitative texture analysis will become possible. In addition, it is an easy experimental step from Bragg-edge transmission to small angle scattering, which has become a well-established method to determine the size and shape of precipitations or voids in materials.

It is certainly correct to state that neutron transmission with Bragg edges opens both scientific and technological perspectives still to be exploited.

Acknowledgements

Discussions with, as well as original contributions from, many colleagues are gratefully acknowledged. Thanks go to R. Bless, C. D. Bowman, M. A. M. Bourke, G. Bruno, L. Edwards, E. Lynn, S. Matthies, K. Meggers, J. Santisteban, M. Stalder, A. Steuwer, W. Trela, S. Vogel, D. Wang, C. Windsor, P. Withers and V. Yuan.

References

[1] Fermi E. *et al.*, The transmission of slow neutrons through micro crystalline materials, *Phys. Rev.* **71**, 589–594 (1947).

[2] Cassels J. M., The scattering of neutrons by crystals, in *Progress in Nuclear Physics 1*, O. R. Frisch (ed.), London, Butterworth-Springer Ltd, 1950.

[3] Bless R., Eine elektronisch gesteuerte Zugvorrichtung zur Untersuchung elastischer Werkstoff-kennwerte mit Hilfe der Neutronen-Transmissions-Diffraktometrie, Thesis, University of Kiel, 1997.

[4] Strunz P., Lukas P., Mikula P., Wagner V., Kouril Z. and Vrana, M, Data evaluation procedure for energy-dispersive neutron-transmission–diffraction geometry, *Proceedings of the Fifth International Conference on Residual Stress*, Linköping, Sweden, 1997, pp. 688–690.

[5] Windsor C., private communication.

[6] Johnson R. G. and Bowman C. D., High resolution powder diffraction by white source transmission measurements, in *AIP Conference Proceedings 89, Neutron Scattering – 1981*, J. Faber (ed.), American Institute of Physics, 1982.

[7] Santisteban J., 1999, private communication.

[8] Priesmeyer H. G. *et al.*, *Fast Transient Neutron Diffraction at LANSCE*; Los Alamos Memo 1989 (unpublished), 1989.

[9] Bowman C. D. *et al.*, Experiments using single neutron pulses, in *ICANS-XI KEK Report 90-25 March 1991, AMRD* Vol. II, M. Misawa, M. Furusaka, H. Ikeda and N. Watanabe (eds), 1991, 840ff.

[10] Priesmeyer H. G., Transmission Bragg-edge measurements, in *Measurement of Residual and Applied Stress using Neutron Diffraction, Proceedings NATO ARW Oxford*, M. T. Hutchings and A. D. Krawitz (eds), Dordrecht, Kluwer, 1992, pp. 380–394.

[11] Priesmeyer H. G., Larsen J. and Meggers K., Neutron diffraction for non-destructive strain/stress measurements in industrial devices, *J. Neutron Res.* **2**, 31–52 (1994).

[12] Bowman J. D. *et al.*, Current-mode detector for neutron time-of-flight studies, *Nucl. Instrum. Methods* **A297**, 183 (1990).

[13] Knudson J. N. *et al.*, A high-rate detection system to study parity violation with polarized epithermal neutrons, *J. Neutron Res.* **4**, 209–213 (1996).

[14] Vogel S., Rietveld approach for the analysis of neutron time-of-flight transmission data, PhD thesis, University of Kiel, 2000.

[15] Steuwer A., Withers, P. J., Santisteban J. R., Edwards L., Bruno G., Fitzpatrick M. E., Daymond, M. R., Johnson M. W. and Wang D, Bragg edge determination for accurate lattice parameter and elastic strain measurement, *Phys. Stat. Sol. A*, **185**, 221–230 (2001).

[16] Meggers K., Neutronendiffraktometrische Dehnungsmessung an einer rotierenden Stahlprobe, Diplomarbeit Christian-Albrechts-Universität Kiel, 1991.

[17] Wang D. Q., Strain measurement using neutron diffraction, PhD thesis, The Open University, 1996.

[18] Meggers K. *et al.*, Transmissions-Kurzzeitmessungen mit Neutronen, *Verhandlungen der Deutschen Physikalischen Gesellschaft*, Hannover, 1992.

[19] Meggers K. *et al.*, Investigation of the austenite–bainite transformation in gray iron using real-time neutron transmission, *Mater. Sci. Eng.* **A188**, 301–304 (1994).

[20] Fowler P. H. and Taylor A. D., Temperature imaging using epithermal neutrons, LA 11393-C Conference (UC 414), 1987, pp. 46–80.

[21] Mayers J., Baciocco G. and Hannon A. C., Temperature measurement by neutron resonance radiography, *Nucl. Instrum. Methods Phys. Res.* **A275**, 453–459 (1989).

[22] Stalder M., Neutronentransmissionsuntersuchungen an Stahl und Temperaturbestimmung mit Hilfe der resonance-doping-Methode, Exp.-Phys. Diplomarbeit, Christian-Albrechts-Universität Kiel, 1996.

[23] Priesmeyer H. G. *et al.*, Progress in single shot neutron transmission diffraction, in *5th International Conference on Application of Nuclear Techniques*, Neutrons in Research and Industry, Crete 1996, *SPIE* **2867**, 164–167 (1997).

[24] Yuan V. W., Asay B. W., Boat R., Bowman J. D., Funk D. J., Hixson R., Hull L., Laabs G., London R., Morgan G. L., Rabie R. and Ragan C. E., Dynamic temperature and velocity measurements using neutron resonance spectroscopy, in *Nuclear Data for Science and Technology*, G. Reffo, A. Ventura and C. Grandi (eds), Italian Physical Society Conference Proceedings, Vol. 59 Part II, 1997.

[25] Steuwer A., Withers P. J., Santisteban J. R., Edwards L., Fitzpatrick M. E., Daymond M. R., Johnson M. W. and Bruno G., Neutron transmission spectroscopy: a solution to the d_0-problem? paper presented at *ICRS 6*, 2000.

[26] Matthies S., Lutterotti L. and Wenk H. R., Advances in texture analysis from diffraction spectra, *J. Appl. Crystallogr.* **30**, 31–42 (1997).

[27] Lutterotti L., MAUD: Material analysis using diffraction, web downloading from: http://www.ing.unitn.it/~luttero/maud/maud.html.

[28] Santisteban J. R., Edwards L., Fitzpatrick M. E., Steuwer A., Withers P. J., Daymond M. R., Johnson M. W., Rhodes N. and Schooneveld E. M., Strain imaging by Bragg edge neutron transmission. *Nuclear Instr. and Methods in Phys. Res.* **A481**, 255–258 (2002).

[29] Weiss R. J. and Clark J. R., Neutron diffraction studies of cold-worked brass, *J. Appl. Phys.* **23**, 1379–1382 (1952).

Part 4

Areas of study

12 Strain mapping

P. J. Webster

12.1 Introduction

Neutron strain scanning techniques have been developing since the early 1980s [1–3] and synchrotron X-ray techniques rapidly since the 1990s [4, 5]. The first neutron measurements used single detector instruments at medium flux reactor sources, 'gauge volumes' that were relatively large and coarsely defined, manual positioning of samples and computer-aided but labour-intensive interactive Gaussian peak fitting routines. Statistical data quality was often high but spatial resolution, data collection rates and data processing speeds were all relatively low. Even in comparatively simple experiments, for example, scans through thin sections of ferritic steel test samples with moderate residual strain gradients, only a handful of measurements could be made and processed in a day. Consequently most neutron strain investigations were restricted to linear scans at a dozen or so points, usually in three orthogonal directions so that stresses could be derived. Area mapping, which typically might require orthogonal measurements at a hundred or more points was not then generally a practical proposition due to time and resource cost considerations.

With the introduction of high flux neutron sources, multidetectors, automated sample positioning, optimisation of data acquisition, fast and cheap computers, fully computerised peak fitting and commercial surface fitting software, the situation has changed dramatically. The data for an adequately defined neutron diffraction peak might now be collected in a few minutes, or even less, and be processed near on-line, so that multipoint neutron area strain scanning is now practicable [6, 7].

Synchrotrons produce near-parallel X-ray beams with extremely high photon fluxes, typically billions of times the equivalent neutron flux of even the most powerful nuclear reactors. On the other hand X-ray attenuation lengths, at similar wavelengths to those used with neutron strain scanning ($\lambda \approx 1.5$ Å), tend to be several orders of magnitude smaller than the corresponding neutron attenuation lengths. Consequently, at these wavelengths the attenuation of X-ray beams is generally much higher than for neutron beams, especially through thick samples. However, for harder more penetrating X-rays ($\lambda \approx 0.3$ Å), attenuation lengths can be similar ($\approx$mm), for light element materials, to those for neutrons. Scattering parameters and attenuation lengths for neutrons and for X-rays of typically used wavelengths are shown for elements that comprise the majority component of common engineering alloys in Table 12.1. In favourable circumstances, such as when measuring through relatively thin samples made of lower attenuating light element materials, synchrotron X-ray count rates, even with single detector instruments, can be orders of magnitude greater than is typical for neutron count rates. Furthermore, synchrotron technology is advancing rapidly, producing higher fluxes at even higher energies. At wiggler and undulator locations synchrotron X-ray

Table 12.1 Neutron and X-ray scattering parameters for four elements which comprise the majority component of many common engineering alloys

Element	Al	Ti	Fe	Ni
Atomic number	13	22	26	28
Neutron scattering length (fm)	3.45	−3.44	9.45	10.3
Neutron attenuation length (mm)	100	20	8.3	5.6
1.5 Å X-ray attn. length (mm)	0.083	0.012	0.004	0.003
0.3 Å X-ray attn. length (mm)	6.71	1.06	0.38	0.27
0.15 Å X-ray attn. length (mm)	52.6	8.47	3.03	2.19

Note

The attenuation length L is the thickness of material required to attenuate a beam by the factor 1/e. It is the reciprocal of the linear attenuation coefficient μ, so the intensity I after attenuation of a beam of incident intensity I_0 through a thickness x is $I = I_0 e^{-x/L} = I_0 e^{-\mu x}$.

fluxes are particularly high. When used with area detectors, data from several reflections, at all orientations within a nearly flat cone, may be simultaneously recorded in seconds or less. Consequently, although synchrotron strain scanning is a relatively new technique, synchrotron area strain mapping is developing rapidly. As developments proceed it is probable that volume strain mapping will also become routine.

The aim of this chapter is to outline the methods used in neutron and synchrotron strain mapping and to illustrate by examples the potentials of the two techniques.

12.2   Optimisation and efficient strain scanning

12.2.1   *Resource cost*

Neutron and synchrotron strain scanning both require large, high capital cost, central facilities that are often used on a timeshare basis. Generally both the demand for and the cost of beam-time are relatively high. As beam-time is usually a large fraction of the total resource cost, it is essential to make efficient use of it. Time for academic investigations is usually restricted and competitively allocated, and commercial use is cost sensitive. The time required for an experiment, using an incrementally stepped single detector, is proportional to the scan range and inversely proportional to the sample volume. In a routine scientific neutron diffraction experiment to determine the structure of a powdered material, for example, the incident and detected beam cross-sections might be about 40 mm high and 10 mm wide giving a scattering volume about 4000 mm³. The detector scan range to obtain a full diffraction pattern might be ≈100°. Strain scanning is particularly profligate in its use of neutrons or X-ray photons. In neutron strain scanning the beam cross-section might be reduced to 2 × 2 mm giving, with a 2 mm wide detector slit, a 'gauge volume' of 8 mm³. The scan range required to define a single peak might be 2°. For the powder diffraction experiment the time required would be proportional to 100/4000 and for strain scanning 2/8. Hence, it would typically be necessary to count about 10 times longer to collect the data for one peak profile when measuring strain than is needed for a complete powder diffraction experiment, if similar peak counts are required. Additionally, strain scanning requires multipoint measurements, usually in three orientations, and precise sample positioning. Consequently, unless appropriate measures are taken, the time required for strain scanning could be hundreds or thousands of times longer than for routine powder diffraction experiments, which would be impractical.

12.2.1.1 Multidetectors

It is now common, when neutron strain scanning, to use multidetectors or position sensitive detectors subtending a few degrees so that data from across the profile of a selected peak can be collected at the same time rather than sequentially. Using a multidetector might typically provide a gain in speed by a factor of $10\times$, but can introduce potentially severe instrumental aberrations at surfaces and interfaces. Multidetectors can also, of course, provide similar gains for routine powder diffraction experiments so that, although there is a real gain in efficiency when strain scanning, there may be no improvement in competitive advantage relative to powder diffraction.

12.2.1.2 Setting-up and positioning time

Initial setting-up of the scanner and precise positioning of the gauge relative to the 'reference point' can be time-consuming, as can be the initial positioning of the sample. Subsequently the time taken to automatically move the sample when scanning can also be a significant proportion of the total time required, particularly when counting times are short.

In well-designed dedicated strain scanners, setting-up times should be short and the gauge volume should, after initial calibration, be automatically adjustable and be precisely positioned at a reference point on the scanner axis. Initial sample positioning is facilitated if the *XYZ* sample translation and rotation equipment has a precise positioning, orienting and fixing system, preferably conforming to a common standard. This should enable a sample to be positioned off-line on a standard matching mount to within 0.1 mm and aligned to better than $0.1°$, before incurring beam-time costs. Efficient rapid mounting, and final position refinement as necessary, is then possible at minimum beam-time cost on-line on the scanner.

The time taken to move the sample when scanning depends upon the inertia of the system and the mechanics, electronics and control software. Time can be saved if the system is designed so that movements can be rapidly, smoothly and precisely executed without significant delay due to over- or under-shooting. Generally, it should be possible to reset the system, when scanning in small steps, in not more than a few seconds. The reduction of dead-time is likely to become an increasingly important factor, especially for synchrotron experiments, as data collection rates increase.

12.2.1.3 Gauge size

Gauge volume sizes should be as large as is practicable considering the instrumental optics, the spatial resolution that is required and attenuation lengths. Minimum gauge sizes tend to be limited by beam divergence when neutron strain scanning, and by grain size considerations when using near-parallel synchrotron X-radiation.

12.2.2 Peak profile determination

The peak profile parameters generally required are: peak centre to determine macrostrain, peak width which is a measure of plastic deformation and microstrain, and peak height and intensity which are indicators of texture and grain size.

The longer the count time the better the statistical data quality, which is a function of $\sqrt{N}$, where N is the neutron or X-ray photon count, but the fewer the number of points that can be sampled in a given time. The number of data points is linearly related to the time available, but the data quality improves as the square of the time. To optimise the utilisation of counting time, it is necessary to compromise between data quality and quantity. Usually, once the data are adequate to enable a specified statistical uncertainty in strain to be attained, counting time is more efficiently utilised if additional points, rather than statistically improved points, are obtained, especially in area and volume strain mapping. Optimisation of data collection and processing is discussed in Ref. [8]. Particular factors to be considered are the following.

Peak angular scan range – This should be large enough to define the peak profile and background but not much more. A practical range for a centred Gaussian peak is $3W$ or 7σ, where W is the full width at half maximum (FWHM) and σ is the standard deviation ($W = 2.35\sigma$ for a Gaussian, so that $3W \sim 7\sigma$).

Number of points/bins in a peak profile – It is generally found that 20–30 or so points is near-optimum to visually define a peak with a near-Gaussian single peak profile and flat background.

Counts in the peak – Assuming that the background is low, the statistical uncertainty U_C in a near-Gaussian peak position is:

$$U_C \sim (2/3)W/\sqrt{C_T} \tag{1}$$

where C_T is the total neutron or photon count in the peak. The corresponding uncertainty in strain U_ε is:

$$U_\varepsilon \sim (1/3)W \cot\theta/\sqrt{C_T} \tag{2}$$

where θ is the Bragg angle of the peak and the peak width W is in radians.

Hence a total count of $C_T \sim 4500$ (equivalent to a 21-step, $W/7$ increment, $3W$ range, peak of height 600 with a low background) will result in an uncertainty $U_C \sim 1\%W$. For neutron strain scanning at a detector angle $2\theta = 90°$ and typical peak width $0.5°$, the corresponding uncertainty in strain is $U_\varepsilon \sim 40 \times 10^{-6}$. For the same count when synchrotron strain scanning at a typical detector angle $2\theta = 10°$ and peak width $0.04°$ the uncertainty in strain is approximately the same. If a strain uncertainty $U_\varepsilon \sim 100 \times 10^{-6}$ is sufficient, as is often the case, counting times could be reduced by a factor $\times 6$ ($C_T \sim 700$, peak height ~ 100) and the number of measurement locations could be correspondingly increased.

Optimisation of peak parameters – To optimise counting times the required strain uncertainty should be specified. Peak parameters, such as counting time, step size and angular range, should then be chosen so as to enable the specified uncertainty value to be attained. Shorter counting times can be used if instrumental peak widths are low and detector angles are high.

12.2.3   *Mapping*

The number and spacing of points required adequately to define the required detail in a residual strain line scan depends upon whether the strain gradient is rapidly or gradually varying. The spacing problem is similar to that of other signal sampling and reconstituting situations so it is possible to use sampling theory to estimate an optimum step resolution [9].

If the general form of the strain variation is known in advance, step sizes may be chosen so that spatial resolution is appropriate. A common practical approach, if the form of the strain pattern is unknown, is to make a coarse step scan which is then fitted by an appropriate function that interpolates between neighbouring points and to fill in with finer steps if it appears to be necessary. A similar approach is used when area mapping but with a function that interpolates in two dimensions. In this case, data of somewhat lower quality can be used as the interpolation will effectively be between four or more neighbouring points, rather than just two when line scanning. However, care should be taken with commercially available fitting routines to ensure that the fitting algorithm is appropriate. In some cases routines fit exactly at experimental data points and generate spurious oscillations in between. The appropriateness of a fitting routine can be judged by superimposing the locations of data points on the map and observing whether the map is overly distorted by the mapping matrix. If it is, another fitting routine or scanning matrix should be used.

12.2.3.1 *Variable step mapping*

In regions of low strain gradient variation the step size between points can be much greater than in regions of high strain gradient variation. However, if substantially varying step sizes are used, it can lead to apparent features on the strain map which are related to the variations in the mapping matrix rather than to the strain field. The distorting features appear as a result of the statistical uncertainties associated with any experimental measurement. If experimental points are close together, rapid uncertainty fluctuations will be observed on the map. If the points are well spaced out, the fitting routine will interpolate a smooth variation between the distant points.

The visually distorting effect of uneven mapping is illustrated in Figures 12.1 and 12.2. Figure 12.1 shows two mapping matrices, one a regular square matrix and the other a more complex pattern in which the spacing varies by a factor $\times 10$. Figure 12.2 shows the effects produced on a random number array pattern by using the two mapping matrices. Figure 12.2(a) is the full regular map with the expected randomly uneven pattern. Figure 12.2(b) uses the same array of random numbers but is based on the partial irregular matrix obtained by omitting some of the points. Two of the six small square regions fully mapped in both figures have been outlined to permit a direct visual comparison. Within the small squares both maps are essentially identical. In the intervening rectangular areas, similar to the one shown outlined, streaked patterns are generated as the periodicity in one rectangular direction is $10\times$ that in the other. In the other areas of Figure 12.2(b), a variety of smoother larger scale patterns related to the more spaced mapping matrix and random variations are evident.

When a strain map is generated from data collected at a matrix of points, the uncertainties in each data point will cause a spurious matrix-related pattern to be superimposed upon the 'true' strain map. If an area strain pattern is generated from an irregular matrix combination of coarse and fine steps between points, such as shown in Figure 12.1(b), the effect can be particularly visually misleading. In general, if a strain field contains distinct high and low strain gradient regions in which high and low density mapping is respectively appropriate, it is advisable to present the results on several maps rather than just one. For example, if the matrix in Figure 12.1(b) was used to collect the data, the results could be presented as six regular 10×10 small squares and one regular 10×10 large square. The former would give detailed pictures of each of the six small regions. The latter would only use a small proportion of the data collected but would give an undistorted low-resolution large-scale view.

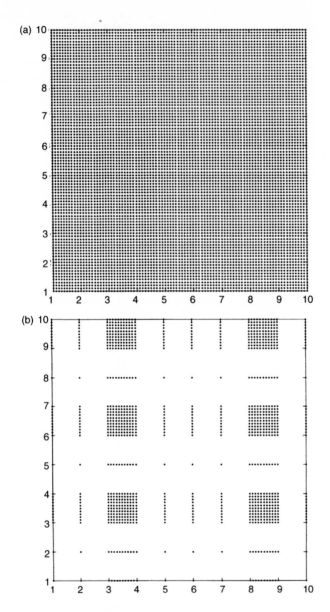

Figure 12.1 (a) Regular square matrix of 100 × 100 points. (b) Irregular matrix comprising
six small square and linear arrays of parts of the 100 × 100 matrix shown in (a).
(These matrices were used to generate the corresponding interpolated maps shown
in Figure 12.2.)

12.3   Neutron strain mapping

Figure 12.3 shows one of the first residual stress maps to be derived from two-dimensional
neutron strain data. The data were collected using D1A at the ILL, Grenoble, in its single
detector strain scanning mode [9]. Measurements were made over the head of a transverse
railway rail section 12 mm thick using the 211 reflection in the three symmetry orientations,

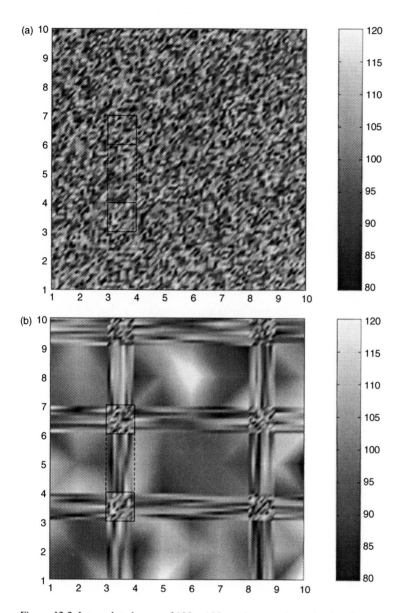

Figure 12.2 Interpolated maps of 100×100 numbers, each nominally of value 100 but with a randomly generated Gaussian uncertainty with a standard deviation 10: (a) arranged on the matrix shown in Figure 12.1(a); (b) arranged on the matrix shown in Figure 12.1(b) with the non-marked data points omitted.

longitudinal, transverse and vertical using a near cubic sampling volume of side 2 mm. The scans were made in nine parallel lines from $z = 0$ at the top of the head to a depth of $z = 40$ mm at transverse locations $y = 0, \pm 8, \pm 16, \pm 24$ and ± 30 mm. Increments between points were varied from 2 mm near the top surface to 6 mm near $z = 40$ mm as strain gradients were known to be substantially higher near the top surface.

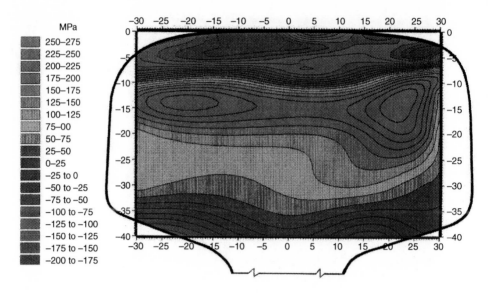

Figure 12.3 Transverse residual stresses in a BR rail head measured using D1A at ILL, Grenoble. (See Colour Plate VI.)

The rail, which had been used, was taken from a straight section of British Rail (BR) track. The 'running line', the line of contact between wheel and rail, was centred at $y = +6$ mm. The map shows the computed interpolated transverse residual stress pattern which varies from $+275$ MPa in the central tensile band to -225 MPa in the compressive near-top-surface region. The residual stress pattern is generally beneficial as the compressive outer layer inhibits the propagation of fatigue cracks from surface defects and the internal balancing tensile region does not reach the surface. The pattern shows some asymmetry due to the running line being 6 mm off-centre.

There is no apparent evidence of spurious features due to the scanning matrix. This is because the statistical data quality was relatively high and the uncertainties small, typically ± 25 MPa, the step sizes were almost constant in y, changed smoothly in z and were matched to the strain gradients. The first interpolation attempted showed distinct nodes at the line locations $y = 0, \pm 8, \pm 16, \pm 24$ and ± 30 mm and oscillations in between because it was constrained to fit exactly at the measured points and to change smoothly in between. The interpolation routine was subsequently revised so that fitting at the measurement points was allowed to vary within a standard deviation. The resulting smoother pattern may be considered to be a good representation as it shows no apparent evidence of features related to the measuring matrix but does reveal significant detail in the residual stress pattern.

12.4   Synchrotron X-ray strain mapping

Figure 12.4 shows results from a synchrotron X-ray study using the high-resolution diffractometer BM16 at the ESRF, Grenoble, in its single detector with analyser strain scanning mode [10]. The sample was a 10 mm thick transverse cross-section cut from a double-V multipass weld made in an aluminium alloy plate 44 mm thick. Scans were made in the transverse orientation using an incident beam 1 mm² and a 1 mm high detector slit. The reflection used

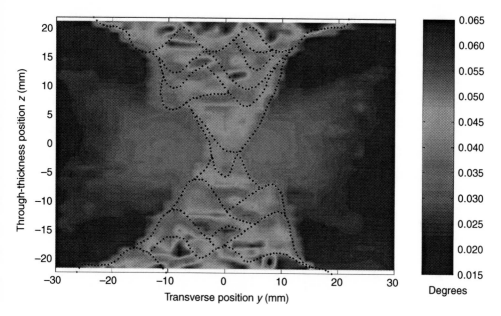

Figure 12.4 Peak width variation over a cross-section of an aluminium alloy double-V multipass weld measured using BM16 at ESRF, Grenoble. The weld bead pattern is superimposed. (See Colour Plate VII.)

was 311, which at a wavelength 0.32 Å gave a gauge ≈ 8 mm long in the longitudinal direction at a detector angle $\approx 15.1°$. The measurements were made at mid-plane over a regular square matrix of points spaced at 1.333 mm intervals. In this particular case, as the longitudinal stresses had been substantially relaxed by the sectioning and were not expected to change significantly through the thickness, the elongated shape of the gauge was advantageous as it gave a better average, a higher count and less grain size variation. The 1 mm y and z gauge dimensions, and the matrix spacing, were chosen to be significantly smaller than the weld bead dimensions so that intra- and inter-bead, as well as fusion zone, boundary variations could be observed.

Figure 12.4 shows the multipass weld bead pattern superimposed upon the measured peak widths which are a measure of microstrain. The passes were made in layers alternately on top and bottom sides so that the weld was as symmetrical as possible. The central layers are thus partially annealed by subsequent passes but the surface passes are not. The bead pattern was not entirely regular and the fusion zone boundaries were neither straight nor symmetrical. The step size was sufficiently small, and the gauge was so well defined, that the map reveals several significant features. There is a clear correlation between the peak widths and details of the weld such as the edge of the fusion zone and the individual weld beads. At transverse distances $y > \approx 20$ mm from the weld the peaks are uniformly $\approx 0.015°$ wide, which corresponds to the inherent instrumental resolution and low microstresses in the plate prior to welding. The increases in peak widths elsewhere are mainly a measure of microstrains and, to a small extent, macrostrain gradients resulting from the welding process and plastic deformation. The maximum peak broadening, $0.065°$, corresponds to a microstrain of $\approx 3 \times 10^{-3}$. The fusion zone boundary is outlined in detail by a narrow layer, about one matrix step wide, in which peak widths are ≈0.03°. This indicates that the microstrain change at the boundary is generally very sharp and is less than or equal to the spatial resolution. Outside the weld, roughly within the band $-20 < y < 20$ mm, $-5 < z < 5$ mm, there is another region in which peak widths are similarly ≈0.03°. In this region there has probably been some plastic deformation as it would have been initially in tension as the first beads were laid down, but finally became a balancing compressive region as later beads were laid. Inside the weld, peak widths vary from ≈0.04° in the partially annealed near mid-thickness region to ≈0.065° in the last deposited surface beads where microstrains are highest.

12.5   Conclusions

Two-dimensional strain mapping is becoming increasingly common using both neutrons and synchrotron X-radiation. This development is a result of demand from engineers, who are now beginning to insist on detailed residual strain and stress information so as to be able to refine their designs and to validate their modelling codes, and the increasing ability of strain scanners now to supply large amounts of repeatable data at acceptably low unit cost. It is anticipated that demands will continue substantially to increase and that standardisation and scanner design, measurement techniques, automated data processing and analysis, will develop so that non-destructive residual strain and stress mapping of components will become as routine and essential in engineering as body scanning is now in medicine.

References

[1]   Allen A., Andreani C., Hutchings M. T. and Windsor C. G. Measurement of internal stress within bulk materials using neutron diffraction, *NDT Int.*, **14**, 249–254 (1981).

[2]   Pintschovius L., Jung V., Macherauch E. and Vohringer O. Residual stress measurements by means of neutron diffraction, *Mater. Sci. Eng.*, **61**, 43–50 (1983).

[3]   Stacey A., MacGillivray H. J., Webster G. A., Webster P. J. and Ziebeck K. R. A. Measurement of residual stresses by neutron diffraction, *J. Strain Anal.*, **20**, 93–100 (1985).

[4]   Webster P. J., Mills G. and Wang X. D. Residual strain scanning of engineering samples, *SRS Annual Report* 21/40 A65 (1992/93).

[5]   Webster P. J. Strain scanning using X-rays and neutrons, *Mater. Sci. Forum*, **228–231**, 191–200 (1996).

[6]   Webster P. J. Neutron strain scanning, *Neutron News*, **2**, 19–22 (1991).

[7]   Webster P. J., Mills G., Wang X. D., Kang W. P. and Holden T. M. Neutron strain scanning of a small welded austenitic stainless steel plate, *J. Strain Anal.*, **30**, 35–43 (1995).

[8]   Webster P. J. and Kang W. P. Optimisation of neutron and synchrotron data collection and processing for efficient Gaussian peak fitting. *Engineering Science Group Technical Report* ESG01/98, The Telford Institute of Structures and Materials Engineering, University of Salford, March 1998.

[9]   Kang W. P. Application of numerical analysis to neutron strain scanning. PhD Thesis University of Salford, August 1996.

[10] Webster P. J., Vaughan G. B. M., Mills G. and Kang W. P. High-resolution synchrotron strain scanning at BM16 at the ESRF, *Mater. Sci. Forum*, **278–281**, 323–328 (1998).

13 Study of stress gradients using synchrotron X-ray diffraction

A. Pyzalla and W. Reimers

13.1 Introduction

By definition, residual stresses σ_{ij} are self-equilibrating stresses within a body. They fulfil the equilibrium condition

$$\sigma_{ij,j} = 0, \quad \text{e.g. } \sigma_{11,1} + \sigma_{12,2} + \sigma_{13,3} = 0 \tag{1}$$

within each point of the body. This implies that regions with tensile residual stresses always have to be balanced by regions containing compressive residual stresses and that for reasons of consistency residual stress gradients are present in between the stress extrema. Furthermore, a gradient in one residual stress component, for instance in σ_{11}, that is $\sigma_{11,1} \neq 0$, always means that there is also a gradient in at least one other residual stress component, for example, $\sigma_{12,2} \neq 0$ or $\sigma_{13,3} \neq 0$ [1]. Thus, the residual stress state of a sample or a component cannot be characterised by a single value but only by a residual stress distribution and hence residual stress gradients.

Steep residual stress gradients typically are present in those areas where strong deformation inhomogeneities exist, for example, in the vicinity of notches or at interfaces between different materials. Also thin films due to the boundary condition

$$\sigma_{ij} n_j = 0 \tag{2}$$

at the free surface often are subject to steep residual stress gradients. In equation (2), n denotes the normal to the surface.

The residual stress distribution and its gradients within a body are of major practical relevance, for example, with respect to distortion of components during further machining or cutting. Near-surface stress gradients are known to play an important role with respect to fatigue properties, for instance, in case of shot-peened surfaces, where compressive residual stresses often are required in a certain magnitude and depth. In the case of thin films or layers, the residual stress gradient near the interface often significantly influences the adhesion on the substrate.

In order to analyse the residual stress distribution and its gradients in a sample or a component by using diffraction methods, the choice of the radiation and the technique have to be done adequately with respect to the in-depth resolution and the lateral resolution as well as with respect to the angular resolution or the resolution in energy required [2]. With regard to the in-depth resolution three regions can be roughly distinguished: the region very near the surface up to approximately $20\,\mu m$ depth, the near-surface-region up to approximately $200\,\mu m$ depth and the bulk (Figure 13.1). The very-near-surface region is accessible to X-rays

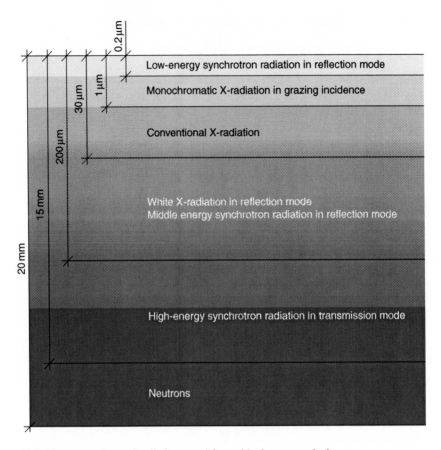

Figure 13.1 The range of use of radiations used for residual stress analysis.

from laboratory sources and low-energy X-rays from synchrotron sources, for example, in the grazing incidence mode. The near-surface region is accessible using conventional X-ray tubes with a white spectrum or medium energy X-rays from synchrotron sources. The bulk residual stresses are the field of neutrons (see Chapter 1 and Part 4 of this book) and high-energy X-rays from a laboratory or a synchrotron source.

13.2   Analysis of very-near-surface stress gradients

Very-near-surface stress gradients have to be expected in thin coatings, but also as a consequence of surface working processes, for example, grinding. The presence of a stress gradient normal to the surface direction manifests itself in a curvature of the $d_{\varphi\psi}^{hkl}$ versus $\sin^2\psi$ graph. This is due to the fact that the penetration depth of the radiation varies with the inclination angle ψ as a consequence of the absorption of the radiation. In fact, X-radiation yields volume averages of the strain and the stress which are weighted with respect to the absorption of the radiation [3]:

$$\sigma_{ij}(\tau) = \frac{\int \sigma_{ij}(z)\varepsilon^{-z/\tau}\,\mathrm{d}z}{\int \varepsilon^{-z/\tau}\,\mathrm{d}z} = \frac{1}{\tau}\varepsilon_{ij}(1/\tau) \tag{3}$$

τ denotes the penetration depth, $\varepsilon(1/\tau)$ is the Laplace transform of $\tau(z)$ and $\sigma(\tau)$ is called the Laplace stress. The penetration depth τ depends on the Bragg angle θ, on the inclination angle ψ, the absorption coefficient μ and the diffraction geometry. For measurements in the Ω-mode where the sample is tilted so the scattering vector remains in the $\theta - 2\theta$ plane, τ is given by [4]

$$\tau = \frac{1}{2\mu} \frac{(\sin^2\theta - \sin^2\psi)}{\sin\theta\cos\psi} \tag{4}$$

whereas in case of measurements in the ψ-mode [4] where the sample is tilted about an axis in a plane normal to the $\theta - 2\theta$ plane

$$\tau = \frac{\sin\theta}{2\mu}\cos\psi \tag{5}$$

Several methods have been developed in order to determine the residual stress gradients by a systematic variation of the penetration depth: see, for example, Refs [4–7]. A variation of the depth by using different Bragg angles can be performed by using different reflections close to grazing incidence in Seemann–Bohlin geometry [8–12] (Figure 13.2). This allows for the analysis of the very-near-surface layers. Here, synchrotron radiation due to the small divergence of the beam is exceptionally beneficial for the analysis of regions less than 1 μm beneath the surface. From the discontinuous residual stress values obtained in the Laplace space the residual stress in the real space can be obtained, assuming $\sigma_{33} = 0$, by fitting the stress values versus penetration depth and subsequent inverse Laplace transformation [12].

A similar method is the multi-wavelength method, which is based on the fact that the absorption depends on the energy and thus the wavelength of the radiation. Thus, the use of different radiations and reflections enables the determination of discrete residual stress values by the $\sin^2\psi$ method in different depths [13]. The adaptation of this method from conventional X-rays (where different sources, for example, CuK_α, CrK_α are used) to synchrotron radiation is very advantageous since, due to the possibility of tuning the wavelength due to using a white beam, a far more continuous sequence of residual stress values versus depth can be made available [14].

These approaches of varying diffraction angles, wavelengths or energies and using several reflections, have been summed up in the so-called Universal-plot method [15–18], where the $\sigma_{11}(\tau)$, $\sigma_{12}(\tau)$, and $\sigma_{22}(\tau)$ curves obtained in Ω and/or ψ geometry for different radiations and reflections are integrated into one curve. Figure 13.3 shows an application of this method

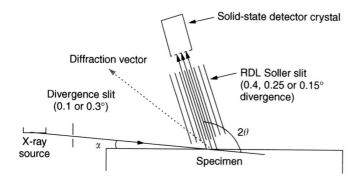

Figure 13.2 Principle of residual stress analysis in the grazing incidence mode [10].

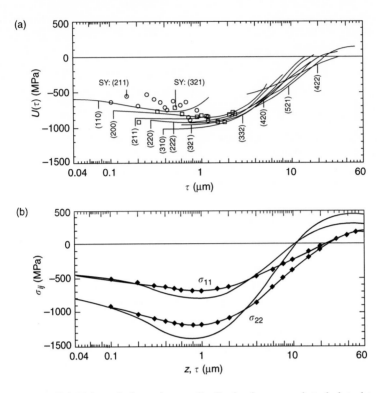

Figure 13.3 Universal-plot and stress distribution in a ground steel plate determined using white synchrotron radiation [15].

to the determination of Universal-plot $U(\tau) = (1/2)[\sigma_{11}(\tau) + \sigma_{22}(\tau)]$ of the residual stress gradient in a ground steel plate by energy dispersive diffraction [17].

Apart from that, methods such as the so-called polynomial method [13, 19] have been developed, which are a consequent further development of the $\sin^2 \psi$ method.

A promising new procedure for analysing stress gradients in the near-surface zone using X-rays or synchrotron radiation is based on the systematic variation of the penetration depth at defined φ, ψ-measuring directions [5, 6, 20, 21]. Therefore, the information depth τ is varied by rotating the sample by the angle η (Figure 13.4) around the measuring direction $\vec{g}_{\varphi\psi}$ (scattering vector).

$$\tau_\eta = \frac{\sin^2\theta - \sin^2\psi + \cos^2\theta \sin^2\psi \sin^2\eta}{2\mu \sin\theta \cos\psi} \qquad \mu - \text{linear attenuation coefficient} \qquad (6)$$

The scattering vector technique is furthermore useful for analysing stress gradients in textured materials (for instance, thin PVD coatings) because in this case the d-depth profile of the intensity poles can be followed.

An example of the application of the scattering vector technique is the analysis of the stress gradients in thin layers deposited by the PVD or CVD methods. As an example the residual stress gradient obtained by synchrotron diffraction [22] of a TiN layer on a steel substrate deposited by a bias voltage on the substrate of -200 V up to a layer thickness of $2.5\,\mu\text{m}$, and then reduction of the bias voltage to 0 V, is shown in Figure 13.5. Obviously, the steep residual stress gradient obtained on both reflections agrees well.

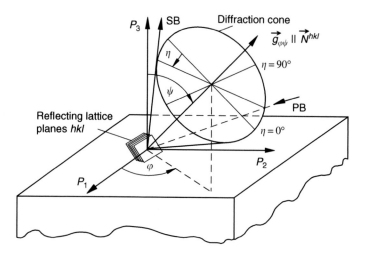

Figure 13.4 Principle of the scattering vector method [21].

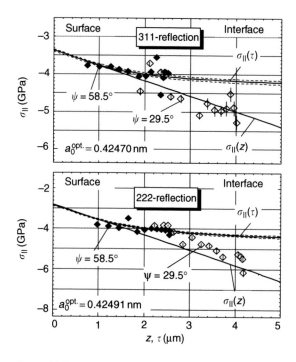

Figure 13.5 Residual stress gradient in a TiN-layer determined by the scattering vector method using synchrotron radiation ($\lambda = 0.13$ nm).

13.3   Analysis of near-surface stress gradients

Whereas the information depth of conventional X-ray diffraction measurements using the characteristic wavelength of the typical anode materials Co, Cu or Cr is limited to some 10 µm in case of metallic materials, the information depth can be extended to approximately

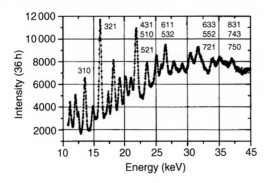

Figure 13.6 Spectrum of a steel sample obtained using white X-radiation at $2\theta = 60°$ [14].

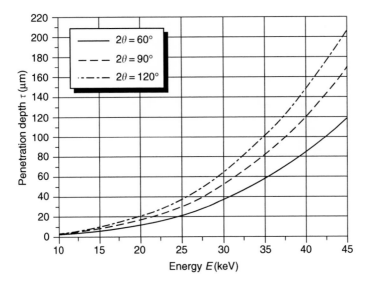

Figure 13.7 Penetration depth as a function of the energy and the diffraction angle.

200 μm using the white X-radiation sources, which allow for the use of higher photon energies [14, 23–26]. For analysing the energy value corresponding to the reflecting lattice planes hkl and their d^{hkl} values, an energy-dispersive detector is required. The measurements are performed in reflection geometry at a constant 2θ angle. Figure 13.6 shows the intensity–energy spectrum of a steel sample [14]. It reveals that the energy-dispersive analysis allows the simultaneous recording of several reflections. Each of the reflections is registered at a different energy level and therefore represents a certain information depth. The information depth can further be varied by the choice of the reflecting angle θ (Figure 13.7).

Thus, stress depth profiles over a scale of several hundred micrometres, which are typical for surface treatments such as rolling, shot peening and case hardening, are often obtainable by measuring only one spectrum. This is illustrated by Figure 13.8, which shows the results of an energy dispersive analysis of the residual stress gradient in a shot-peened steel spring. The

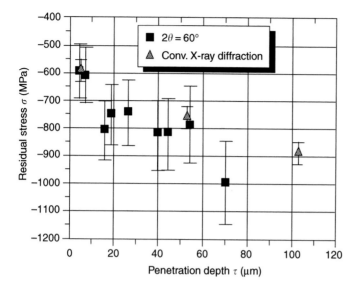

Figure 13.8 Residual stresses in a shot-peened steel spring.

residual stress values are compared to the results of conventional X-ray diffraction measurements after repeated etching of the surface. The results obtained by the two different methods are in good agreement. Further application examples for white beam high-energy X-ray diffraction residual stress analyses of near-surface stress gradients can be found, for example, in Refs [27–29]. In the case of lightweight materials such as aluminium, monochromatic radiation of medium energy can often also be used [30].

13.4   Analysis of bulk residual stress gradients

Whereas in the case of residual stress analyses in the very-near-surface region and the near-surface region the information depth is governed by the penetration depth of the radiation, in the case of high-energy synchrotron radiation residual stress analysis, similar to neutron diffraction, the volume element is defined by slits. But the diffraction angle, in contrast to neutron diffraction, has to be small, usually less than $2\theta = 10°$ (see Ref. [31]). Thus the volume element has the shape of an elongated diamond, whose small dimension in two directions allows for the analysis of the strain gradients with a resolution of several micrometres [32]. This analysis can be performed with extreme accuracy in the strain values by using monochromatic synchrotron radiation [33]. However, since for the analysis of the residual stress rotation of the sample is necessary, due to the low 2θ angles, different parts of the sample are illuminated when, for example, the strain components in radial and axial directions of a cylindrical sample are determined. Thus, the assumption of the homogeneity of the stress state within the gauge volume has to be checked carefully. White synchrotron radiation is especially useful in those cases where the residual stress gradients are accompanied by other gradients such as composition gradients or texture gradients (Figure 13.9), since in those cases a simultaneous determination of the residual stresses and the volume fractions of the different phases and/or the texture is possible. A few examples of the determination of residual

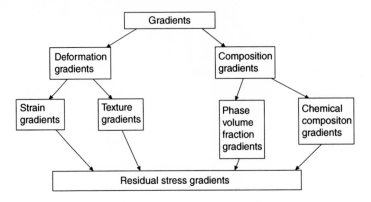

Figure 13.9 Microstructural and residual stress gradients.

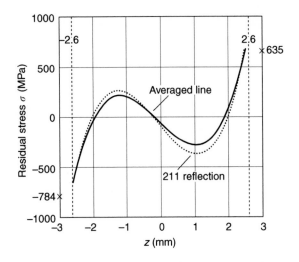

Figure 13.10 The residual stress distribution obtained across the sample thickness and the results of conventional X-ray residual stress analysis at the surface of the sample.

stress fields in the bulk by energy dispersive high-energy synchrotron diffraction are given below.

13.4.1 *Analysis of the load stress gradient in a plastically deformed steel*

In order to prove the residual stress analysis method developed at the high-energy beam line ID15A at the ESRF [27, 28], the load stress distribution of a plastically deformed steel sheet (German steel grade 42CrMo4, dimensions: 150 mm width × 40 mm height × 6 mm thickness) sample was determined *in situ* in a three-point bending device. Figure 13.10 illustrates the residual stress distribution obtained across the sample thickness and the results of conventional X-ray residual stress analysis at the surface of the sample. A comparison of the full line representing the average of five independent reflections and the dotted line which corresponds to the 211 reflection reveals that the scatter of the data is small. Furthermore, the results are in good agreement with the calculations. They reveal the high local resolution obtainable by this method, which allows for the determination of steep local stress gradients near the residual stress extrema.

13.4.2 Simultaneous analysis of residual stress and composition gradients

Functionally graded materials (FGMs), e.g. NiCr80/20 reinforced with 8Y–ZrO$_2$ meet the demand for adjusting defined property gradients within technical parts. Due to the different coefficients of the microstructural constituents, strong phase-specific residual micro stresses can develop during manufacturing or under service conditions. These residual microstresses can be superimposed on macro residual stress fields caused, for example, by temperature gradients during cooling. High-energy synchrotron diffraction measurements of the phase-specific residual stresses (Figure 13.11) determined in the FGM reveal that the out-of-plane and the in-plane residual stress values are of the same amount; thus the residual stress state in the bulk of the sample is almost hydrostatic [34]. Obviously, the high compressive residual stresses in the ZrO$_2$ decrease with increasing volume fraction of the ZrO$_2$, which is partially compensated by the NiCr80/20 matrix. Due to the small size of the gauge volume, even the analysis of the residual stress distribution in individual layers seems possible (Figure 13.12).

13.4.3 Simultaneous analysis of residual stress and texture gradients

An example of the simultaneous analysis of the texture and the residual stress analyses in a semi-finished product was done on a full forward-extruded sample. Within this sample residual stress and texture gradients (Figures 13.13 and 13.14) arise due to strong plastic deformations, which are inhomogeneously distributed across the sample diameter. The residual stress values determined in the radial and the hoop direction in the centre of the sample agree well, which proves that the experimental error is small. Furthermore, the difference

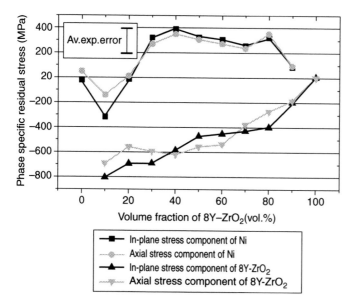

Figure 13.11 Phase-specific residual stresses in the bulk of a functional gradient material.

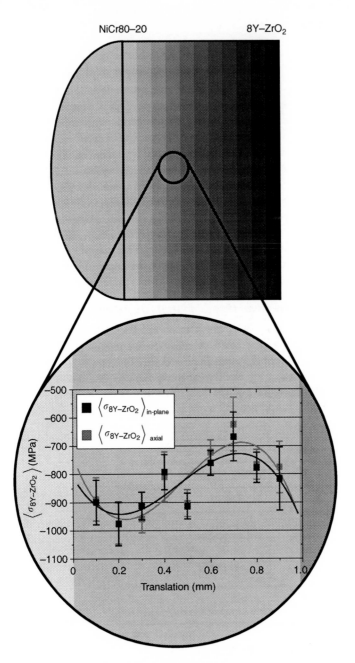

Figure 13.12 Phase-specific residual stresses in the ceramic component within an individual layer.

between the axial and the radial or hoop component in the centre of the sample determined by neutron diffraction compares well to the difference of these values determined by high-energy synchrotron residual stress analyses. Neutron diffraction and synchrotron diffraction, in very good agreement, reveal that in the inner part of the specimen the residual stresses in

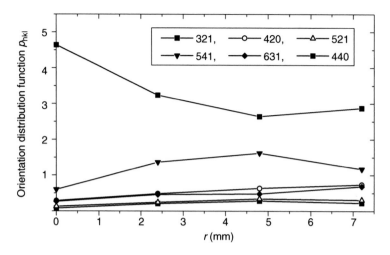

Figure 13.13 Texture distribution in an extruded steel sample.

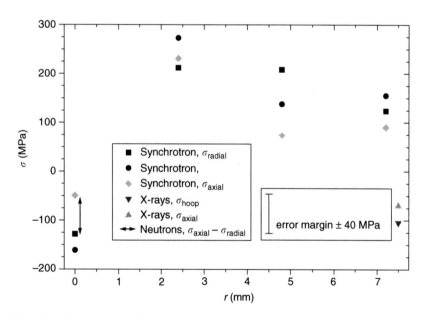

Figure 13.14 Residual stress distribution in an extruded steel sample. $r = 0$ is at the centre of the rod.

radial, hoop and axial direction are compressive (Figure 13.14). These compressive residual stresses are balanced by tensile residual stresses in the outer part of the sample. The quantitative stress values also fulfil, within an experimental error margin of ± 80 MPa, the mechanical equilibrium condition.

13.5   Future prospects

The use of synchrotron radiation can be expected to have a large impact on the analysis of residual stress gradients, since the low divergence of the beam and the high spatial resolution obtainable will facilitate residual stress analyses with extreme local resolution in the order of some micrometres.

But this high local resolution is bound to bring about new challenges. One of them is the problem of coarse grains. In the case of coarse-grained materials, the variety of orientations within the gauge volume is too small with respect to the statistics necessary for a smooth curve of the reflection. Since these single grains are situated at different positions within the gauge volume (for example, at the opposite sharp points of the elongated diamond), it is evident that the reflections arise from positions that both are slightly off-centre and the line positions are therefore shifted. This shift superimposes on the line position shift induced by the residual stresses. The grain size d_C above which the material has to be classified as coarse grained can be estimated using the image that the Debye–Scherrer ring should be completely occupied:

$$d_C = \sqrt[3]{\frac{J \cdot \Delta\theta_0 \bullet ms}{360° \cdot 360°} V} \tag{7}$$

The bold dot symbolises a convolution between the beam divergence $\Delta\theta_0$ and the mosaic spread 'ms' of the crystals. J is the multiplicity factor. In the case of synchrotron diffraction, due to the small beam divergence, this convolution can be roughly substituted by the mosaic spread alone. The critical grain sizes with respect to coarse grain effects for a gauge volume V of $0.12\,mm^3$ are plotted in Figure 13.15 versus the mosaic spread of the crystal. As expected, the critical grain size decreases with decreasing gauge volume and decreasing mosaic spread of the crystals and multiplicity factor. The critical grain sizes estimated indicate that coarse grain effects can be expected to occur especially for ceramics, which usually have small mosaic spreads and often structures of low symmetry, but that for metals with large mosaic

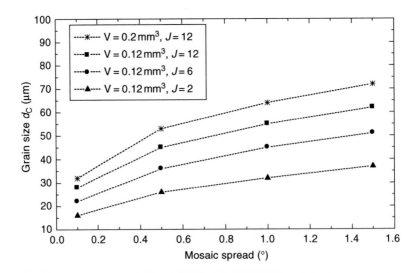

Figure 13.15 Critical grain size with respect to coarse grain effects [35].

spreads the critical grain size is nearly in the order of the critical grain size of conventional X-ray analyses. However, the problem of large grains can in some cases be overcome by small oscillations of the sample, preferably in the angle ψ.

Another challenge for synchrotron residual strain scanning is the accuracy in positioning required. Due to the smallness of the volume elements, which is in the order of micrometres, this is also the error margin required, for example, in the position of the gauge volume after turning the sample.

Future developments anticipated by people involved in synchrotron residual stress experiments are the completion of dedicated instruments, such as the second diffractometer at the beamline ID11 of the ESRF in Grenoble and an instrument at the APS in Chicago, as well as the further development of the diffractometer at the ID15A at the ESRF and the high-energy X-ray diffractometer at the HASYLAB, Hamburg. Instrumental improvements such as energy-dispersive detectors with improved energy resolution will be beneficial, for example, with respect to the analysis of strain gradients in composite materials. Finally, the increase in availability of synchrotron radiation will allow for the development of new methods of taking advantage of the high photon flux and small divergence of synchrotron radiation.

Acknowledgements

The authors would like to thank the instrument scientists working at the beamline ID15A at the ESRF Grenoble, particularly Dr Klaus-Dieter Liss, Dr Agnès Royer and Dr Veijo Honkimäkki, for their support during the experiments and we acknowledge beamtime at ID15A to the ESRF Grenoble.

References

[1] Ruppersberg H., Stress fields in the surface region of pearlite, *Mater. Sci. Eng.* **A224**, 61–86 (1997).

[2] Webster P. J., Wang X. D. and Mills G., Strain scanning using neutrons and synchrotron radiation, *Proc. ECRS 4* **1**, 127–134 (1996).

[3] Dölle H. and Hauk V., Der theoretische Einfluß mehrachsiger tiefenabhängiger Eigenspannungszustände auf die röntgenographische Spannungsermittlung, *HTM* **34**, 272–277 (1979).

[4] Hauk V., *Structural and Residual Stress Analysis by Nondestructive Methods: Evaluation, Application, Assessment*, H. Behnken *et al.* (eds), Elsevier Science B. V., Amsterdam (1997).

[5] Genzel Ch., Non-destructive analysis of residual stress gradients $\sigma_{ij}(z)$ in the near-surface region by diffraction methods – problems and attempts at their solution, *Proc. ICRS 5*, 514–521 (1997).

[6] Genzel Ch., X-ray stress gradient analysis in thin layers – problems and attempts at their solution, *Phys. Stat. Sol. (a)* **159**, 283–296 (1997).

[7] Hauk V. and Krug W. K., Der theoretische Einfluß tiefenabhängiger Eigenspannungszustände auf die röntgenographische Spannungsermittlung II, *HTM* **39**, 273–279 (1984).

[8] Zedehroud J., Wieder T., Thoma K. and Gärtner H., Tiefenaufgelöste röntgenographische Dehnungsmessungen an TiN-Schichten in Seemann–Bohlin-Geometrie, *HTM* **48**, 41–49 (1993).

[9] Ligen Y., Halin S., Kewei X. and Jiawen H., A correction of the Seemann–Bohlin method for stress measurements in thin films, *J. Appl. Crystallogr.* **27**, 863–867 (1994).

[10] Predecki P., Ballard B. and Zhu X., Proposed methods for depth profiling of residual stresses using grazing incidence X-ray diffraction (GIXD). *Adv. X-ray Anal.* **36**, 237–245 (1993).

[11] Ballard B., Zhu X., Predecki P. and Braski D. N., Depth-profiling of residual stresses by asymmetric grazing incidence X-ray diffraction (GIXD). *Proc. ICRS 4*, 1133–1143 (1994).

[12] Ballard B. L., Predecki P. K. and Braski D. N., Stress–Depth profiles in magnetron sputtered Mo films using grazing incidence X-ray diffraction (GIXD). *Adv. X-ray Anal.* **3**, 189–196 (1994).

[13] Eigenmann B., Röntgenographische Analyse inhomogener Spannungszustände in Keramiken, Keramik–Metall-Fügeverbindungen und dünnen Schichten. Dissertation, Karlsruhe, 1990.

[14] Brusch G. and Reimers W., Evaluation of residual stresses in the surface region and in the bulk of materials by high energy X-ray diffraction, *Proc. ICRS 5*, 557–562 (1997).

[15] Ruppersberg H., Detemple I. and Krier J., $\sigma_{xx}(z)$ and $\sigma_{yy}(z)$ Stress-fields calculated from Diffraction Experiments performed with Synchrotron Radiation in the Ω- and Ψ-mode techniques, *Z. Kristallogr.* **195**, 189–203 (1991).

[16] Ruppersberg H., Formalism for the evaluation of pseudo-macro stress fields $\sigma_{33}(z)$ from Ω- and Ψ-mode diffraction experiments performed with synchrotron radiation, *Adv. X-ray Anal.* **35**, 481–486 (1992).

[17] Ruppersberg H., Detemple I. and Bauer C., Evaluation of stress fields from energy dispersive X-ray diffraction experiments. In: *Residual Stresses*, V. Hauk, H. P. Hougardy, E. Macherauch and H.-D. Tietz (eds), DGM Informationsgesellschaft Verlag 171–178 (1993).

[18] Ruppersberg H., Complicated average stress-fields and attempts at their evaluation with X-ray diffraction methods, *Adv. X-ray Anal.* **37**, 235–244 (1994).

[19] Leverenz T., Eigenmann B. and Macherauch E., Das Abschnitt-Polynom-Verfahren zur zerstörungsfreien Ermittlung gradientenbehafteter Eigenspannungszustände in den Randschichten von bearbeiteten Keramiken, *Z. Metallkde.* **87**, 616–625 (1996).

[20] Dupke R. and Reimers W., Residual stress evaluation and damage characterisation of machined silicon wafers using X-ray diffraction, *Proc. ICRS 4*, 1097–1105 (1994).

[21] Genzel Ch., Formalism for the evaluation of strongly non-linear surface stress fields by X-ray diffraction in the scattering vector mode. *Phys. Stat. Sol. (a)* **146**, 629–637 (1994).

[22] Wroblewski T., Claus O., Crostack H.-A., Fandrich F., Genzel Ch., Hradil K., Ternes W. and Woldt E., The new diffractometer for materials science and imaging at HASYLAB Beamline G3. *Nucl. Instrum. Methods A* **428**, 570–582 (1999).

[23] Bechthold C. J., Placious R. C., Boettinger W. J. and Kuriyama M., X-ray residual stress mapping in industrial materials by energy dispersive diffractometry, *Adv. X-ray Anal.* **25**, 329–338 (1982).

[24] Black D. R., Bechtoldt C. J., Placious R. C. and Kuriyama M., Three-dimensional strain measurements with X-ray energy dispersive spectroscopy, *J. Nondestr. Eval.* **5**, 21–25 (1985).

[25] Shibano J., Ukai T. and Tadano S., Measurement method of the stress distribution along a depth by polychromatic X-rays, *Proc. ICRS 4*, 207–213 (1994).

[26] Lavelle B. and Jaud J. Transmission X-ray stress determination using a polychromatic radiation on aluminium alloy and stainless steel, *Proc. ICRS 5*, 563–567 (1997).

[27] Reimers W., Broda M., Brusch G., Dantz D., Liss K.-D., Pyzalla A., Schmackers T. and Tschentscher T., Evaluation of residual stresses in the bulk of materials by high energy synchrotron diffraction, *J. Nondestr. Eval.* **17**, 129–139 (1998).

[28] Reimers W., Pyzalla A., Broda M., Brusch G., Dantz D., Liss K.-D., Schmackers T. and Tschentscher T., The use of high energy synchrotron diffraction (HESD) for residual stress analyses, *J. Mater. Sci. Lett.* **19**, 581–583 (1999).

[29] Reimers W., Analysis of residual stress, *Mater. Sci. Forum* **321–324**, 66–74 (2000).

[30] Webster P. J., Mills G., Wang X. D. and Kang W. P., Synchrotron strain scanning through a peened aluminium alloy plate, *Proc. ICRS 5*, 551–556 (1997).

[31] Withers P. J., Use of synchrotron X-ray radiation for stress measurement, this book, pp. 170–189.

[32] Daymond M. R. and Withers P. J., A synchrotron radiation study of internal strain changes during the early stages of thermal cycling in an Al/SiC$_w$ MMC, *Scripta Mater.* **35**, 1229–1234 (1996).

[33] Lorentzen T., Clarke A. P., Poulsen H. F. and Garbe S., Local strain contours around inclusions in wire-drawn Cu/W composites, *Composites A* **27** (1997).

[34] Dantz D., Genzel Ch. and Reimers W., Analysis of macro and micro residual stresses in functionally graded materials by diffraction methods, *Mater. Sci. Forum* **308–311**, 829–836 (1999).

[35] Pyzalla A., Methods and feasibility of residual stress analysis by high energy synchrotron diffraction in transmission geometry using a white beam, *J. Nondestr. Eval.* **19** (2000).

14 Near-surface stress measurement using neutron diffraction

L. Edwards

14.1 Introduction

The measurement of near-surface residual strains is of significant technological importance since many industrial surface engineering processes improve surface properties by the introduction of relatively shallow residual stress fields. The determination of near-surface stresses has historically been undertaken by X-rays. X-ray diffraction can measure surface stresses to a depth of $\sim 20\,\mu$m non-destructively. Furthermore when used with the destructive layer removal technique, near-surface stress profiles up to several hundred μm in depth can be achieved, although it has been suggested that such measurements should be limited to a depth of $\sim 100\,\mu$m at most [1]. Current developments in synchrotron strain scanning suggest that this may also be a technique suitable for near-surface stress measurements.

Neutron diffraction, however, uniquely possesses the ability to measure accurately and non-destructively residual stresses from the surface to deep within the bulk of engineering components. When using neutron diffraction to measure near-surface strains, the gauge volume is often only partially immersed in the material and so special analyses are required to allow for the changing diffraction geometry. Furthermore, as many surface engineering treatments produce large stress gradients near the surface of components, precise near-surface strain measurement is required, which involves both an accurate determination of measurement position and a small systematic error in strain measurement. This chapter will describe the requirements for precise near-surface strain measurements by neutron diffraction and will then show how these requirements can be met for both monochromatic wavelength measurements (reactor neutron sources) and time-of-flight spectrometry measurements (pulsed neutron sources). Finally, the advantages of near-surface measurement using vertical scanning (Z-scanning) will be illustrated. Examples of near-surface measurements will be given where appropriate.

14.2 Requirements for near-surface measurement

There are three basic requirements for the accurate measurement of near-surface strains. First, the position of the surface with respect to the neutron beam must be accurately known. This may be achieved by extremely careful optical alignments using theodolites and other optical instruments. Although accurate optical alignment of both the instrument and specimen is essential, determining the precise location of the specimen surface with respect to the incoming neutron beam is best achieved by fitting intensity variations measured by fast surface scanning to an attenuation based model of the resulting entrance curve. Using this method an alignment precision of better than 0.05 mm can be achieved [2].

Second, the 'centre of gravity', or centroid, of a near-surface gauge measurement volume must be calculated, as the centroid of the diffracting material clearly differs from the centre of the volume defined by the collimated neutron beam as soon as the latter volume straddles the specimen surface. It may also be necessary to consider the effects of attenuation for some materials as this may change the apparent position of the measurement centroid.

Finally, a correction may be required to allow for any pseudo-strain effects caused when the gauge volume straddles the surface or where the phase being measured is not homogeneously distributed within the gauge volume, resulting in a changing diffraction geometry. This surface effect due to the changing diffraction geometry is instrument dependent and must be characterized for a given collimator and detector setup.

Early determinations of near-surface strains in surface engineered components [3–6] were made on single-detector angular-sensitive neutron diffractometers that, fortunately, possessed relatively low surface aberrations, for example, D1A at ILL, Grenoble. However, with further development of the neutron diffraction technique, most stress diffractometers were fitted with multi-detector position-sensitive instrumentation to increase the efficiency of data collection. It became clear that such instruments produced pronounced surface effects and several authors analysed the principal causes of these effects for both reactor-based [7–11] and pulsed source-based instruments [2, 12].

The most succinct treatise on near-surface measurements using reactor sources is given by Webster *et al.* [8]. They show that the three main effects seen during monochromatic wavelength measurements on reactor-based instruments are:

(i) *A wavelength effect.* At reactor neutron sources, a monochromatic beam of neutrons is produced by using a crystal monochromator to select a given neutron wavelength from a 'white' polychromatic neutron beam by Bragg diffraction. In order to obtain usable neutron intensities, the monochromators may be bent and the collimators have finite width. As a consequence there is a gradient of wavelengths across the beam. When the gauge volume is immersed in the specimen the average wavelength of the diffracted beam remains constant, but as the gauge volume moves through the surface the occupied portion of the gauge volume changes position relative to the incident beam, so a different average wavelength applies at each point in the scan.

(ii) *A geometrical effect.* As the diffracting gauge area changes shape, its centroid moves relative to the detector system. If a multi-detector such as a position sensitive detector (PSD) is used, this displacement of the centroid causes a change in diffraction angle, causing a shift of the peak on the detector.

(iii) *A peak clipping effect.* When a multi-angle detector is used, the detector slit used to define the gauge volume can also limit the angular divergence of the beam diffracted from a part of the gauge volume. This can cause peak clipping. When the gauge volume is immersed in the specimen, this effect is symmetrical and the position of the peak is not affected. However, when scanning through a surface so that the gauge volume is only partially filled, the clipping is asymmetric and can lead to an offset in the measured peak position.

Several attempts have been made to produce both analytical and numerical analyses [9–12] of the first two of these effects. All authors agree that the above surface effects can be minimized but not eliminated. It is clear, however, that they can be determined experimentally – by passing the gauge volume through a stress-free surface such as a powder material – or calculated, allowing corrections to be made to raw data.

The author and co-workers have studied the near-surface measurement effects using the time-of-flight ENGIN diffractometer on the ISIS pulsed neutron spallation source. Spallation source stress diffractometers do not suffer from wavelength or peak clipping effects and possess relatively low geometrical effects; but the radial collimators now commonly used at pulsed source instruments require specific analyses of their geometrical effect to be undertaken.

This chapter will use an analysis of ENGIN to illustrate the requirements for precise near-surface stress measurement. However, the principles are generic and the method could be modified to deal with other neutron diffraction conditions. Many of the basic analyses are also applicable to synchrotron near-surface strain scanning.

14.3   A methodology for near-surface strain measurement

14.3.1   Precise determination of measurement position

To perform near-surface strain scanning using neutron diffraction, the surface of the component must be established to a precision of at least 0.1 mm. This is particularly important if the specimen contains large stress gradients near its surface. It is difficult and sometimes impossible to determine the precise specimen surface position using only theodolites and telescopes, for example, owing to the large size or complicated geometry of some components. The location of the sample position with respect to the neutron beam can be obtained by observing the increase in neutron counts as the gauge volume enters the sample. A quick neutron scan (about 20 mins)[1] can be performed by moving the gauge volume from a point off the surface to a point inside the surface, and researchers have been using this technique for a number of years [13]. One could then obtain a plot of the integrated neutron intensity against position relative to the surface.

The case where the gauge volume possesses a rhombic prism shape has been considered in detail by Brand and Prask [14]. However, in order to obtain a good fit to the experimental data a Levenberg–Marquardt technique had to be used in which the background, attenuation and position and size of gauge volume were all adjustable. Larsen *et al.* [15] have also considered this problem, but their model required the use of a non-standard attenuation coefficient, that is, $0.152 \, mm^{-1}$ for iron when a value of $0.112 \, mm^{-1}$ is typically quoted in the literature [16]. A general mathematical description of this problem for the ENGIN diffractometer at the ISIS spallation neutron source has been developed which utilizes two opposing collimators with axes normal to the incident beam [2, 12]. Other authors have addressed the same problem, showing that the approach could be extended to other diffractometers [9–11].

The traditional horizontal scanning geometry for making all neutron measurements is shown in Figure 14.1. In the general case the gauge volume is not cubic. This is the usual situation at ENGIN, where the geometry of the focusing collimators is fixed and the shape of the gauge volume is varied by changing incident beam slits [17]: as may be seen from Figure 14.2, which illustrates the basic geometry of ENGIN.

The gauge volume is defined by the dimensions of the slits provided by the incident beam collimation and by the dimension of the focusing function of the collimating detectors. The integrated neutron intensity during surface scanning is governed by two principal factors (though other phenomena, such as multiple scattering, and texture [2] can also be involved in

---

[1] This time estimation is given on the basis of data collection for seven points at ENGIN, ISIS, UK.

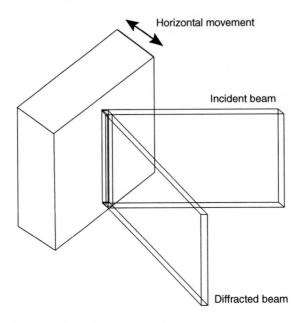

Figure 14.1 The traditional horizontal surface scanning geometry.

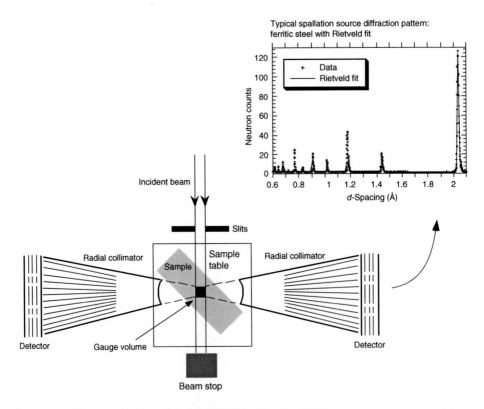

Figure 14.2 Schematic drawing of the ENGIN diffractometer at ISIS.

some circumstances). The integrated neutron intensity initially increases as the gauge volume enters the sample, so increasing the diffracted volume. It is also attenuated by absorption and diffraction within the sample. So, in practice, the integrated neutron intensity reaches a maximum, the position of which depends on the shape of the gauge volume and the material's attenuation. For the case of ENGIN, the intensity is additionally dependent on the focusing function of the collimators, so throughout this chapter the term 'detected intensity' is used in preference to integrated neutron intensity.

The mathematics of the modelling can be obtained from Ref. [2] and only a few relevant points will be made here. Firstly, it is important to include both attenuation and beam divergence when fitting intensity curves to define the position of the surface.

Figure 14.3 plots typical entrance curves for both aluminium and iron for a nominal 2×2 mm incident beam at ENGIN – the third dimension is fixed by the radial collimator. It can be seen that the position of the surface relative to the maximum intensity observed varies significantly with attenuation.

This approach has been applied to a variety of materials on ENGIN. The specimen surface is accurately located by using a least-squares method in which only the background and sample

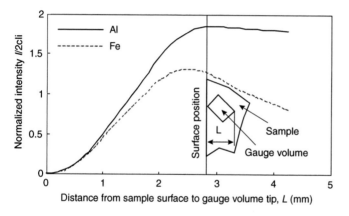

Figure 14.3 Intensity variation as a $1.4 \times 2.5 \times 2.5$ mm^3 gauge volume enters a sample.

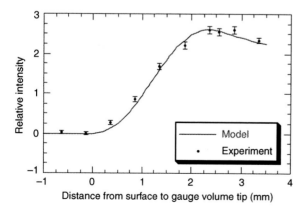

Figure 14.4 The model fitted to experimental data obtained from an iron powder in a vanadium can.

position are variables to fit experimental data to the model. The result for a 60 µm iron powder (used to eliminate any texture effects) in a vanadium can is plotted in Figure 14.4. Incident beam slits of 2×5 mm^2 were used, giving an effective gauge volume of $1.4 \times 2.2 \times 5.2$ mm^3. It can be seen that the model fits the experimental data extremely well, and the surface is located to better than 0.05 mm.

14.3.2　Calculation of gauge volume centroid

When measuring components that contain high strain gradients at the surface, a relatively high number of near-surface measurements need to be performed. As typical gauge volume dimensions are large compared to the distances over which strain gradients occur, measurements are usually taken with the gauge volume only partially immersed in the component. Under these conditions it is important to know the precise position of the 'centre of gravity' of the diffracting volume is, as it is not at the centre of the gauge volume. A similar condition can occur when the gauge volume is completely immersed in the sample if the material possesses significant attenuation.

This is achieved by calculating the centroid of the diffracting volume – the effective point of strain measurement. If the gauge volume is completely immersed in the specimen, provided the attenuation is not too high, the centroid is simply its geometrical centre, provided that the spatial distribution of the diffracted intensity possesses a point symmetry at the geometrical centre. However, when the gauge volume partially enters the specimen, the position of the centroid relies on two functions. One is the function of the diffracted volume, $V(x, y, z)$. The other is the spatial distribution function of the detected intensity, $I(x, y, z)$. The term 'detected intensity' is used because for ENGIN the spatial distribution of the intensity depends not only on the spatial intensity distribution of the incident beam and the diffraction homogeneity of the specimen, but also on the focusing effect of the collimators.

To calculate the position of the centroid, we use the fact that the total diffracted neutron intensity in both halves of the gauge volume separated by the centroid position should be equal, so that:

$$\int_0^L I(x, y, z)\, dV = 2 \int_0^r I(x, y, z)\, dV \tag{1}$$

where the left-hand side integrates over the whole diffracted volume; the right-hand side either over the half of the volume from the sample surface to the centroid, or from the centroid position to the tip of the gauge volume; L is the distance from the tip of gauge volume to the sample surface; and r is the distance from the sample surface to the diffraction centroid.

Considering the typical geometry shown in Figure 14.1, if the spatial distributions of the detected intensity in the three orthogonal directions are rectilinear and the specimen possesses homogeneous diffraction properties, then the position of the geometrical centroid can be calculated as described in Ref. [2]. In practice, the spatial diffracted intensity function may contain some instrumental or material broadening and so it may have extremities that look like, for example, one half of a Gaussian. In such cases it is assumed that the volume can be approximated by a rectilinear gauge volume of dimensions given by the FWHM of the spatial function. The centroid position will also be affected by the distribution of diffracted intensity within the gauge volume, which is affected by both attenuation and the spatial intensity distribution of the incident beam. For iron this produces a shift of 0.13 mm for a gauge volume of $1.4 \times 2.5 \times 2.5$ mm^3.

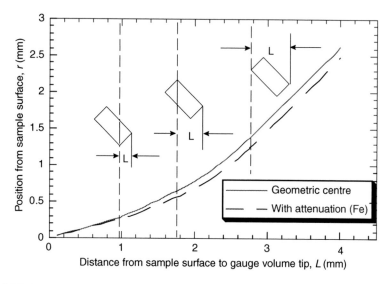

Figure 14.5 Centroid position relative to L, the distance from the sample surface to the gauge volume.

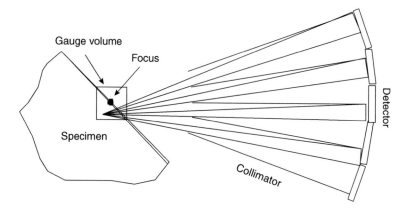

Figure 14.6 Differences in average diffraction length for an individual detector on ENGIN [12].

Again I shall give only illustrative results here. The results for such a $1.4 \times 2.5 \times 2.5 \, \text{mm}^3$ gauge volume are plotted in Figure 14.5. The solid line assumes no attenuation, and the dashed line uses the attenuation coefficient for iron. It can also be seen that the change in the centroid position as the gauge volume enters the sample surface is larger when attenuation is present.

14.3.3 *Correction of near-surface effects due to changing diffraction geometry*

As noted earlier, an anomalous strain can be measured when the gauge volume straddles the sample surface or an internal interface between two phases, or where the distribution of the phase being measured is not homogeneously distributed within the gauge volume. This can be quite large at ENGIN, where the effect results from differences in the average diffraction

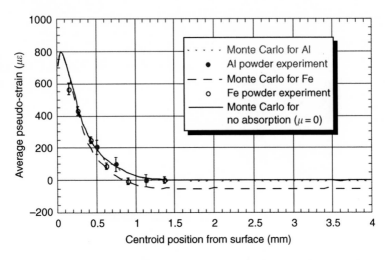

Figure 14.7 Comparison of measured and simulated pseudo-strains in iron and aluminium powders. (See Colour Plate IX.)

length and visible gauge volume for individual detectors within the focusing collimator, as is schematically shown in Figure 14.6. The radial focusing collimator possesses 40 vanes and is backed by three banks of 45 detectors [17]. A dual-pronged strategy was taken to characterize this pseudo-strain effect on ENGIN. A numerical model of the diffraction geometry of ENGIN was produced using the Monte Carlo method [12]. The results of this simulation were then compared to pseudo-strains detected when passing the gauge volume through a series of strain-free surfaces created by placing metal powders in vanadium cans. It was assumed that no type I strain or texture exists in the powder specimens. Two types of powder, iron and aluminium (particle size: $60\,\mu m$), were put into containers consisting of an aluminium frame with two 0.1 mm vanadium windows. The results are shown in Figure 14.7 where it can be seen the simulation compares well with the experimental results. It can be seen that the pseudo-strain measured in both iron and aluminium is similar, showing that the effect of attenuation is small when compared to the geometry effect.

14.4 Z-scanning – a better method for surface strain measurement

Of the three essential requirements for precise near-surface measurement detailed above by far the most difficult and awkward to deal with are the pseudo-strains described in section 3.3. These can be very large, often bigger than the physical strain in the component surface, and conventionally can only be dealt with by either a specific calibration measurement or Monte Carlo type simulation of each measurement geometry used. Furthermore, the magnitude of the effect is unique to each specimen and diffractometer used.

However, if the incident beam is scanned vertically out of a horizontal surface as shown in Figure 14.8, there is no change in diffraction angle and, hence, no pseudo-strains are generated as the gauge volume moves out of the surface [12]. I shall term this geometry Z-scanning.

Determination of the surface location from entrance curves and determination of the gauge volume centroid is also easier for this geometry where one side of the gauge volume is

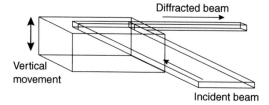

Figure 14.8 The Z-scan geometry for surface strain measurement.

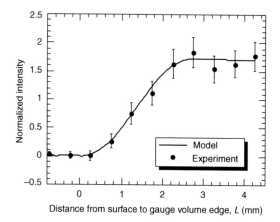

Figure 14.9 The model fitted to experimental data for the geometry shown in Figure 14.3.

essentially parallel to the surface. Indeed, for low absorption materials there is a linear change in both centroid position and diffracted intensity as the gauge volume enters the surface [2].

Figure 14.9 shows the Z-scan entrance curve for an aluminium alloy, measured with a $2 \times 2 \, \text{mm}^2$ incident beam and a slit to gauge volume distance of 250 mm. Taking the beam divergence (3×10^{-3} rad for ENGIN) into account, then the extremities of the gauge volume are $2 \times 3.5 \times 3.5 \, \text{mm}^3$. Following the method suggested in Ref. [2], the FWHM of the detected intensities can be used to define an equivalent gauge size in the three orthogonal directions so that the effective gauge volume is $1.4 \times 2.75 \times 2.75 \, \text{mm}^3$.

This expected relatively simple entrance curve behaviour is, however, not seen in all cases. In particular, when high attenuation material is measured with 90° detectors that typically subtend a larger angle (up to ±10°) of the Debye cone, then 'wings' are seen to the entrance curve as shown in Figure 14.10, which plots a Z-scan entrance curve for a 5 mm thick shot-peened IN 718 Nickel alloy plate.

The explanation for this unusually shaped Z-scan emergence intensity curve is illustrated in Figure 14.11. The relatively large angle typically used (~±10°) when a detector is at 90° to the incoming neutron beam allows diffracted beam paths to the top and bottom of the detector to spend differing lengths in the sample when the gauge volume is near the surface. So, as shown in Figure 14.11, when the gauge volume is deep in the sample the specimen path length is roughly the same for neutrons entering any part of the detector. However, when the gauge volume approaches the surface, then neutrons entering the top half of the detector reduce their path length in the specimen as they pass through the top surface of the specimen

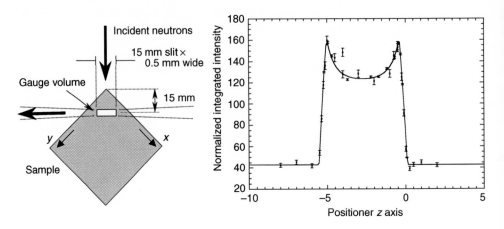

Figure 14.10 Diffraction geometry and Z-scan entrance curve through Ni plate.

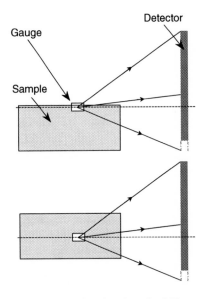

Figure 14.11 Explanation for 'decaying' Z-scan entrance curve seen in high absorption materials.

and thus are less attenuated than one would expect. This gives a larger signal than expected when the gauge volume is straddling the surface and produces the Z-scan entrance curve shown in Figure 14.10.

This effect is strongest for high absorption materials. It is distinctly different to the decaying entrance curves seen in horizontal surface scanning geometries and illustrated in Figures 14.3 and 14.4, in that once the volume is fully within the material then there is no further decrease in diffracted intensity.

There are fewer complications to centroid determination when Z-scanning. For a simple orthogonally-sided gauge volume, then, the centroid position for this geometry may be simply

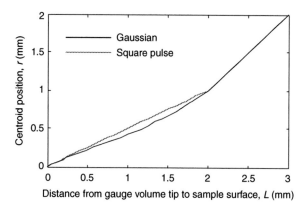

Figure 14.12 Centroid position for square and Gaussian shaped gauge volumes in Z-scanning.

deduced as occurring half way across the actual diffracting volume. The deviation from this behaviour for a Gaussian shaped gauge volume is small, as may be seen from Figure 14.12.

14.5 Example: grinding stresses

Grinding can be a major cause of residual stress at the surface of engineering components and can critically influence their service life. Relatively mild grinding induces low work-piece temperatures and compressive residual stresses are typically produced owing to mechanical indentation effects. More severe grinding conditions increase work-piece temperature until thermal stresses dominate, owing to expansion of the material closest to the surface, leading eventually to the production of tensile stresses at energy levels even below those at which damage such as burn, tempering or phase changes occur [18].

As part of a research programme on Controlled Stress Grinding, the onset temperatures and conditions for the generation of tensile stresses in ground surfaces of a series of commercial steels has been characterized. A predictive model has been developed which has been validated by experimental determination of temperature from embedded thermocouples and near-surface residual stress profiles by X-ray and neutron diffraction [19]. A comparison of the residual stresses measured in a specimen of EN31 is given in Figure 14.13. Excellent agreement can be seen between the two diffraction stress measurement methods.

The material studied was EN31 steel, quenched and tempered to a hardness of 850 HV. Samples ($16 \times 8 \times 50\,\text{mm}^3$) were ground with different conditions (wheel speed, work-piece speed and depth of cut). The X-ray diffraction measurements were performed using Cr radiation ($\lambda = 2.2897\,\text{Å}$) on a Bruker D5005 diffractometer equipped with a single detector, sampling from the (211) plane in the ferrite phase at $2\theta = 156.3°$. Stresses were calculated by measuring 11 psi-angles. The high carbon content of the martensitic EN31 produces very wide diffraction peaks, and thus a sliding gravity method was used to determine accurately peak positions. The near-surface measurements were performed by electrolytic surface removal of an area of $0.5\,\text{cm}^2$ using steps between 10 and $50\,\mu\text{m}$. Of course, using this method the X-ray technique is, at best, a semi-destructive method of measuring near-surface stresses.

The neutron diffraction surface stress measurements were performed using the Z-scan method on the ENGIN diffractometer. A sampling volume of $20 \times 1.7 \times 0.5\,\text{mm}^3$ was scanned vertically out of the sample surface, in increments of $25\,\mu\text{m}$. The strain-free reference

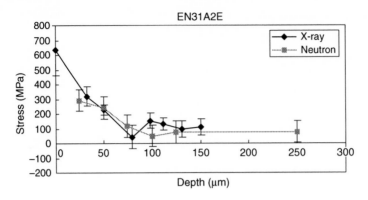

Figure 14.13 Near-surface grinding stresses in an EN31 sample as measured by X-ray (layer removal) and neutron (Z-scan) diffraction on ENGIN.

Figure 14.14 Cylindrically ground EN31 component sample.

lattice spacing was taken from a measurement 4 mm from the ground surface (middle of the specimen). Strains were obtained after fitting all the peaks of the diffracted spectrum using a Rietveld analysis program. The fitting was performed using two phases (ferrite and austenite) due to the presence of retained austenite (estimated at around 7%). Stresses were calculated assuming a zero principal stress perpendicular to the ground surface.

Of course, the two techniques are complementary in that neutron diffraction cannot make measurements at the surface and X-ray diffraction cannot make near-surface measurements non-destructively. This complementarity was used in the Controlled Stress Grinding investigation to completely non-destructively determine the near-surface residual stress profile in several large cylindrical grinding components (Figure 14.14).

Figure 14.15 compares the residual stresses in two EN31 samples ground under similar conditions (work-piece speed 40 m/s; wheel speed 0.3 m/s; depth of cut 8 μm) with two different grinding wheels. Sample 1 was ground with an alumina wheel whilst sample 2 was ground with a cubic boron nitride (CBN) wheel. The results show two distinctly different profiles, one highly tensile and one highly compressive, and illustrate the benefit of using CBN if tensile residual stresses are to be avoided. This is probably due to the greater thermal

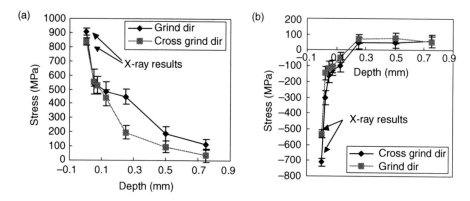

Figure 14.15 Stress profiles from the cylindrically ground samples. (a) Sample 1; (b) Sample 2.

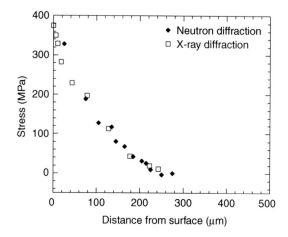

Figure 14.16 Near-surface grinding stresses in an EN9 sample as measured by X-ray (layer removal) and neutron (Z-scan) diffraction on the G5.2 diffractometer at LLB, Saclay, Paris (courtesy Dr M. Fitzpatrick).

conductivity of the CBN wheel. However, it should be noted that grinding with alumina usually produces a substantially better surface finish than grinding with CBN.

The neutron Z-scan measurements illustrated above were performed on the ENGIN diffractometer at ISIS, UK, which is particularly suited for Z-scan measurements as the vertical divergence of the neutron beam is relatively low (3×10^{-3} rad). Although the higher neutron fluxes available at constant wavelength reactor-based diffractometers should enable the use of smaller gauge volumes, the intrinsically higher vertical divergence used at most of these instruments enlarges the vertical gauge volume size substantially even if narrow slits are used. Thus, there has been little work using the Z-scan geometry to measure near-surface stresses on constant wavelength instruments. However, the technique can be used on such instruments, as can be seen from Figure 14.16 which compares X-ray results with neutron diffraction residual stress measurements obtained using Z-scanning on the G5.2 diffractometer at LLB, Saclay, Paris. It is, however, interesting to note that despite the use of a 0.3 mm slit to define the vertical gauge volume size (nominal sampling volume $= 10 \times 10 \times 0.3 \, mm^3$) on this ground EN9 sample, the effective gauge volume was nearly double that found on ENGIN when using a 0.5 mm slit. So unless constant wavelength diffractometers can be configured with lower vertical divergence, then the advantages of Z-scanning is likely to be most utilized on pulsed spallation source instruments.

14.6   Example: shot peening

Shot peening produces compressive residual strain (stress) in the surface region of components to improve their resistance to fatigue. Quantification of the residual stress distribution near shot-peened surfaces is desirable both to monitor the shot peening procedure and to predict the fatigue life of components.

The in-plane stress distribution in a nickel-based superalloy IN 718 plate measured by a Z-scan on ENGIN is shown in Figure 14.17. This specimen was a VAMAS round robin specimen used in the development of an international standard of the measurement of stress using neutron diffraction and was shot-peened using 230 R at an intensity of 0.0083 A. The entrance curve intensity profile for this specimen is given in Figure 14.10.

14.7   Conclusions

Near-surface strain measurement by neutron diffraction requires the measurement location to be precisely determined. This can be achieved by comparing the intensity distribution generated by a rapid neutron scan with a model of the expected behaviour, and calculating the gauge volume centroid position taking into account both gauge volume size and attenuation. In addition, for measurement of strains normal to a surface, the pseudo-strains

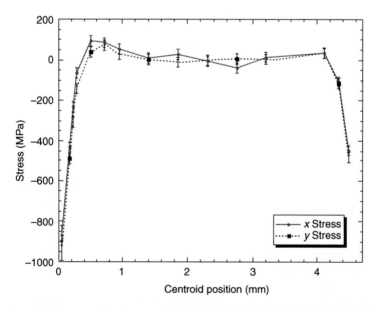

Figure 14.17 In-plane stress distribution near a Nickel superalloy shot-peened surface.

arising from changes in the diffraction geometry must be allowed for. Strain scanning in the direction perpendicular to the incident and diffracted beams (Z-scanning) produces no changes in diffraction geometry and is the preferred method of measuring near-surface strains when physically possible. However, strain measurement must typically occur both perpendicular and parallel to a surface if the full stress tensor is to be determined so that accurate models of surface pseudo-strains are usually required. Used in this way, neutron diffraction can uniquely measure the near-surface stress within a component and complements X-ray measurements which can only make measurements deeper than 50 μm if a destructive layer removal technique is used.

Acknowledgements

The author would like to thank David Wang, Mark Daymond, Mike Fitzpatrick, John Wright, Ahmed Bouzina, Phil Withers, Hans Priesmeyer and Mike Johnson for their considerable help in this area of study over a number of years. This work would not have been possible without the support of the UK Engineering and Physical Sciences Research Council (EPSRC).

References

[1] Ruppersberg H. and Detemple I., Evaluation of the stress field in a ground steel plate from energy-dispersive X-ray diffraction experiments, *Materials Science and Engineering*, **A161**, 41–44 (1993).

[2] Wang D. and Edwards L., Precise determination of specimen surface position during near-surface strain scanning by neutron diffraction, *Proceedings of the 4th European Conference on Residual Stresses*, eds., Denis, S *et al.*, Société Française de Métallurgie et de Matériaux, **1**, 135–144 (1998).

[3] Ezeilo, A. N., Webster, P. S., Webster, G. A. and Webster, P. J., Development of the neutron diffraction technique for the determination of near-surface residual stresses in critical gas turbine components, in *Measurement of Residual and Applied Stress Using Neutron Diffraction*, NATO Advanced Research Workshop Series, eds., Hutchings M. T. and Krawitz A. D., Kluwer Academic Publishers, London, pp. 545–553 (1992).

[4] Ezeilo A. N., Webster P. J., Webster G. A. and Webster P. S., Comparison of shot peening residual stress distributions in a selection of materials *Proceedings of 5th International Conference on Shot Peening*, Oxford, pp. 273–281 (1993).

[5] Webster P. J. and Wang X. D., Neutron strain scanning of metal components, *Surface Engineering*, **10**, 287–291 (1994).

[6] Webster G. A. and Ezeilo A. N., Neutron scattering in engineering applications, *Physica B*, 949–955 (1997).

[7] Webster P. J., Spatial resolution and strain scanning, in *Measurement of Residual and Applied Stress using Neutron Diffraction*, NATO Advanced Research Workshop Series, eds., Hutchings, M. T. and Krawitz, A. D., Kluwer Academic Publishers, London, pp. 545–553 (1992).

[8] Webster P. J., Mills G., Wang X. D., Kang W. P. and Holden T. M., Impediments to efficient through-surface strain scanning, *Journal of Neutron Research*, **3**, 223–240 (1996).

[9] Spooner S. and Wang X.-L., Diffraction peak displacement in residual stress samples due to partial burial of the sampling volume, *Journal of Applied Crystallography*, **30**, 449–455 (1997).

[10] Lorentzen T., Numerical analysis of instrumental effects on strain measurement by diffraction near-surfaces and interfaces, *Journal of Neutron Research*, **5**, 167–180 (1997).

[11] Wang X.-L., Spooner S. and Hubbard C. R., Theory of the peak shift anomaly due to partial burial of the sampling volume in neutron diffraction stress measurements, *Journal of Applied Crystallography*, **31**, 52–59 (1998).

[12] Wang D, Harris I. B., Withers P. J. and Edwards L., Sub-surface strain measurement by means of neutron diffraction, *Proc. Fourth European Conference on Residual Stresses*, eds., Denis S. *et al.*, Société Française de Métallurgie et de Matériaux, Vol. 1, pp. 69–78 (1998).

[13] Ezeilo A. N., Residual stress determinations by neutron and X-ray diffraction methods, PhD thesis, Imperial College, University of London (1992).

[14] Brand P. C. and Prask H. J., New methods for the alignment of instrumentation for residual-stress measurements by means of neutron diffraction, *Journal of Applied Crystallography*, **27**, 164–173 (1998).

[15] Larsen J., Priesmeyer H. G., Ohms C., Harris I. B. and Withers P. J., Investigation of the vibrational stress relief method using neutron diffraction on ENGIN, *PREMIS Report*, RAL (1994).

[16] Bacon G. E., *Neutron Diffraction*, Clarendon Press, Oxford, Chaps. 2 and 3 (1962).

[17] Johnson, M. W., Edwards, L. and Withers, P. J., ENGIN – a new instrument for engineers, *Physica B*, **234**, 1141–1143 (1997).

[18] Peters J., Snoeys R. and Maris M., Thermally induced damage in grinding, *Annals of the CIRP*, **27**, 1–15 (1978)

[19] McCormack D. F., Row W. B., Chen X., Bouzina A., Fitzpatrick M. E. and Edwards L. Characterizing the onset of tensile residual stresses in ground components, *Proceedings of the 6th International Conference on Residual Stresses*, IoM London, pp. 225–234 (2000).

Part 5

Applications

15 Shot peening

A. Ezeilo

15.1 Introduction

Structural failures of engineering components are usually characterised by the initiation and growth of fatigue cracks at surfaces. This is because surfaces usually experience the highest applied stresses, and are susceptible to manufacturing defects and environmental attack. Residual stresses also influence fatigue lifetimes, particularly levels of residual stress which are present at the surface of a component. Tensile surface residual stresses tend to promote failure while compressive surface residual stresses offer resistance to the initiation and growth of fatigue cracks [1].

Shot peening is a mechanical surface treatment procedure which introduces compressive residual stresses at the surface of a component and is widely used to improve fatigue performance [2–4]. The improvement in fatigue life is attributed to the introduction of near-surface compressive residual stresses, a complex distribution of dislocations, and by work hardening of the material at the surface. Figure 15.1 shows the improvement in fatigue life achieved by applying a shot peening operation [3]. The greater the magnitude and depth of the compressive stress layer the greater the improvement in fatigue life. However, excessive peening operations may introduce cracks.

It is therefore desirable to know the magnitude and depth of the compressive stress layer, as this can be used to assess the potential improvement in fatigue life and can also be used to determine the optimal shot peening parameters. Neutron diffraction provides a non-intrusive procedure for measuring shot peening residual stresses provided the necessary precautions are taken. It is the purpose of this chapter to introduce shot peening and describe how the neutron diffraction technique can be used to establish an accurate and reliable profile of the residual stress distribution in shot peened specimens and components. Examples are given of shot peening residual stresses generated in a variety of materials over a range of shot peening intensities. An example is also given of how neutron diffraction can be used to measure the evolution of a shot peening residual stress state. By improving the fatigue life of engineering components there is greater scope for design optimisations and weight reductions, which ultimately translate to cost savings.

15.2 The shot peening process

Controlled shot peening involves bombarding the surface to be treated with shot [3–5]. Typical shot is made of steel, ceramic or glass beads with sizes ranging from 50 μm to 6 mm in diameter and having velocities at impact of up to 150 m s^{-1}. Each individual shot produces local plastic deformation in the target material and subsequent multiple indentations subject the surface layer to local cyclic plastic loading. The resulting effect is that the surface layer is

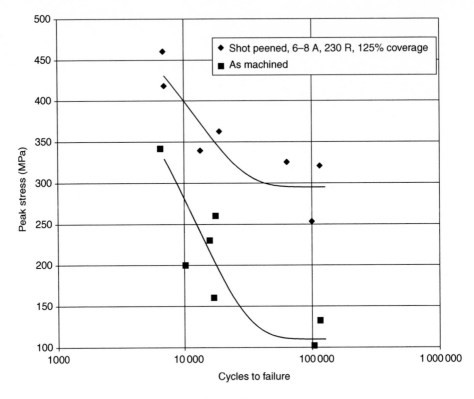

Figure 15.1 The effect of shot peening on fatigue life.

permanently stretched in-plane with the elastic subsurface attempting to return to its former shape (Figure 15.2). This results in the generation of elastic compressive residual stresses at the surface. The magnitude and distribution of the residual stresses, produced by a shot peening operation, are dependent on the target material properties and the shot peening parameters. Material properties such as yield stress, hardening rate and cyclic softening/hardening all influence the final residual stress profile. Characteristics of the shot, such as size, material, velocities and shape, also influence the residual stresses introduced.

A typical shot peening residual stress distribution is known to have its maximum compressive stress value just below the surface in the region $0 < z < 10\,\mu$m. Duplex peening involves using two separate peening operations in an attempt to increase the magnitude of the surface compressive stresses while maintaining a good overall depth of compressive residual stresses [6].

The most widely used specification for a shot peening operation is the Almen Intensity which is the curvature experienced by a standard strip of steel exposed to identical peening conditions as the treated component. For example, 18–20 A is the deflection in thousands of an inch of an A-gauge Almen strip. Other parameters usually quoted are the size and hardness of the shot and the amount of coverage. These specifications are empirically based and aim to give the design engineer some indication of the improvements in mechanical and surface properties as a result of a peening operation. There is therefore a need to know the relationship between peening conditions, target material and residual stress levels introduced.

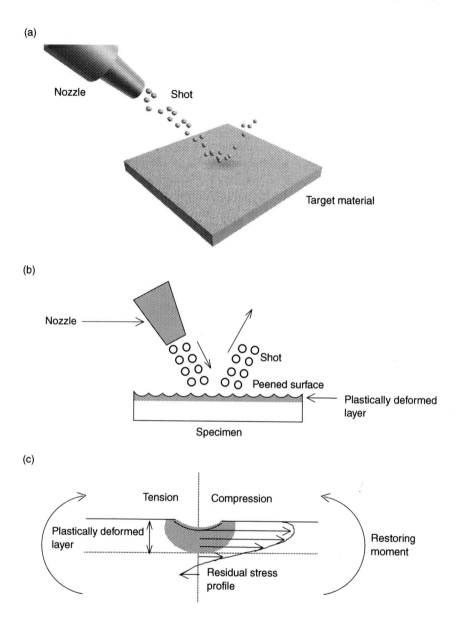

Figure 15.2 The shot peening process. (a) Shot peening of a plate. (b) Schematic of the shot peening process. (c) Generation of residual stresses from a single indentation.

15.3 The evaluation of shot peening residual stresses

Shot peening residual stress distributions are characterised by a shallow depth of compressive stresses, with a maximum value of the order of the compressive yield stress of the material. A steep stress gradient and plastically deformed material exist at the surface of the component. These conditions must be considered when measuring shot peening residual stresses.

There are a variety of established techniques available for measuring residual stresses, such as X-ray diffraction, sectioning, incremental hole drilling, crack compliance and neutron diffraction [7]. Most of these have been applied to shot peening situations with various degrees of success. The X-ray diffraction technique has been used for measuring shot peening residual stress distributions [5]. The technique is non-destructive if surface-only measurements are required, and laboratory sources are also quite cost effective and easy to use. The main disadvantage is that depths of examinations are restricted to typically only 30–50 μm beneath the surface. If depth profiles of residual stresses are required, material removal is necessary, usually using an electrochemical etching procedure. In this case, it is necessary to ensure that the measuring surface remains flat, and also that there is minimal grain boundary erosion, for the measurements to be reliable. Correction algorithms are often used to account for residual stress redistribution caused by the material layers removed. In addition, because a very small volume of material is sampled, grain size, texture and prior machining stresses influence the results and need to be interpreted properly in an analysis. High-energy synchotron X-rays produced in accelerators are orders of magnitude more intense than laboratory X-ray sources and can also be used for residual stress measurements [8]. In light metals (aluminium, titanium, etc.) it is possible to penetrate several millimetres [8].

The incremental hole drilling technique has been applied to the measurement of shot peening residual stresses [9]. Particular precautions are necessary for the interpretation of results where the local stresses are above half the yield stress of the material. The crack compliance technique has been successfully used to measure shot peening residual stresses [10].

The neutron diffraction technique has also successfully been applied to the measurement of shot peening residual stresses [11–15]. The technique is non-destructive and does not disturb the original residual stress field. Because of the relatively weak scattering of neutrons, the volume of material required to produce an acceptable neutron count is significantly larger than the volume required for X-rays. As such, the closest measurement point to the surface is typically 0.1 mm. There are particular precautions associated with using neutrons for residual stress measurements which are discussed in the next section. Measurements have been made successfully at reactor [11–13] and pulsed sources [14].

15.4   Neutron diffraction measurement of shot peening residual stresses

Near-surface residual stress measurements using neutrons usually involve partially immersed sampling volumes and measurements in regions of steep stress gradients and plastically deformed material. It is necessary to consider these factors in the interpretation and analysis of the diffraction data. A good overview of near-surface residual stress measurements using neutrons, including precautions necessary to achieve reliable measurements, is presented by Edwards in Chapter 14. Also, VAMAS (The Versailles Project on Advanced Materials and Standards) under the Technical Working Area TWA20 (Measurement of Residual Stress) is developing procedures for accurate and reliable near-surface residual stress measurements using neutrons.

Partially immersed sampling volumes are necessary when measuring shot peening residual stresses because the stresses of interest are in the near-surface region (0–1 mm). High-intensity neutron sources have an advantage for this application, as very small sampling volumes are necessary for stress measurements close to the surface. Partially immersed sampling volumes have the tendency to introduce shifts to diffraction peaks, unrelated to the levels of strain present. If these spurious peak shifts are not correctly interpreted, errors will be introduced into the analysis. This occurs both at reactor and pulsed sources whether single or position sensitive detectors are used [11–14]. These spurious peak shifts are largely the result of a shift in the centre of gravity of the diffracting volume and the geometry and type of detectors used. Many investigations have been carried out to try and model these effects and produce correction algorithms. Smaller slit sizes, the use of sollers and the quality of the monochromator all reduce this effect.

To correct for these spurious peak shifts it is suggested that calibration measurements are made on a powder cell in the shape of the specimen. Alternatively this effect can be eliminated by re-orientating the specimen in the neutron beam as shown in Figure 15.3(b)

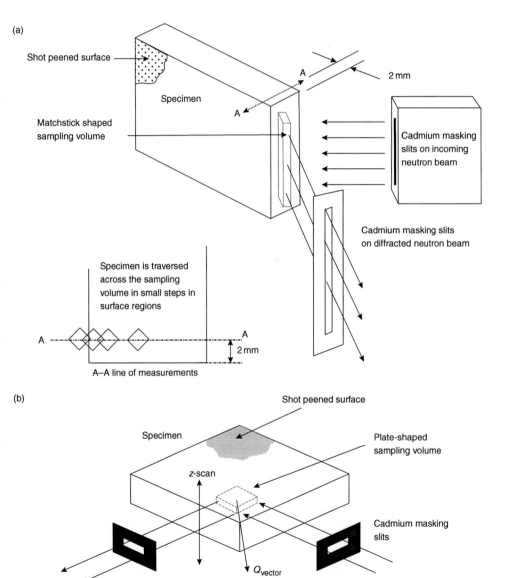

Figure 15.3 Definitions of neutron sampling volumes: (a) xy-scan; and (b) z-scan.

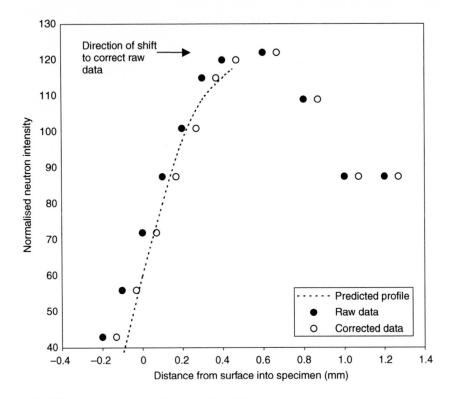

Figure 15.4 Typical entering curve of neutron intensities.

which eliminates changes in the centre of gravity. This is sometimes described as the z-scan method. In the case of symmetrical specimens it has been suggested that self-calibration can be achieved by taking the average of the strain measurements made on each face. Entering curves show the intensities of scattered neutrons as the sampling volume is traversed from a point outside of the specimen to a point inside the specimen. This intensity scan across the surface can be used to precisely determine the position of the surface of the specimen relative to the neutron beam (Figure 15.4).

To obtain the stress at a point, the component must be rotated and repositioned in the neutron beam during the course of a single stress measurement to obtain components of the strain tensor. Because shot peening introduces steep stress gradients, it is important that a point in the component can be positioned to within 0.1 mm during a course of strain measurements. Particular attention must therefore be paid to the sample positioning device and the neutron beam optics. It is essential that these are set up properly. The translators that position the specimen in the neutron beam must be able to do this to within 0.1 mm at every stage of the measurements. It must also be ensured that the sampling volume is always measuring in the plane of the peened layer. It is necessary to centre the incoming neutron beam over the centre of the rotation table using slits of 0.2–0.5 mm to ensure no movement of the gauge volume when the table is rotated. It is recommended that small masking slit sizes are used to improve the resolution, particularly in the direction of the stress gradient. Figures 15.5(a) and (b) illustrate the effects of using different size sampling volumes on the

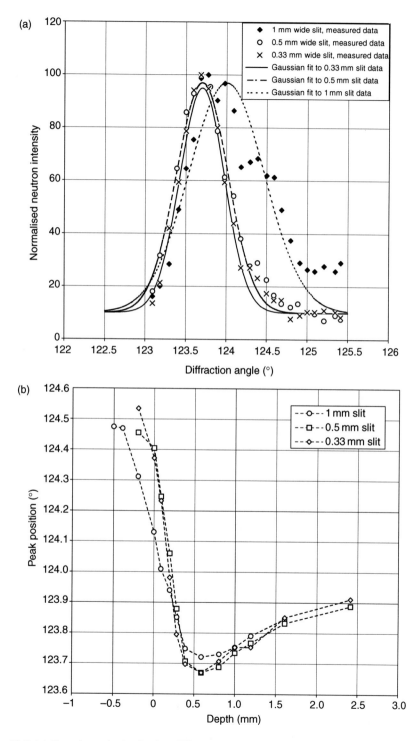

Figure 15.5 (a) Gaussians obtained using different slit sizes. (b) Effect of different size volumes on measured strain distribution.

Table 15.1 Details of diffraction measurements

Alloy	Base element	Crystal structure	Neutron spectrometer and wavelength (nm)	(hkl)	Sampling volume (mm³)	Scan method
Udimet 720	Ni	FCC	D1A, Grenoble, France, 0.19	(311)	1 × 1 × 25	xy-scan
Waspaloy	Ni	FCC	HB4, Petten, Holland, 0.19	(311)	1 × 1 × 25	z-scan
Jethete	Fe	BCC	G5.2, LLB Saclay France, 0.29	(211)	0.5 × 10 × 10	z-scan
IMI 834	Ti	HCP	D1A, Grenoble, France, 0.19	(11$\bar{2}$2)	1 × 1 × 25	xy-scan

Table 15.2 Shot peening parameters

Specimen no.	Alloy	Thickness (mm)	Shot size	Intensity (A)	Coverage (%)
1	Udimet 720	12.75	MI 330 R	18–20	200
2	Udimet 720	12.75	M 110 H	6–8	125
3	Waspaloy	5.00	MI 230 R	14–16	200
4	Jethete	5.00	MI 230 R	14–16	200
5	IMI 834	12.75	MI 330 R	18–20	200

Figure 15.6 Photograph of a specimen in a neutron spectrometer.

Gaussian data and the measured peak shift with depth, indicating that a 0.5 mm wide slit is about the practical limit for slit width to obtain a good resolution.

As a result of the shot peening operation, the surface layer will have different mechanical properties to the elastic base material. For example, for work hardening material it is possible for the residual stresses generated at the surface to be greater than the static compressive stress of the material. For work softening material the opposite effect may be observed.

15.5 Examples of measured profiles

Residual stress measurements have been made at a number of neutron sources in specimens made from a variety of engineering materials, shot peened over a range of peening conditions, to demonstrate the application of the neutron diffraction technique to the measurement of shot peening residual stresses. Residual stress measurements have also been made before and after fatigue cycling to establish how shot peening residual stresses re-distribute during loading. All the measurements described here were made at reactor sources using a monochromatic beam of neutrons (Table 15.1). A description of the specimens and shot peening conditions is given in Table 15.2.

Each specimen was shot peened on both of its surfaces with the same intensity. Specimen 1 had an additional peen on one of its faces (duplex peening). Two approaches have been used to measure the in-plane residual stresses; the xy-scan method and z-scan method (Figure 15.3). Specimens were mounted in the neutron beam using optical alignment. Figure 15.6 shows a typical specimen in the neutron beam. Entering curves were used to establish the specimen surface position to 0.1 mm. Sampling volume dimensions are given in Table 15.1. The residual strains were obtained and converted to stresses using the diffraction elastic constants (DECs) given in Table 15.3.

The measured stress profiles are shown in Figures 15.7–15.9. Figure 15.7 shows the effect of different shot peening intensities on the residual stress distributions in a nickel base superalloy. It is observed that increased peening intensities result in increased depths of the compressive residual stresses in conjunction with higher maximum subsurface tensile stresses. The duplex peening operation used shows that for depths >0.1 mm there is no significant advantage over the single peening operation. X-ray residual stress measurements will tend to show the benefits of a duplex peening operation in the region $0 < z < 0.1$ mm. However, the duplex peened surface shows slightly higher compressive stresses and higher tensile subsurface stresses when compared with the single peened surface.

Figure 15.8 shows the effects of different target materials subjected to similar shot peening conditions on the residual stress distributions produced. Nickel, titanium and steel have been investigated. The nickel and steel alloys used have similar mechanical properties and for the similar peening conditions used show very similar residual stress distributions. In contrast the titanium alloy has similar yield properties to the nickel and steel but a significantly lower Young's modulus. The depth of compression introduced is not as significant as that produced in the other materials.

Figure 15.9 shows the re-distribution of residual stresses that occur due to fatigue loading in bending, in a nickel base superalloy. The specimen was peened on both sides and residual

Table 15.3 Diffraction elastic constants

Material (hkl)	E (hkl) (GPa)	v (hkl)	E (mechanical) (GPa)	v (mechanical)
Steel (211)	–	–	210	0.3
Nickel (311)	209	0.36	225	0.3
Titanium (11$\bar{2}$2)	126	0.26	119	0.32

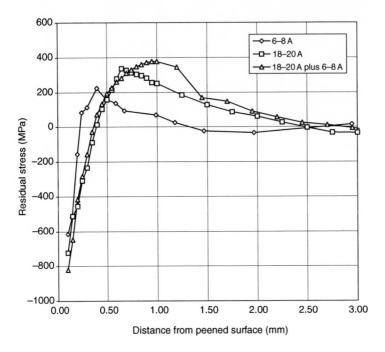

Figure 15.7 In-plane shot peening residual stresses in nickel superalloy resulting from different peening conditions.

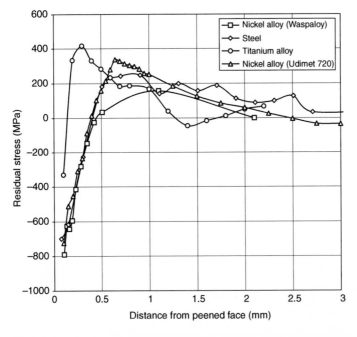

Figure 15.8 Shot peening residual stresses produced in different materials.

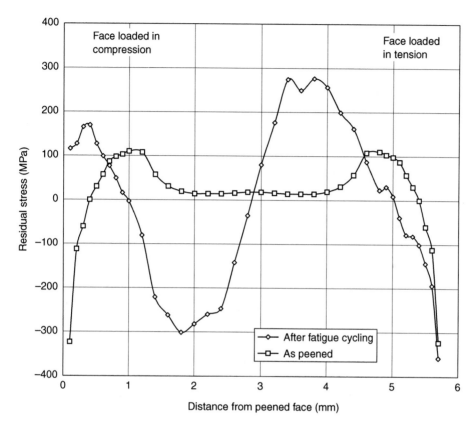

Figure 15.9 Re-distribution of shot peening residual stresses.

stress measurements were made before and after fatigue cycling, before the initiation of any fatigue cracks. The fatigue loading conditions were 36 000 cycles at 1.13% total strain (0−max). This is in the low cycle fatigue (LCF) regime as the strain at yield for this material is 0.35%. It is observed that the fatigue loading has significantly altered the original residual stress profile by superimposing the stress field expected from plastic bending of a bar. The initial compressive residual stresses limited the levels of tensile stresses generated at the surface of the face loaded in tension. This is observed in Figure 15.9, where greater compressive residual stresses are obtained at the surface after fatigue cycling than are introduced from the peening process. The benefits of the peening process are therefore evident as the compressive residual stresses inhibit the initiation of fatigue cracks.

15.6   Conclusions

Residual stresses have successfully been measured using neutrons in a variety of shot peened specimens. It has been demonstrated how attractive the neutron diffraction technique is for the non-destructive measurement of near-surface and subsurface residual stresses. Surface compressive stresses have been measured as well as subsurface tensile stresses. The necessary precautions required for measurements have been presented.

Acknowledgements

The author would like to thank Rolls-Royce and the UK Engineering and Physical Sciences Research Council (EPSRC) for the provision of material to test and for financial support respectively. He would also like to thank Steven Brook for his assistance with the illustrations. MSX International are thanked for their support during the production of this paper.

References

[1] Webster G. A. and Ezeilo A. N. Neutron scattering in engineering applications, *Physica B*, **234–236**, 949–955, 1997.

[2] Townsend D. P. and Zaretsky V. Effect of shot peening on the surface fatigue life of carburized and hardened AISI 9310 spur gears, *NASA Technical Paper 2047*, 1982.

[3] Marsh K. J. (ed.) *Shot Peening: Techniques and Applications*, Chamelion Press Ltd, ISBN 0-947817-64-6, 1993.

[4] Tange A. and Ando K. Study on shot peening processes of coil springs, *Proc. 6th Int. Conf. on Residual Stresses*, Oxford, UK, pp. 897–904, 2000.

[5] Fuchs H. O. Shot Peening, *Mechanical Engineers' Handbook*, John Wiley & Sons, pp. 941–951, 1986.

[6] Ishigami H., Matsui K., Jin Y. and Ando K. Study on stress, reflection and double shot peening to increase fatigue limit, *Proc. 6th Int. Conf. on Residual Stresses*, Oxford, UK, pp. 667–673, 2000.

[7] Ericsson T., Odén M. and Andersson A. (eds), *Proc. 5th Int. Conf. on Residual Stresses*, Linkoping Inst. of Tech., Linköpings Universitet, Sweden, 1997.

[8] Webster P. J. Synchotron strain scanning through a peened aluminium alloy plate, *Proc. of the 5th Int. Conf. on Residual Stresses*, eds T. Ericsson, M. Odén and A. Andersson, Inst. of Tech., Linköpings Universitet, Sweden, pp. 551–556, 1997.

[9] Niku-Lari A., Lu J. and Flavenot J. F. Measurement of residual stress distribution by the incremental hole-drilling method, *J. Mech. Working Technol.*, **2**(2), 1985.

[10] Cheng W., Finnie I., Gremaud M., Rosselet A. and Streit R. The compliance method for measurement of near surface residual stresses – aplication and validation for surface treatment by laser and shot peening, *J. Eng. Mater. Technol.*, **116**, 556–560, 1994.

[11] Garot-Piant A., Lodini A., Holden T. M., Rogge R. and Braham C. Stress measurement in shot peened plates by neutron and X-rays diffraction, *Proc. of the 5th Int. Conf. on Residual Stresses*, eds T. Ericsson, M. Odén and A. Andersson, Inst. of Tech., Linköpings Universitet, Sweden, Vol. 2, 244–249, 1996.

[12] Garot-Piant A. Thesis, Champagne-Ardenne University, France, 1996.

[13] Ezeilo A. N., Webster G. A., Webster P. J. and Webster P. S. Comparison of shot peening residual stress distributions in a selection of materials, *Proc. of the 5th Int. Conf. on Shot Peening*, Oxford University, pp. 274–281, 1993.

[14] Wang D. Q. and Edwards L. *Pulsed Neutron Strain Scanning – Technical Report FRG/97/3/1*, Faculty of Technology, The Open University, Milton Keynes MK7 6AA, 1996.

[15] Ezeilo A. N., Webster P. S., Webster G. A. and Webster P. J. Development of the neutron diffraction technique for the determination of near surface residual stresses in critical gas turbine components, in *Measurement of Residual and Applied Stress Using Neutron Diffraction*, NATO Advanced Research Workshop Series, eds M. T. Hutchings and A. D. Krawitz, Kluwer Academic Publishers, Netherlands, pp. 535–543, 1992.

16 Composite materials

M. E. Fitzpatrick

16.1 Introduction

Multi-phase materials have been, and continue to be, a key area of study for diffraction-based stress measurement techniques. The reason is that diffraction methods are among a very small number of methods by which through-thickness strain profiles can be measured unambiguously in each phase of such a material, and are essentially unique for the non-destructive study of metal and ceramic composites. Bulk monitoring techniques such as strain gauging, or destructive techniques such as sectioning or hole drilling, reveal only the total stress within the material, and not how that stress is partitioned between the individual phases.

The development of neutron diffraction as a technique for stress analysis has run roughly concurrently with the development of many new metal- and ceramic-based composite materials for engineering applications. Inorganic composites have, as a result, been studied extensively using the technique, both for the acquisition of understanding of the fundamental properties of these materials, and also for the study of specific problems relating to their use in engineering components. As composites begin to be used in targetted engineering applications, greater emphasis is beginning to be placed on the study of residual stress in real components. The development of synchrotron X-ray techniques, as both a strain measurement method and for tomographic scanning, offers exciting possibilities for future studies of composite materials [1, 2].

There have been detailed reviews published of the application of neutron diffraction to the study of composite systems [3, 4]; this chapter will attempt to indicate how neutron and synchrotron diffraction methods have been applied to the study of composites, and the type of information which can be obtained.

Most of the studies that have been undertaken on composite materials have focussed on metal matrix composites. A lesser number have looked at ceramic matrix composites, whilst polymer matrix composites have been avoided owing to the near-impossibility of obtaining sensible diffraction data from a hydrogen-containing material with its very high level of incoherent scattering, although future studies on polymer composites may be undertaken using deuterated materials. This review is confined therefore to inorganic composites.

The predominance of metal matrix composites (MMCs) in stress studies reflects the volumes of these materials that are in production relative to ceramic matrix composites (CMCs). MMCs have found quite diverse applications in engineering, from structural and load-bearing components such as bicycle frames and gear selector forks, to applications where the tailoring of thermal properties is important, such as brake discs and packaging for microelectronic devices [5]. CMCs, although candidate materials for high temperature applications where higher toughness than conventional monolithic ceramics is required, have yet to show the same degree of market penetration. I will concentrate primarily on metal-based composites in this article: CMCs are covered in detail elsewhere in this book [6].

16.2   Internal stress in composites

The stress state in a composite, and indeed in any two-phase material, is more complicated than in a homogeneous material. If the second phase is present in an appreciable volume fraction (say 1% or greater) then components of stress will be present in both phases owing to differences in the physical and mechanical properties of the two phases. At lower volume fractions there will be significant stress components in the second phase, whilst the stress in the matrix will average to near-zero.

The advantage of diffraction-based stress analysis for composite materials is that stresses are obtained separately in each phase, and, importantly, the total stress in each phase is determined.

The total stress in a composite comprises contributions from several different components. These components are dealt with individually in the following sections, but they are: the macrostress (type I stress); the elastic mismatch stress, and thermal and plastic misfit stresses (type II stresses). A diffraction measurement supplies the mean field stress in the diffracting volume: this may have contributions from all these components. By using a large gauge volume which floods the sample, it is possible to 'average out' any macrostress variation. This allows the misfit and mismatch strains to be measured with a well-designed experiment. Using model data, in conjunction with experimentally determined mean strains, it is possible to separate, by calculation, the contribution of misfit stresses from macrostresses in the case where the gauge volume is embedded within the material in a stress gradient.

Many studies have been constructed to look at the effects of individual stress components in composites, whereas others have been concerned solely with the total stress in each phase and its effect on the overall properties of the composite. Indeed, the majority of internal stress studies performed on composite systems have had the object of drawing a link between the internal stress states and the macroscopic mechanical behaviour of the material. An example of this type of study is the work of Lewis and Withers [7], where particle cracking in an Al/ZrO_2 composite was related to the measured stress levels in the zirconia reinforcement. The technique has also been applied to both metal and ceramic systems to investigate reinforcement phase changes; either by temperature effects [8] or by the effects of imposed strain [9].

Neutron and synchrotron diffraction has proved to be useful for the study of phenomena such as creep [10, 11], where there is a time-dependent evolution of the phase strains in the composite, and thermal cycling, again where the phase strains evolve over time [12–14].

Some ceramic reinforcements pose problems for diffraction measurements, in that they are not pure single-phase materials. Silicon carbide (SiC) is an example of this: Figure 16.1 shows a spectrum from an aluminium-matrix MMC reinforced with 17% by volume of 3 μm SiC particles.

In Figure 16.1, the reinforcement data has been analysed assuming that the SiC is a cubic phase (the cubic phase of β-SiC is a polytype with $F\bar{4}3M$ space group). In reality, it can be seen that there is an additional peak present above the SiC {111} peak, indicating the presence of an additional phase. This is hexagonal α-SiC, with a $P63MC$ space group; the peak in question is the {10$\bar{1}$} reflection. There is also the possibility of having an orthorhombic $R3M$ phase. It is a good idea to be aware of the crystallographic structure of the reinforcement.

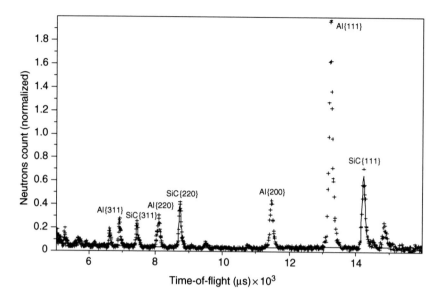

Figure 16.1 Time-of-flight neutron diffraction spectrum obtained from an Al/SiC composite material, showing a modified Pawley fit to the SiC peaks with the assumption that the SiC is a cubic phase. Higher-time-of-flight corresponds to higher lattice spacing. For reference, the Al{111} peak is at a *d*-spacing of ~2.4 Å. The unfitted peak to the right of the SiC{111} reflection arises from a non-cubic polytype of the SiC that is present in the composite.

Some polytype peaks may overlap with matrix diffraction peaks, and errors in strain may arise if an analysis is conducted assuming an incorrect crystal structure: such errors are not large if there is no great distortion between the crystal structures of the different polytypes, and for SiC fibres may be only of the order of ~5% [15].

16.3 The measured stress variation

When a diffraction measurement is performed on a composite material, the total stress determined in each phase is the sum of three contributions:

$$\bar{\sigma}_{s,i}(\mathbf{r}) = \sigma_s^{macro}(\mathbf{r}) + \langle\sigma\rangle_{s,i}^{mE}(\mathbf{r}) + \langle\sigma\rangle_{s,i}^{mSh}(\mathbf{r}) \tag{1}$$

Here the macrostress within the gauge volume is denoted as σ^{macro}. The total, or average, stress component in the s direction $\bar{\sigma}_{s,i}$, in the ith phase of the composite, therefore comprises σ_s^{macro}, which is the same in all phases; plus the 'mean' elastic mismatch stress $\langle\sigma\rangle_{s,i}^{mE}$ which acts as an offset to the macrostress; and finally the 'mean' shape misfit stress $\langle\sigma\rangle_{s,i}^{mSh}$, which is comprised of thermal and plastic misfits between the two phases. Any one of these components of stress may be of interest for a particular experiment, and typical results are summarized later in this review. However, it is the total or 'average' stress which is measured (in fact, calculated from the measured strains) within the gauge volume during a neutron diffraction experiment.

The strain variation measured in a composite is shown schematically in Figure 16.2 [16]. In Figure 16.2, the macroscopic stress variation is offset by the thermal shape misfit stress

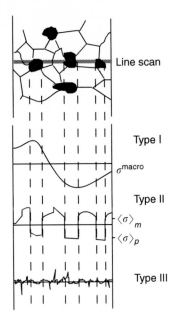

Figure 16.2 Schematic stress variation in a composite material.

in each phase; and there will also be type III stresses arising from local stress fields around point defects, etc.

Once the stress has been measured in both phases of a composite, the macrostress can be calculated by summing the phase stresses in proportion: this is possible as the elastic mismatch and shape misfit stresses will sum to zero over a volume of material which is several times the interparticle spacing in dimension [16, 17].

$$f \langle \sigma \rangle_{s,p}^{\mathrm{mTh}}(\mathbf{r}) + (1 - f) \langle \sigma \rangle_{s,m}^{\mathrm{mTh}}(\mathbf{r}) = 0 \qquad (2)$$

This is essentially the limit of the information that can be calculated directly from the measured phase strains. The phase strains allow calculation of the phase stresses, from which the macrostress can be evaluated. Obtaining additional information from the results requires some modelling to be undertaken, based on the known physical and mechanical properties of the composite phases.

Using a model such as that of Eshelby [18–20], the elastic stress in each phase arising from the (calculated) macrostress can be evaluated. This requires a knowledge of the elastic properties of each phase, along with the volume fraction of the reinforcing phase, and its aspect ratio if it is an aligned whisker or fibre composite [5, 16]. This gives the $\langle \sigma \rangle_{s,i}^{\mathrm{mE}}(\mathbf{r})$ elastic misfit term from equation (2). It is then clear that subtracting the macrostress and elastic misfit stress from the total measured stress will give the shape misfit stress component $\langle \sigma \rangle_{s,i}^{\mathrm{mSh}}(\mathbf{r})$. A full treatment of this method is given in Fitzpatrick *et al.* [16].

The shape misfit is itself comprised of two components: a thermal misfit and a plastic misfit. In many studies, for example, where the composite has been subjected to an anneal or other high temperature heat treatment, it can be assumed that the shape misfit is primarily of thermal origin, with any plastic effects removed during the anneal.

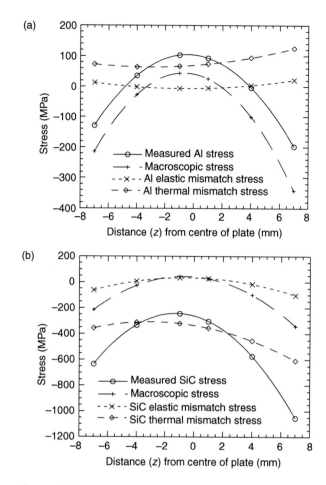

Figure 16.3 Separation of macrostress, elastic mismatch and thermal misfit stresses in the (a) matrix and (b) reinforcement of an Al/SiC quenched plate [21].

Using an approach like this, the individual stress terms on the right-hand side of equation (1) can be obtained from a set of measurements made in the two phases of a composite. This can be seen from the results in Figure 16.3, where the total measured stress in the components of an Al/SiC composite have been separated to give the macrostress, thermal misfit and elastic mismatch stresses [21].

In that case, as the samples were examined following a heat treatment and quench, it is reasonable to assume that any shape misfit has a thermal origin. In other cases, the misfit stress may have contributions from plastic effects [22, 23].

I will now consider the individual contributions to the measured stress variation.

16.3.1 Thermal misfit stress

Firstly, a stress will exist in the composite material if there is a difference in the coefficient of thermal expansion (CTE) between the phases. Fabrication processing of composites is

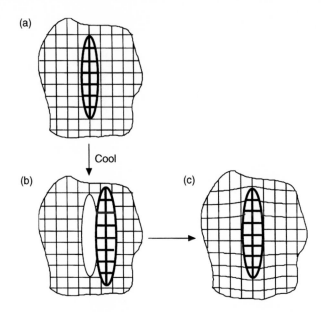

Figure 16.4 Generation of a thermal misfit in a composite material where the matrix has a higher
CTE than the reinforcement: (a) shows the initial condition with no misfit between
the phases; (b) shows the thermal misfit generated by a temperature drop: as the
composite is cooled, the reinforcement contracts less than does the matrix. This leads
to a misfit strain as shown in (c), which is tensile in the matrix and compressive in
the reinforcement. The shape change is exaggerated for clarity. Adapted from Clyne
and Withers [5], reproduced with permission.

undertaken at elevated temperatures, and a *thermal misfit* stress will be generated as the
composite is cooled to room temperature. For metal/ceramic systems, the metal matrix generally has a higher CTE than the ceramic reinforcement, and this leads to the matrix being
in net tension and the reinforcement being in net compression after cooling. This is shown
schematically in Figure 16.4.

Changes in the misfit stress with temperature can be predicted using a model such as the
equivalent inclusion model of Eshelby [18, 20, 24]; it is usually not possible, though, to
predict precisely what thermal misfit stress will be present in a particular MMC at room
temperature, as there will be relaxation of the stress which is generated during the cooling
process by diffusional or plastic flow within the matrix.

An example of this is demonstrated by measurements that have been made on Al/SiC
particulate composites. Eshelby calculations using the method of Withers [5, 20, 24] show that
the matrix stress becomes more tensile by 0.293 MPa, and the reinforcement stress becomes
more compressive by 1.423 MPa, per °C cooled, for an Al–17vol% SiC composite with
particles having an aspect ratio of 1. Such a composite typically experiences a temperature
drop of ∼500 K from processing temperature to room temperature, implying that a residual
stress of around 147 MPa would be expected in the matrix, and around −710 MPa in the
reinforcement. In reality, experiments on these composites have shown values of around
half this [16, 25, 26], indicating that there is relaxation during or after the cooling process,
although results from Maeda *et al.* [27], on slightly different Al/SiC systems, show stresses

that are essentially unrelaxed compared to predicted values. Ohnuki *et al.* [28], in a study on whisker-reinforced MMC, indicated that the final thermal misfit residual stress in the two phases appears to depend on the temperature from which the material is quenched. If a bulk stress measurement, with the gauge volume encompassing the entire sample, is made on a composite using a diffraction technique, the results from each phase give the thermal misfit stress, as any *macrostress* variation will average to zero within the sampling volume (any plastic misfit stress will be included in the measured value of the misfit stress, but cannot be separated from the thermal effects). Bulk stress measurement studies performed on MMC materials using neutron diffraction [16, 29–33] produce, as expected, the results of tensile matrix stresses and compressive reinforcement stresses. Several similar studies on CMCs [34, 35] show clearly the presence of thermal misfit between the different ceramic phases.

Several studies have shown how the thermal misfit stresses evolve as the temperature is altered, for MMC [14, 29, 31] and CMC [34, 36–38] systems. As the misfit is thermally generated, there exists a 'stress-free' temperature at which there is no shape misfit between the phases, and this may be calculated or measured directly by varying the temperature of the composite sample *in situ* on the diffractometer. As with metal systems [16, 28], the thermal misfit in CMCs can be seen to be dependent on the precise prior thermal history of the material [38].

Much of the preceding discussion deals with measurements on particulate-reinforced materials; however, whisker-reinforced systems have also seen considerable interest. Whisker reinforcement offers improved elastic moduli for similar volume fractions than particulate reinforcements, at the expense of introducing anisotropy to the mechanical properties of the composite. The development of whisker-reinforced systems has, however, been hampered by the health hazards present in handling of the whisker material [39].

When measuring strains in perpendicular directions for aligned, high reinforcement aspect ratio composites, higher strains are found in the alignment direction than in the transverse directions [31, 40]; even when the whiskers are not perfectly aligned, but are oriented in one plane as a consequence of the processing operations [30]. Of course, for materials with an aligned structure, the precise level of the average phase strains will depend on the orientation distribution of the reinforcement [28], that is, how well the whiskers are aligned. As with particle-reinforced systems, the thermal misfit is tensile in the matrix and compressive in the reinforcement [41].

An extension of the effects seen in whisker-reinforced systems is observed in systems which use long-fibres or continuous fibres as the reinforcement. Such systems, particularly those based on Ti/SiC are being studied increasingly owing to the possibility of incorporating them into aero-engines as replacement for conventional Ti-based disc materials, giving weight savings by allowing a change in component design from a disc to a ring [42].

Once again, the thermal misfit stress results are much as expected, showing tensile matrix strains and compressive fibre strains [43], with larger stresses longitudinal to the fibres than transverse to them [44, 45].

16.3.2 *Plastic misfit stress*

The thermal misfit stress discussed in the previous section is an example of a stress which arises because of a shape misfit between the two phases of the composite. In that case, the shape misfit occurs because of the differential contraction of the two phases following cooling. Another source of shape misfit, primarily in MMCs, is plasticity. In this case, a shape change can be accommodated by plastic flow in the matrix, whilst the reinforcement remains only elastically-deformable. This is shown schematically in Figure 16.5.

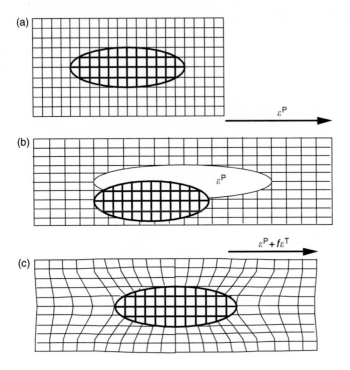

Figure 16.5 Generation of a plastic misfit stress in a composite: (a) shows the initial state of the composite, with a non-deformable particle in a matrix that can deform plastically; in (b), the composite has been subjected to a plastic strain of ε^P. As the particle is deformed only elastically, there is a shape misfit between the two phases, which is represented in (c) as a transformation strain ε^T, that depends on the elastic properties and volume fraction of reinforcement. Adapted from Clyne and Withers [5], reproduced with permission.

Such deformation leads to the generation of a residual misfit stress which is compressive in the matrix and tensile in the reinforcement, for a tensile plastic deformation [5]. Like the thermal misfit stress, the plastic misfit stress component cannot be measured independently. However, if the stress state in a composite can be assumed not to have a plastic misfit component – for example, after the composite has been subjected to an anneal or a solution heat treatment – than the effect of subsequent plastic flow can be assessed.

The misfit generated by plastic deformation has been seen to reduce the thermal misfit stress in Al/SiC composites [22, 23], as shown in Figure 16.6, and even to reverse their sense at large plastic deformations, of several percent [25]. In these cases, modelling and calculation is required to extract the misfit components from the total measured stress field, as discussed earlier.

16.3.3 Elastic mismatch stress

In a composite material, the different elastic response of the two phases gives rise to a partitioning of the load within the material. The stiffer phase will bear a higher proportion of the applied load. In most metal/ceramic systems it is the ceramic which has the higher Young's modulus, and there is hence a transfer of load from the matrix to the reinforcement

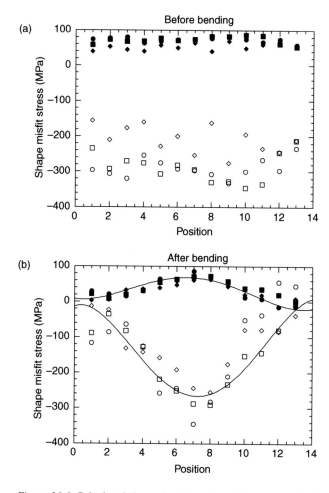

Figure 16.6 Calculated thermal and plastic misfit stress (a) before and (b) after bending of an MMC bar [22]. The closed symbols represent the Al matrix stresses; the open symbols are the SiC reinforcement stresses. Different symbol shapes correspond to different principal stress directions.

as the composite is loaded. The precise value of the elastic misfit depends on the elastic properties of the two phases (Young's modulus and Poisson's ratio), the volume fraction and aspect ratio of the reinforcement, and also on any anisotropy of the reinforcement properties, particularly if the reinforcement is in the form of aligned whiskers or fibres. The elastic mismatch stress can be calculated more easily than the thermal misfit stress, and again an Eshelby-based model is one method whereby this can be achieved. The generation of the elastic mismatch is shown schematically in Figure 16.7.

The effect of the elastic mismatch can be seen in studies where both phases of a composite are measured as the load is increased. Figure 16.8 shows results from Withers *et al.* [19], where an Al/SiC whisker-reinforced composite was loaded *in situ* in the neutron beam, and strain measurements made in both phases during loading. The strains are shown relative to the initial unstressed state of the composite.

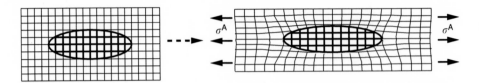

Figure 16.7 Elastic mismatch between matrix and reinforcement. The stiffer reinforcement deforms less under an applied load than does the matrix. The result is a transfer of load from the matrix to the reinforcement, and a higher stress in the reinforcement than in the matrix. From Clyne and Withers [5], reproduced with permission.

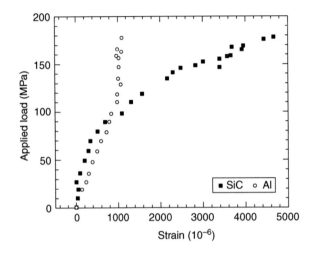

Figure 16.8 Measured longitudinal strains in an Al/SiC whisker-reinforced composite, with 5% SiC whiskers. The composite was loaded *in situ* in the neutron beam [19].

The results show clearly that, in the initial elastic region (applied load ~80 MPa) the SiC reinforcement is the stiffer phase, as it shows less strain than the Al matrix for a given load. Once the matrix becomes plastic, there is a reduction in load transfer from the matrix to the reinforcement owing to the development of plastic misfit between the two phases. Similar results have been reported also by Allen *et al.* [30].

Results from an experiment are shown in Figure 16.9 for an Al/SiC particulate composite which was loaded and unloaded *in situ* in the beam [46]. However, in this case a stress-free reference was measured for each phase, and strain measurements were made both longitudinally and transverse to the load axis, so allowing calculation of the absolute values of stress.

These results show the effect of the elastic misfit stress, which is proportional to the applied stress. The slope of the SiC line is clearly greater than that of the Al line as the load in increased, reflecting the transfer of load from the matrix to the reinforcement by virtue of the reinforcement's higher stiffness. The degree of load transfer is in good agreement with that predicted by Eshelby-based modelling [16], which predicts that for this composite the reinforcement should experience an elastically derived stress ~120% greater than the matrix at a given load.

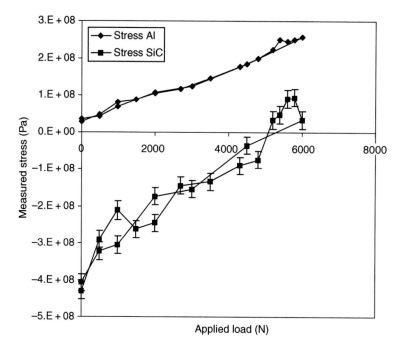

Figure 16.9 Measured longitudinal total stress in the two phases of an Al/SiC$_p$ composite during loading and unloading [46]. No plastic flow occurs in the matrix during loading, so the loading and unloading data lie on top of each other. The error in the Al data is small, $\sim \pm 4$ MPa, so error bars are not shown for Al.

Similar results of the load partitioning owing to the elastic mismatch stress have been obtained by other workers. Withers and Clarke [45] showed that, for a Ti/SiC fibre-reinforced system, the measured strains in each phase during elastic loading agreed well with the load partition predicted from simple rule-of-mixtures and Eshelby modelling.

Another composite system which has been studied using neutron diffraction is the Al/TiC system, which is well suited to study using this technique owing to the greater accuracy in strain determination in TiC relative to SiC [47]. Bourke *et al.* [48], from measurements made on each phase of this composite during applied loading, showed that very good agreement could be obtained between the measured stress and that expected from finite element predictions.

Not all composite studies are restricted to metal–ceramic systems. There is interest in metal–metal composite systems also, both from the point of view of load transfer, and also for monitoring the initiation and development of plasticity in composites where both phases are plastically deformable. Carter and Bourke [49] showed that in a beryllium–aluminium composite, the yield points of the two phases were different: 100 MPa for Al and 200–250 MPa for Be. Synchrotron X-ray diffraction has also been useful for these materials, as indicated by a recent study of load partitioning effects in a copper–molybdenum composite [50].

It should be recalled that measurements of the thermal and elastic misfit stresses can be made only by diffraction-based stress measurement methods, as it is necessary to sample the strain in each phase separately. Material removal methods such as sectioning, layer removal or hole drilling can sample the macroscopic stress profile only.

16.3.4   Macroscopic residual stress–strain profiling

The macroscopic stress field – the type I stress or *macrostress* – is the stress component which varies typically over several millimetres or centimetres within a material. A macrostress usually arises in practice from a thermal or mechanical treatment which causes non-uniform plasticity: examples of this are a quench from a solution treatment temperature, surface treatments such as shot peening, and room temperature non-uniform plastic deformation by bending, for example.

For the purpose of this discussion, I will treat the macrostress as being a stress field which varies with position. Of course, the measurements described in the preceding section, which give information on the elastic mismatch, at the same time indicate the macrostress in each phase.

Macrostresses in composite materials partition between the matrix and reinforcement, with the level of partitioning being governed by the elastic misfit stress. Whilst it is possible to measure the average type I macrostress at a point in a composite using destructive techniques such as layer removal [51] or hole drilling [52], these techniques are destructive and will not provide information on the phase mismatch strains in the material [53].

In a composite material, knowledge of the stress distribution in a single phase should generally give an indication of the form of the macrostress variation, if it can be assumed that any thermal misfit stress is relatively uniform. However, the precise value of the macrostress cannot be evaluated even if stress balance is invoked, unless knowledge of the elastic misfit stress is available. Evaluation of the stress in both phases allows more accurate evaluation of the macrostress.

Measurements of misfit stress, or elastic mismatch during loading, can be undertaken by using large neutron gauge volumes that encompass most of the sample being studied. For macrostress measurements, it is usual to use a restricted gauge volume, so that a point-to-point profile of strain (and hence stress) can be determined. This requires accurate sample positioning if measurements are to be made successfully.

The first measurements of macrostress variations in composites using neutrons were made by Hutchings and Windsor [16, 54, 55], measuring the quench stress field in Al/SiC_p MMC plates. The same system was used for a study of the macrostress variations around a fatigue crack in the same composite system [26], and for the effects of prior plastic deformation on the macrostress and misfit stress fields [23].

The study of plastic deformation of particulate MMCs, as mentioned earlier, requires measurement of the total stress, which generally shows the macrostress variations in each phase, before calculation of the misfit stresses is possible. In effect, a strain profiling measurement must be made if it is the intention to compare the distribution of macrostress and misfit stress at different positions within a deformed sample. Measurements made on plastically deformed composite bars [22, 56] allowed information on the macrostress changes caused by misfit stress changes to be determined. Recently, Korsunsky *et al.* [57–59] have shown that synchrotron X-ray diffraction can be used successfully to perform strain profiling experiments in MMCs during and after bending, with good spatial resolution. Their results show excellent correlation with similar results that were obtained from MMCs using neutron diffraction. It is likely that the synchrotron X-ray technique will see increased application in the future for such studies.

Studies of macrostress profiles in whisker- and fibre-reinforced composites have been relatively scarce, partially because of the relatively smaller volumes of these materials that are in production, but also because fewer studies have progressed to the stage of requiring strain profiles to be conducted. One exception is the recent work of Withers *et al.* [60], where an SiC fibre bundle incorporated into a titanium block prior to machining of a prototype ring [42], introduces long range stress variations that are detectable using neutron diffraction. The synchrotron technique also shows great promise for the study of fundamental mechanisms in fibre-reinforced materials. Maire *et al.* [2] were able to monitor the strain evolution in individual fibres of an SiC fibre-reinforced titanium composite sample by virtue of the small sampling volume (typically 100 μm) that can be used. Korsunsky and Wells [59], using the synchrotron method, were able to map the strain distribution around an indent in a similar fibre-reinforced sample. Such a measurement would not be practical using neutron diffraction, and shows the utility of synchrotron X-ray diffraction in opening up new areas of study for composite materials.

16.4   Conclusion

Inorganic composite systems have been studied widely using diffraction stress measurement techniques, and it is likely that such techniques will continue to be of importance in determining the mechanisms of stress transfer in these materials.

Many aspects of the fundamental properties of composites are still not well-understood; such as the evolution of plastic misfits during fatigue cycling. At the time of writing (2002), processing technology for composites is still developing rapidly, and the anticipated improvements in interfacial bonding and homogeneity of reinforcement distribution should have beneficial effects in the attainable properties of the material.

Development of strain measurement methods, particularly in the field of synchrotron X-ray diffraction, will give enhanced tools for the study of these materials in the near future.

Acknowledgements

Thanks are due to Prof. Phil Withers for his comments on this paper and permission to reproduce some of his figures. I wish to thank all of the researchers who have participated in my contribution to this field of research: particularly Dr M. R. Daymond, Dr M. Dutta, Dr L. Edwards, Dr M. T. Hutchings, Dr R. Levy, Prof. A. Lodini and Prof. P. J. Withers.

References

[1] Withers, P. J. Use of synchrotron X-ray radiation for stress measurement. Chapter 10, this volume.

[2] Maire, E., Owen, A., Buffiere, J.-Y. and Withers, P. J. A syncrotron X-ray study of a Ti/SiC$_f$ composite during in-situ straining. *Acta Mater.* **49**, 153–163 (2001).

[3] Kupperman, D. S. Neutron diffraction determination of residual stresses in advanced composites. *Annu. Rev. Mater. Sci.* **24**, 265–291 (1994).

[4] Withers, P. J. Neutron strain measurement of internal strain in metal and ceramic matrix composites. *Ceramic Matrix Composites*, Key Engng Mater., 108–110, eds G. M. Newaz, H. Neber-Aeschbacher and F. H. Wöhlbier. Trans Tech Publications, Switzerland, pp. 291–314, 1995.

[5] Clyne, T. W. and Withers, P. J. (1993) *An Introduction to Metal Matrix Composites.* Cambridge University Press, Cambridge.

[6] Todd, R. I. Residual stresses in ceramic materials. Chapter 20, this volume.

[7] Lewis, C. A. and Withers, P. J. Weibull modelling of particle cracking in metal-matrix composites. *Acta Metall. Mater.* **43**(10), 3685–3699 (1995).

[8] Wang, X.-L., Fernandez-Baca, J. A., Hubbard, C. R., Alexander, K. B. and Becher, P. F. Transformation behaviour in Al_2O_3–ZrO_2 ceramic composites. *Physica B* **213/214**, 824–826 (1995).

[9]   Lewis, C. A. The Internal Stress State and Related Microstructural Changes During Deformation of Al/ZrO₂ Metal Matrix Composites. Thesis, University of Cambridge, 1994.

[10] Daymond, M. R., Lund, C., Bourke, M. A. M. and Dunand, D. C. Elastic phase-strain distribution in a particulate-reinforced metal-matrix composite deforming by slip or creep. *Metall. Mater. Trans.* **30A**(11), 2989–2997 (1999).

[11] Winand, H. M. A., Whitehouse, A. F. and Withers, P. J. An investigation of the isothermal creep response of Al-based composites by neutron diffraction. *Mater. Sci. Engng* **A284**, 103–113 (2000).

[12] Daymond, M. R. and Withers, P. J. A synchrotron radiation study of transient internal strain changes during the early stages of thermal cycling in an Al/SiCw MMC. *Scripta Mater.* **35**(10), 1229–1234 (1996).

[13] Daymond, M. R. and Withers, P. J. A new stroboscopic neutron diffraction method for monitoring materials subjected to cyclic loads: thermal cycling of metal matrix composites. *Scripta Mater.* **35**(6), 717–720 (1996).

[14] Daymond, M. R. and Withers, P. J. In situ monitoring of thermally cycled metal matrix composites by neutron diffraction and laser extensometry. *Appl. Composite Mater.* **4**(6), 375–393 (1997).

[15] Tomé, C. N., Bertinetti, M. A. and MacEwen, S. R. Correlation between neutron diffraction measurements and thermal stresses in a silicon carbide/alumina composite. *J. Am. Ceram. Soc.* **73**(11), 3428–3432 (1990).

[16] Fitzpatrick, M. E., Hutchings, M. T. and Withers, P. J. Separation of macroscopic, elastic mismatch and thermal expansion misfit stresses in metal matrix composite quenched plates from neutron diffraction measurements. *Acta Mater.* **45**(12), 4867–4876 (1997).

[17] Winholtz, R. A. Separation of microstresses and macrostresses. In: *Measurement of Residual and Applied Stress using Neutron Diffraction*, eds M. T. Hutchings and A. D. Krawitz, Kluwer Academic, London, pp. 131–145 (1992).

[18] Eshelby, J. D. The determination of the elastic field of an ellipsoidal inclusion and related problems. *Proc. Roy. Soc.* **A241**, 376–396 (1957).

[19] Withers, P. J., Lorentzen, T. and Stobbs, W. M. A study on the relation between the internal stresses and the external loading response in Al/SiC composites. *ICCM VII*, eds W. Yunshu, G. Zhenlong and W. Renjie, Pergamon Press, pp. 429–434 (1989).

[20] Withers, P. J., Stobbs, W. M. and Pedersen, O. B. The application of the Eshelby method of internal stress determination to short fibre metal matrix composites. *Acta Metall.* **37**(11), 3061–3084 (1989).

[21] Fitzpatrick, M. E., Hutchings, M. T. and Withers, P. J. The determination of the profile of macrostress and thermal mismatch stress through an Al/SiC$_p$ composite plate from the average residual strains measured in each phase. *Physica B* **213**, 790–792 (1995).

[22] Fitzpatrick, M. E., Withers, P. J., Baczmanski, A., Hutchings, M. T., Levy, R., Ceretti, M. and Lodini, A. Effect of plastic flow on the thermal mismatch stress in an Al/SiC$_p$ metal matrix composite. *Fourth European Conference on Residual Stresses*, eds S. Denis *et al.*, ENSAM, Paris, pp. 961–970 (1996).

[23] Fitzpatrick, M. E., Dutta, M. and Edwards, L. Determination by neutron diffraction of effect of plasticity on crack tip strains in a metal matrix composite. *Mater. Sci. Technol.* **14**, 980–986 (1998).

[24] Withers, P. J. The Development of the Eshelby Model and its Application to Metal Matrix Composites. Thesis, Cambridge, 1988.

[25] Fitzpatrick, M. E. The effect of plasticity caused by cold working on the misfit stresses in a metal matrix composite. *Fifth International Conference on Residual Stresses*, eds T. Ericsson, M. Odén and A. Andersson, University of Linköping, pp. 886–891 (1997).

[26] Fitzpatrick, M. E., Hutchings, M. T. and Withers, P. J. Separation of measured fatigue crack stress fields in a metal matrix composite material. *Acta Mater.* **47**(2), 585–593 (1999).

[27] Maeda, K., Wakashima, K. and Ono, M. Stress states in quenched SiC/Al particulate composites examined by neutron diffraction. *Scripta Mater.* **36**(3), 335–340 (1997).

[28] Ohnuki, T., Tomota, Y. and Ono, M. Residual elastic strain measurement in heat-treated SiC whisker A2014 composite by neutron diffraction. *J. Jpn Inst. Metal* **60**(1), 56–64 (1996).

[29] Withers, P. J., Jensen, D. J., Lilholt, H. and Stobbs, W. M. The evaluation of internal stresses in a short fibre metal matrix composite by neutron diffraction. *ICCM VI/ECCM 2*, eds F. L. Matthews *et al.*, Elsevier, Amsterdam, pp. 255–264, (1987).

[30] Allen, A. J., Bourke, M. A. M., Dawes, S., Hutchings, M. T. and Withers, P. J. The analysis of internal strains measured by neutron diffraction in Al/SiC metal matrix composites. *Acta Metall. Mater.* **40**(9), 2361–2373 (1992).

[31] Lorentzen, T., Liu, Y. L. and Lilholt, H. Relaxation of thermal induced internal stresses in metal matrix composites. *Ninth International Conference on Composite Materials*, ed. A. Miravete, University of Zaragoza, pp. 371–378 (1993).

[32] Ceretti, M., Braham, C., Lebrun, J. L., Bonnafé, J. P., Perrin, M. and Lodini, A. Residual stress analysis by neutron and X-ray diffraction applied to the study of two phase materials: metal matrix composites. *Fourth International Conference on Residual Stresses*, Society for Experimental Mechanics, Bethel, CT, pp. 32–39 (1994).

[33] Weisbrook, C. M. and Krawitz, A. D. Thermal residual stress distribution in WC–Ni composites. *Mater. Sci. Engng* **A209**, 318–328 (1996).

[34] Kupperman, D. S., Majumdar, S. and Singh, J. P. Neutron diffraction NDE for advanced composites. *J. Engng Mater. Technol.* **112**, 198–201 (1990).

[35] Wang, X.-L., Hubbard, C. R., Alexander, K. B., Becher, P. F., Fernandez-Baca, J. A. and Spooner, S. Neutron diffraction measurements of the residual stresses in Al_2O_3–$ZrO_2(CeO_2)$ ceramic composites. *J. Am. Ceram. Soc.* **77**(6), 1569–1575 (1994).

[36] Majumdar, S., Kupperman, D. and Singh, J. Determinations of residual thermal stresses in a SiC–Al_2O_3 composite using neutron diffraction. *J. Am. Ceram. Soc.* **71**(10), 858–863 (1988).

[37] Majumdar, S., Singh, J. P., Kupperman, D. and Krawitz, A. D. Application of neutron diffraction to measure residual strains in various engineering composite materials. *J. Engng Mater. Technol.* **113**, 51–59 (1991).

[38] Edwards, L. Estimation of sintering stresses in ceramic matrix composites using neutron diffraction. *Fourth International Conference on Residual Stresses*, Society for Experimental Mechanics, Bethel, CT, pp. 697–705 (1994).

[39] King, J. E. Failure in composite materials. *Metals Mater.* **5**, 720–726 (1989).

[40] Arsenault, R. J. and Taya, M. Thermal residual stress in metal matrix composite. *Acta Metall.* **35**(3), 651 (1987).

[41] Li, J., Lu, J., Perrin, M., Ceretti, M. and Lodini, A. Study of the residual stress in cold-rolled 7075 Al–SiC whisker-reinforced composites by X-ray and neutron diffraction. *J. Composites Technol. Res.* **17**(3), 194–198 (1995).

[42] King, J. E. Composites take off without a parachute. *Mater. World* **5**(6), 324–327 (1997).

[43] Rangaswamy, P., Prime, M. B., Daymond, M., Bourke, M. A. M., Clausen, B., Choo, H. and Jayaraman, N. Comparison of residual strains measured by X-ray and neutron diffraction in a titanium (Ti–6Al–4V) matrix composite. *Mater. Sci. Engng* **A259**(2), 209–219 (1999).

[44] Rangaswamy, P., Bourke, M. A. M., Wright, P. K., Jayaraman, N., Kartzmark, E. and Roberts, J. A. The influence of thermal–mechanical processing on residual stresses in titanium matrix composites. *Mater. Sci. Engng* **A224**, 200–209 (1997).

[45] Withers, P. J. and Clarke, A. P. A neutron diffraction study of load partitioning in continuous Ti/SiC composites. *Acta Mater.* **46**(18), 6585 (1998).

[46] Fitzpatrick, M. E. and Daymond, M. R. Unpublished research (1999).

[47] Shi, N., Bourke, M. A. M., Roberts, J. A. and Allison, J. E. Phase-stress partition during uniaxial tensile loading of a TiC-particulate-reinforced Al composite. *Metall. Mater. Trans.* **28A**, 2741–2753 (1997).

[48] Bourke, M. A. M., Goldstone, J. A., Shi, N., Allison, J. E., Stout, M. G. and Lawson, A. C. Measurement and prediction of strain in individual phases of a 2219Al/TiC/15p-T6 composite during loading. *Scripta Metall. Mater.* **29**, 771–776 (1993).

[49] Carter, D. H. and Bourke, M. A. M. Neutron diffraction study of the deformation behaviour of beryllium–aluminium composites. *Acta Mater.* **48**, 2885–2900 (2000).

[50] Wanner, A. and Dunand, D. C. Synchrotron X-ray study of bulk lattice strains in externally loaded Cu–Mo composites. *Metall. Mater. Trans.* **31A**, 2949–2962 (2000).

[51] Andrews, K. W. *Physical Metallurgy Techniques and Applications.* George Allen & Unwin, London (1973).

[52] Beaney, E. M. Accurate measurement of residual stress on any steel using the centre hole method. *Strain* **12**(3), 99–106 (1976).

[53] Cullity, B. D. Some problems in X-ray stress measurements. *Adv. X-ray Anal.* **20**, 259–271 (1977).

[54] Windsor, C. G. and Hutchings, M. T., *The Through-Thickness Residual Stress in the Metal Matrix Composite Plate A680*, Report AEA-InTec-0886, AEA Industrial Technology, Harwell, 1992.

[55] Hutchings, M. T., *The Through-Thickness Residual Stress in a Number of Al/SiC Metal-Matrix Composite Plates*, Report AEA-InTec-1289, AEA Industrial Technology, Harwell, 1993.

[56] Baczmanski, A., Wierzbanowski, K., Tarasiuk, J., Lodini, A., Levy, R., Ceretti, M. and Fitzpatrick, M. E. Study of second order stresses for single- and two-phase polycrystalline materials. *Fifth International Conference on Residual Stresses*, eds T. Ericsson, M. Odén and A. Andersson, University of Linköping, pp. 208–213 (1997).

[57] Korsunsky, A. M., Wells, K. E. and Withers, P. J. Mapping two-dimensional state of strain using synchrotron X-ray diffraction. *Scripta Mater.* **39**(12), 1705–1712 (1998).

[58] Wells, K. E. and Korsunsky, A. M. Strain analysis in composites using synchrotron X-ray diffraction. *Eleventh International Conference on Experimental Mechanics*, ed. I. M. Allison, A. A. Balkema Publ., Brookfield, VT, pp. 761–766 (1998).

[59] Korsunsky, A. M. and Wells, K. E. High-energy synchrotron X-ray measurements of 2D residual stress states in metal matrix composites. *Mater. Sci. Forum* **321–324**, 218–223 (2000).

[60] Withers, P. J., Rauchs, G., Oliver, E. and Bonner, N., *Development of Stress in Ti/SiC Reinforced Components*, ISIS Experimental Report RB10219, Rutherford Appleton Laboratory, Didcot, UK, 1998.

17 Residual stress analysis in monocrystals using capillary optics

W. Reimers and D. Möller

17.1 Introduction

From the point of view of diffraction methods, polycrystalline materials have to be defined as coarse grained if the diffraction experiment gives evidence for a splitting of the Debye fringes into spatially localised Bragg reflections which can be attributed to individual crystallites. In this case, the usually applied measuring and evaluation procedures for the analysis of residual stresses face two problems. Due to the inhomogeneous intensity distribution in the diffraction pattern, intensity measurements can be performed only at selected sample orientations. Furthermore the single crystal elastic anisotropy leads to a scatter of the experimentally obtained strain data [1] resulting in large error bars when calculating stress values under the assumption of elastic isotropy [2]. Since the number of crystallites contributing to the diffracted intensity is decisive for the observed intensity pattern, the transition from fine grained to coarse grained material depends on the gauge volume under study, the experimental resolution and the grain size.

Applying X-ray diffraction, generally, only the near-surface crystallites contribute to the diffraction pattern. Under usual experimental conditions the Debye fringes are split for crystallite diameters of $\geq 100\,\mu m$. By additional sample movements during the measurement, for example, translation, rotation and inclination, the number of reflecting crystallites can be increased, so that the condition of quasi-isotropic behaviour of the gauge volume may be fulfilled up to grain sizes of approximately $\geq 200\,\mu m$. Using synchrotron radiation, however, separated Bragg reflections may already be observed at significantly smaller crystallite diameters due to the small divergence of the radiation.

In those cases where individual Bragg reflections are observed, the single crystal anisotropy has to be considered when calculating the stresses from experimental strain data. Thus, an evaluation of the orientation of the crystallite under study is necessary. Therefore in a first step the spatial orientation of two independent reflections (*hkl*) of the selected crystallite has to be determined. Using a four-circle diffractometer, this may be done in a straightforward and efficient way. From the knowledge about the spatial orientation of these two reflections the orientation matrix, which relates the crystallographic axes system to the fixed laboratory system, can be established. From there on the diffractometer settings for every reflection (*hkl*) to be studied are calculated and the precise reflection position is determined experimentally by a centring routine. This means that this method is well suited for an automatic measuring routine, which gives the strain tensor based on the experimentally determined diffraction angles 2θ for several reflections of an individual crystallite. Since the orientation of the crystallite has been determined, the stress tensor components are calculated from the experimental strain tensor components by applying Hooke's law for anisotropic materials.

17.2   Evaluation of the orientation matrix

The geometry of a four-circle diffractometer and the definition of the laboratory axes system are illustrated schematically in Figure 17.1. The crystallite selected for the study is located at the instrument centre. Using X-rays, the selection of the crystallite can be done either by collimating the beam or by shielding the sample surface with thin Pb foils. Using neutron diffraction the 90° scattering technique is applied which guarantees that the gauge volume keeps almost constant for all sample orientations.

The incident and diffracted beams are in the horizontal plane. The counter (here: Det) is moved in this plane about the vertical instrument axis ω and makes an angle 2θ with the primary beam direction (here: Col). The whole Eulerian cradle may be rotated around the vertical axis by ω. The χ-axis is in the horizontal plane and makes an angle ω with the primary beam direction. The sample is rigidly attached to the φ-shaft which is supported by the χ-ring. The rotation sense for the rotation movement is defined mathematically positive.

The orientation of the crystallite under investigation is known when it is possible to relate its axes system to the fixed laboratory system. This coincides with the θ-axis system when all instrument angles are zero.

For transforming the system it is convenient to define them in terms of Cartesian axes. A crystal direction $\boldsymbol{h}$ is therefore described by a Cartesian crystal direction $\boldsymbol{h_c}$:

$$\boldsymbol{h_c} = \mathbf{B}\boldsymbol{h} \quad \text{for } \boldsymbol{h} = \begin{pmatrix} h \\ k \\ l \end{pmatrix} \quad \text{and} \quad \mathbf{B} = \begin{pmatrix} b_1 & b_2 \cos\beta_3 & b_3 \cos\beta_2 \\ 0 & b_2 \sin\beta_3 & -b_3 \sin\beta_2 \cos\alpha_1 \\ 0 & 0 & 1/a_3 \end{pmatrix} \quad (1)$$

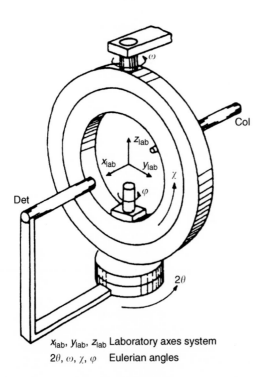

Figure 17.1 Four-circle diffractometer (x_{lab}: parallel to the primary beam, y_{lab}: parallel to the direction of the reflected beam for $2\theta = 90°$, z_{lab}: parallel to the the ω-axis).

a_i, α_i and b_i, β_i are the direct and reciprocal lattice parameters, respectively. Since h_c shall be described in terms of the Φ-axes system also the instrumental angles ω, χ, φ are represented in Cartesian axes system form Ω, X, Φ:

$$\Phi = \begin{pmatrix} \cos\varphi & -\sin\varphi & 0 \\ \sin\varphi & \cos\varphi & 0 \\ 0 & 0 & 1 \end{pmatrix}$$

$$X = \begin{pmatrix} 1 & 0 & 0 \\ 0 & \cos\chi & -\sin\chi \\ 0 & \sin\chi & \cos\chi \end{pmatrix}$$

$$\Omega = \begin{pmatrix} \cos(\omega - \omega_0) & -\sin(\omega - \omega_0) & 0 \\ \sin(\omega - \omega_0) & \cos(\omega - \omega_0) & 0 \\ 0 & 0 & 1 \end{pmatrix} \qquad (2)$$

with $\omega_0 = 2\theta/2$.

The angular set ω, χ, φ gives the diffractometer position where reflected intensity is observed. The corresponding crystal direction can then be described in the Φ-axes system as a unit vector $\boldsymbol{u}_\Phi$:

$$\boldsymbol{u}_\Phi = \tilde{\Phi}\tilde{X}\tilde{\Omega} \begin{pmatrix} 0 \\ 1 \\ 0 \end{pmatrix} \qquad \sim = \text{transposed matrix} \qquad (3)$$

In the case of residual stress measurements, the material investigated and hence the cell parameters are usually known approximately. So the diffracting angle for a reflection h can be calculated and the detector is positioned. The spatial orientation of a reflection h_1 is then searched by systematic variation of φ and χ with fixed ω. The angular position of this reflection is then denoted ω_1, χ_1, φ_1. The plane of a second reflection h_2 is defined by its angular relationship to the first reflection. In cases where the angle between h_1 and h_2 has to be $90°$ for crystallographic reasons, h_2 is found by systematic search using $\Delta\omega$-steps in the plane ($\chi_1 + 90°$, φ_1). The observation of two non-colinear reflections of known indices h_1, h_2 is sufficient to obtain the orientation matrix U [3] which transforms vectors in the crystal Cartesian axes system into the Φ-axes system:

$$h_{1\Phi} = U h_{1c} \qquad h_{2\Phi} = U h_{2c} \quad \text{with } h_{1\Phi}\|u_{1\Phi} \text{ and } h_{2\Phi}\|u_{2\Phi} \qquad (4)$$

Because of experimental errors and uncertainties in the knowledge of the cell parameters, the orthogonal matrix U does not, in general, satisfy both conditions. Therefore, an orthogonal unit vector triple t_{1c}, t_{2c}, t_{3c} in the crystal Cartesian axes system is defined, where t_{1c} is parallel to h_{1c}, t_{2c} lies in the plane of h_{1c} and h_{2c} and t_{3c} is perpendicular to t_{1c} and t_{2c}. In the Φ-axes system another triple t_1, t_2, t_3 is defined, based in the same way on $u_{1\varphi}$ and $u_{2\varphi}$. Since these two unit vector triples are orthogonal by definition they can be superimposed exactly onto each other:

$$t_{i\Phi} = U t_{ic} \quad i = 1, 2, 3 \qquad (5)$$

In matrix notation:

$$\mathbf{T}_\Phi = \mathbf{U}\mathbf{T}_c \qquad (6)$$

Then follows:

$$U = T_\Phi \tilde{T}_c \tag{7}$$

17.3   Angle calculations for any reflection (hkl)

After the evaluation of the orientation matrix by two reflections, the coordinates in the Φ-axes system can be calculated using the matrix **UB** for any reflection (hkl):

$$\boldsymbol{h}_\Phi = \mathbf{UB}\boldsymbol{h} \quad \text{with } \boldsymbol{h}_\Phi = \begin{pmatrix} h_{\Phi 1} \\ h_{\Phi 2} \\ h_{\Phi 3} \end{pmatrix} \tag{8}$$

For the detector position the Bragg equation is used. The interplanar lattice spacing D is given by:

$$D = \frac{1}{\sqrt{h_{c1}^2 + h_{c2}^2 + h_{c3}^2}} \tag{9}$$

The diffractometer setting for the symmetric position ($\omega_0 = 2\theta/2$) is then obtained by:

$$2\theta = 2 \arcsin\left(\frac{\lambda}{2\sqrt{h_{c1}^2 + h_{c2}^2 + h_{c3}^2}} \right)$$

$$\omega_0 = \frac{2\theta}{2}$$

$$\chi_0 = \arctan\left(\frac{h_{\Phi 3}}{\left(h_{\Phi 1}^2 + h_{\Phi 2}^2\right)^{1/2}} \right) \tag{10}$$

$$\varphi_0 = \arctan\left(\frac{h_{\Phi 1}}{h_{\Phi 2}} \right)$$

with λ = wavelength and subscript '0' for the symmetric position defined by $\omega_0 = 2\theta/2$.

17.4   Strain and stress tensor

Since the strain tensor is defined as a symmetric tensor of second rank, at least six reflections have to be analysed concerning their precise reflection position. According to Bragg's law the refined 2θ-position gives the corresponding $D(\boldsymbol{h})$ value so that the strain in $\boldsymbol{h}$ is obtained by:

$$\varepsilon(\boldsymbol{h}) = \frac{D(\boldsymbol{h}) - D_0(\boldsymbol{h})}{D_0(\boldsymbol{h})} \tag{11}$$

with $D_0(\boldsymbol{h})$ = lattice spacing of the unstressed sample.

For expressing the measurements $\varepsilon(\boldsymbol{h})$ in terms of strain tensor components ε_{kl}^c, the Cartesian crystal axes system is chosen as a fixed reference system:

$$\varepsilon(\boldsymbol{h}) = n_k n_l \varepsilon_{kl}^c \quad k, l = 1, 2, 3 \tag{12}$$

In this notation $\boldsymbol{n}$ is a unit vector parallel to the scattering vector whose components n_k are given in the cubic system by:

$$n_k = \frac{h_k}{\sqrt{h_1^2 + h_2^2 + h_3^2}} \tag{13}$$

For non-orthogonal crystal axes system, an orthogonalisation has to be performed:

$$\varepsilon(\boldsymbol{h}) = p_k, \, p_l \varepsilon_{kl}^p \quad k, l = 1, 2, 3 \tag{14}$$

Here the components p_k are defined by:

$$p_k = \alpha_{kl}^{-1} n_l \quad k, l = 1, 2, 3 \tag{15}$$

α_{kl} are the components of the orthogonalisation matrix. In homogeneous media the symmetric strain tensor ε^c can be determined by six measurements in non-coplanar directions. If more information is available, a least-squares refinement can be applied. The stress tensor components σ_{ij}^c are calculated by applying Hooke's law for anisotropic, quasielastic materials. Here, the single crystal elastic constants are inserted:

$$\sigma_{ij}^c = c_{ijkl} \varepsilon_{kl}^c \quad i, j, k, l = 1, 2, 3 \tag{16}$$

Whereas in the triclinic crystal system 21 independent elastic constants are present, their number is reduced to three independent ones ($c_{1111} \neq c_{2222} \neq c_{1212}$) in the case of the cubic system. The resulting stress tensor σ^c is referred to the Cartesian crystal axes system whose orientation is dependent on the crystal orientation. In most cases, especially for comparing the values obtained in crystals which are in direct neighbourhood to each other, it is preferable to transform the stress tensor into the macroscopic sample system. As the first step, the stress tensor σ^c is transferred into the fixed laboratory system (index L). As transformation matrix the **UB** matrix can be used:

$$\sigma^L = \tilde{\mathbf{T}} \sigma^c \mathbf{T} \tag{17}$$

In the second step the stress tensor σ^L has to be transformed into the common reference system. Therefore, the angles describing the orientation of the reference system relative to the laboratory system have to be determined. The transformation matrix $\mathbf{T}$ is then represented by a rotation matrix $\mathbf{R}$, whose components are given by ω, χ, φ.

17.5 Residual stress analysis using a tapered glass capillary

Capillary optics were introduced as a means for generating intense X-ray microbeams from X-ray tubes and other conventional X-ray sources. Many groups demonstrated the new possibilities opened up by these devices in various fields [4–6]. The use of X-ray capillary focusing systems also has an important impact on single crystal measurements since the intensity gain at a beam size of 100 μm now enables residual stress analyses in single grains of not necessarily large grained samples [7, 8].

The monocapillary used was made of borosilicate glass and had a parabolic profile with an entrance diameter of 500 μm and an exit diameter of 100 μm. Due to the different characteristics of the capillary-formed X-ray beam compared to beams formed with conventional collimators, special attention had to be paid to the adjustment of the set-up and the sample positioning during the experiment.

17.5.1   *Residual stress analysis on a steel sample*

As a first step, it was necessary to validate the results obtained with the X-ray optics. The validations were based on the comparison of results obtained with the capillary using the single crystal measurement technique and results from residual stress analyses using the $\sin^2 \psi$ technique and single crystal measurements performed with a conventional collimator. A small sample of a ferritic–perlitic steel (German steel grade 42CrMo4, sample size $8 \times 13 \times 3$ mm^3) was heat treated at 900°C for 3 h under an argon atmosphere to create a microstructure with an average grain size of 20 μm and was cooled under an argon gas flow. Afterwards the sample was ground, polished and etched in order to reveal the ferritic–pearlitic microstructure (Figure 17.2).

17.5.1.1   *Residual stress analyses using the $\sin^2 \psi$ technique*

In a first step, the in-plane stresses were analysed. The results of the d–$\sin^2 \psi$ measurements are $\sigma_{11} - \sigma_{33} = (50 \pm 10)$ MPa along the longitudinal axis (y-axis) of the sample and $\sigma_{22} - \sigma_{33} = (80 \pm 10)$ MPa along the lateral axis (x-axis). However, for the residual stress analysis it was necessary to take the two-phase microstructure of steel into account, which may create a σ_{33} component if the average particle distance of a phase is smaller than the penetration depth of the radiation used [9].

The evaluation of the σ_{33} component was done by determining the lattice distance d^* of the ferritic phase along the strain-free direction $\sin^2 \psi^*$ of the $\sin^2 \psi$ diagram (Figure 17.3) [10]. This was done in a first approximation with the assumption that $\sigma_{33} = 0$. With this assumption it follows for the strain-free lattice parameter a^*:

$$a^* = 2.8679 \, \text{Å}$$

Figure 17.2 Micrograph of the steel sample.

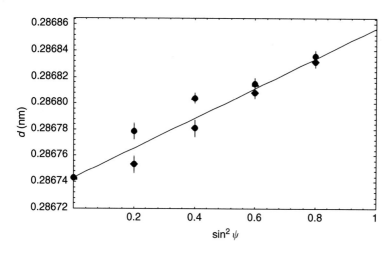

Figure 17.3 d versus $\sin^2\psi$ distribution of a 42CrMo4 steel specimen used for the determination of the strain-free lattice distance.

Accrediting the difference of a^* and the strain-free lattice distance of ferrite ($a = 2.8664\,\text{Å}$) to a non-vanishing σ_{33} component, σ_{33} was determined to be

$$\sigma_{33} = (280 \pm 30)\,\text{MPa}$$

It follows that the phase specific residual stresses for ferrite are

$$\sigma_{11} = (330 \pm 40)\,\text{MPa}$$
$$\sigma_{22} = (330 \pm 40)\,\text{MPa}$$

17.5.1.2   Single grain measurements

The single grain measurements have been performed using a conventional collimator and also using the monocapillary. Both devices formed a microbeam with a diameter of about $100\,\mu\text{m}$. With this beam specification and an average grain size of the sample of $20\,\mu\text{m}$, most of the beam illuminated not only the grain under study, but also a number of grains. However, the beam diameter had to be kept larger than the grain size in order to illuminate single grains completely and as homogeneously as possible.

Due to broad rocking curves of the single grain reflections the absolute peak intensities were low (Figure 17.4). Consequently Fe-Kα radiation was used in order to avoid fluorescence.

The residual stresses of four ferrite grains were analysed. These grains were easy to detect and to locate. Due to the beam characteristics, knowledge of the exact position of the grain under study was necessary.

The stress tensors (sample system) are listed in Table 17.1. The variation of single stress components are in a range of several $100\,\text{MPa}$. Thus, the residual stresses of type II are very large which is one explanation for the observed strong mosaicity. Two stress tensors (Σ_2 and Σ_3) are striking with stress components of 890 and $700\,\text{MPa}$, respectively. Since the yield strength of 42CrMo4 is about $1100\,\text{MPa}$, these stresses are not unrealistic. Furthermore, it

Table 17.1 Comparison of stress tensors determined with the 100 μm capillary (left) and a conventional 100 μm collimator

100 μm capillary

$$\Sigma_1 = \begin{pmatrix} 380 & 30 & -10 \\ & 470 & 0 \\ & & 270 \end{pmatrix} \pm \begin{pmatrix} 30 & 15 & 10 \\ & 30 & 10 \\ & & 30 \end{pmatrix} \text{ MPa}$$

$$\Sigma_2 = \begin{pmatrix} 890 & 190 & 80 \\ & 420 & -30 \\ & & 580 \end{pmatrix} \pm \begin{pmatrix} 50 & 50 & 50 \\ & 50 & 40 \\ & & 50 \end{pmatrix} \text{ MPa}$$

$$\Sigma_3 = \begin{pmatrix} 450 & 80 & -20 \\ & 700 & 90 \\ & & 440 \end{pmatrix} \pm \begin{pmatrix} 70 & 30 & 20 \\ & 50 & 30 \\ & & 60 \end{pmatrix} \text{ MPa}$$

$$\Sigma_4 = \begin{pmatrix} 390 & 0 & 20 \\ & 650 & -20 \\ & & 340 \end{pmatrix} \pm \begin{pmatrix} 60 & 40 & 40 \\ & 50 & 40 \\ & & 60 \end{pmatrix} \text{ MPa}$$

Conventional 100 μm collimator

$$\Sigma_1 = \begin{pmatrix} 410 & 70 & -10 \\ & 510 & 10 \\ & & 310 \end{pmatrix} \pm \begin{pmatrix} 30 & 10 & 15 \\ & 30 & 15 \\ & & 30 \end{pmatrix} \text{ MPa}$$

$$\Sigma_2 = \begin{pmatrix} 950 & 200 & 100 \\ & 510 & -20 \\ & & 450 \end{pmatrix} \pm \begin{pmatrix} 50 & 40 & 50 \\ & 50 & 40 \\ & & 50 \end{pmatrix} \text{ MPa}$$

$$\Sigma_3 = \begin{pmatrix} 540 & 0 & -40 \\ & 800 & 140 \\ & & 420 \end{pmatrix} \pm \begin{pmatrix} 60 & 40 & 30 \\ & 50 & 30 \\ & & 50 \end{pmatrix} \text{ MPa}$$

$$\Sigma_4 = \begin{pmatrix} 320 & -50 & -10 \\ & 540 & -20 \\ & & 220 \end{pmatrix} \pm \begin{pmatrix} 60 & 50 & 40 \\ & 60 & 50 \\ & & 60 \end{pmatrix} \text{ MPa}$$

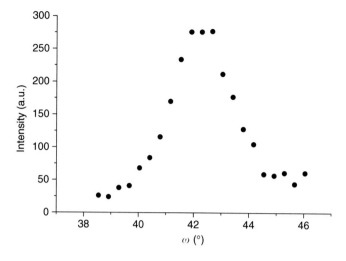

Figure 17.4 Rocking curve of a single grain reflection of ferrite.

is evident that the same stress tensors result, in spite of using different beam collimating techniques.

To describe the residual stresses of type I, the four stress tensors of the capillary have been averaged:

$$\Sigma_{\text{Average}} = \left(\begin{pmatrix} 530 & 80 & 20 \\ & 350 & 10 \\ & & 410 \end{pmatrix} \pm \begin{pmatrix} 210 & 70 & 40 \\ & 120 & 50 \\ & & 120 \end{pmatrix} \right) \text{MPa}$$

Compared with the $\sin^2 \psi$ measurements, the same stress state results from both the single crystal measurements and the measurements on the polycrystal.

17.5.2   Stress analysis on a pure iron bending bar

To describe the macroscopic material behaviour with the anisotropic elastic constants the models of Voigt [11], Reuss [12] and Eshelby–Kröner [13, 14] are commonly used. Thereby xthe models of Voigt and Reuss are limit assumptions based on the idea that each grain shows the same homogeneous strain or stress state as the whole polycrystalline aggregate. Within the scope of the mentioned models the aim of this experiment was to analyse the response behaviour of single grains in a polycrystalline bending bar of pure iron (99.99 + %).

The experiment was performed on a bending bar of pure iron with the dimensions of $40 \times 4.3 \times 3.7$ mm. For creating a microstructure with an average grain size of about 30 μm a heat treatment at 600°C for 6 h under an argon atmosphere was performed. The specimen was cooled under the argon atmosphere as well. In the next step the surface to be studied was ground, polished and etched using the colour etching method of Klemm. By the etching process single crystallites were uncovered, which makes the adjustment of the grain to be studied in the Eulerian cradle easier.

The measurements were performed on four crystallites which were located at a distance of about 1.3 mm from the neutral fibre of the bending bar. The load was applied in several steps ranging from 100 to 400 N corresponding to tensile stresses of 50–200 MPa at the surface of the bar. During the loading process the bar stays elastic, which is a prerequisite of the previously mentioned models, since plastic deformation occurred at a load of 500 N. For the creation of an X-ray microbeam the glass capillary was used.

The unaxial load created by three-point bending corresponded to the σ_{22} component of the stress tensor. The following stress tensors (sample system) of grain number 3 show the stress development with respect to the raised load (the index of the stress tensors denotes the raised load):

$$\Sigma_{100\,N} = \left(\begin{pmatrix} -130 & 20 & 30 \\ & -130 & -10 \\ & & -20 \end{pmatrix} \pm \begin{pmatrix} 30 & 30 & 30 \\ & 40 & 30 \\ & & 30 \end{pmatrix} \right) \text{MPa}$$

$$\Sigma_{200\,N} = \left(\begin{pmatrix} -120 & 30 & 20 \\ & -60 & -20 \\ & & 0 \end{pmatrix} \pm \begin{pmatrix} 40 & 30 & 30 \\ & 30 & 30 \\ & & 30 \end{pmatrix} \right) \text{MPa}$$

$$\Sigma_{300\,N} = \left(\begin{pmatrix} -90 & -10 & 10 \\ & 10 & 0 \\ & & -10 \end{pmatrix} \pm \begin{pmatrix} 20 & 20 & 20 \\ & 20 & 20 \\ & & 20 \end{pmatrix} \right) \text{MPa}$$

$$\Sigma_{400\,N} = \left(\begin{pmatrix} -140 & 10 & -20 \\ & 50 & 10 \\ & & 10 \end{pmatrix} \pm \begin{pmatrix} 50 & 40 & 40 \\ & 50 & 40 \\ & & 40 \end{pmatrix} \right) \text{MPa}$$

Using equation (18), the stress S applied to the bending bar was calculated:

$$S = \frac{3}{2} \frac{Fl}{bh^2} \tag{18}$$

The parameters b and h describe the width and height of the bending bar, l describes the distance of the support points of the bending jig and F the applied load.

For the parameters used in the experiment it follows a stress component of 0.47 MPa/N for the surface under tensile load at the position of the four investigated crystallites.

For the calculation of the stress tensors the elastic constants and the strain-free lattice parameter of pure α-iron were used.

Figure 17.5 shows the stress response of the studied four crystallites in a measured stress vs. applied stress diagram. The internal stress is described by the experimentally obtained σ_{22}-stress component and the external stress is described by the stress which follows from equation (18). Since (in a first approximation) only the slope of each regression line as a measure of the response is of interest, the offsets of the lines were removed. For a better comparison all lines cross the origin.

If each grain responds as Reuss suggested, the slopes of the regression lines would correspond to 1 (MPa/MPa) (dashed line in Figure 17.5). However, due to residual microstresses and the elastic anisotropy of single crystallites, grains number 3 and 4 respond in a different way. The slopes of the regression line of both grains are (1.24 ± 0.11) and (0.7 ± 0.2) MPa/MPa, respectively (see also Table 17.2). Thus, the Reuss model does not describe the material behaviour of each grain. The same is true for the Voigt model (Figure 17.6). However, concerning the average values of the response values from Table 17.2, which represent the macroscopic response of the bar, both the Voigt and the Reuss models describe the material behaviour within the measurement accuracy.

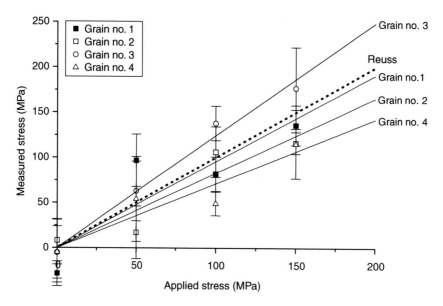

Figure 17.5 Response of the single grains to load/strain development (σ_{22} component approximated by linear regression).

Table 17.2 Experimentally determined response of the σ_{22} and ε_{22} components and the orientation of the soft $\langle 100 \rangle$ direction with respect to the force vector of the load

Grain	Response in σ_{22} (MPa/MPa)	Response in ε_{22} $(10^{-5}/10^{-5})$	Angle of $\langle 100 \rangle$ direction and force vector
1	1.0 ± 0.4	0.9 ± 0.3	49°
2	0.8 ± 0.2	0.8 ± 0.2	63°
3	1.24 ± 0.11	1.18 ± 0.04	36°
4	0.7 ± 0.2	1.0 ± 0.2	45°
Average	0.9 ± 0.2	1.0 ± 0.14	

The slopes in Table 17.2 reflect the elastic anisotropy of the crystallites. Crystallites with soft crystallographic directions (e.g. the $\langle 100 \rangle$ direction), orientated parallel with respect to the force vector of the load, react with large strains and vice versa. Analysing the orientation of the four grains it was found as expected that favourably orientated grains with a small angle between the $\langle 100 \rangle$-direction and the force vector show a large response in their ε_{22} component.

17.5.3 Characterisation of the plastic deformation and residual stress analysis on coronary artery stents

During the placement of metallic coronary artery stents, the material undergoes significant plastic deformations due to the expansion of the stent to the required diameter or due to manual dilating or crimping [15]. The plastic deformation creates residual stresses which

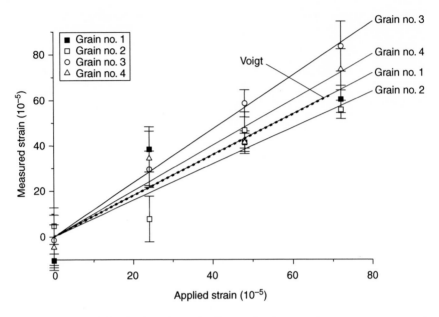

Figure 17.6 Response of the single grains to load/stress development (ε_{22} component approximated by linear regression).

influence the mechanical behaviour of the material and may cause a weakening of the stent. For the characterisation of the plastic deformation, three stents with different degrees of deformation were investigated using single grain reflections which was necessary due to the small size of the stents. Furthermore, residual stress analysis was performed on two stents using the single crystal measuring and evaluation technique. The investigated stents were made of the austenitic steel X2CrNiMo 18 15 3. Depending on the degree of deformation the stents had a length of 10–14 mm and a diameter of 2–4 mm (Figure 17.7). Due to the low surface curvature all measurements were performed at the nodes of the stent design (Figure 17.8). To achieve the required high local resolution maintaining a sufficient X-ray intensity, the formation of an X-ray microbeam with a diameter of 100 μm was done with a tapered capillary.

The first stent studied was non-deformed (as fabricated), the second stent was plastically deformed by dilating and the third stent was plastically deformed by dilating and crimping. The average grain size of the stent material was 20–30 μm.

17.5.3.1   Analysis of plastic deformation

It has been shown elsewhere [8] that the widths of rocking curves are sensitive to plastic deformation processes. Therefore, the widths of single grain reflections were analysed. For the measurements Cr-Kα radiation was used.

Per stent single grain reflections of four different grains were analysed by determining the width of their rocking curves. Figure 17.9 shows selected rocking curves of the three investigated stents. As expected, the effect of reflection broadening depends on the degree of plastic deformation. The quantitative results listed in Table 17.3 confirm this observation.

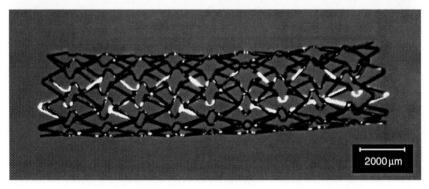

Figure 17.7 Two of the investigated stents. Top: stent without plastic deformation (as fabricated).
Bottom: plastically deformed stent.

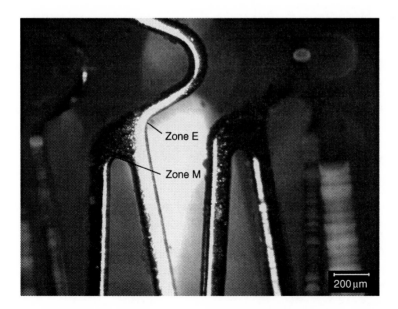

Figure 17.8 Measurement locations. For the plastically deformed stent, 'M' and 'E' denotes zones
where residual stresses were analysed with a high local resolution.

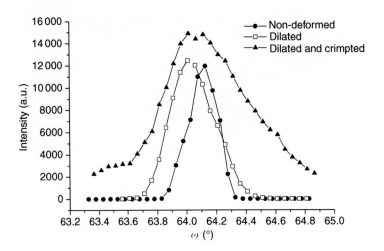

Figure 17.9 Rocking curves of grains of the three investigated stents with different degrees of plastic deformation.

Table 17.3 Widths of single grain reflections

Reflection no.	As-fabricated reflection width (°)	Dilated reflection width (°)	Dilated + crimped reflection width (°)
1	0.19	0.29	0.28
2	0.19	0.23	0.53
3	0.23	0.53	0.38
4	0.22	0.31	0.53
Average	0.21 ± 0.0	0.25 ± 0.04	0.43 ± 0.11

The average reflection width (last row in Table 17.3) of the four reflections studied ascends with the degree of plastic deformation. In addition, by comparing the standard deviation of each stent it becomes evident that the scattering of the widths is also related to the degree of plastic deformation.

17.5.3.2   *Residual stress analysis*

For the measurements an X-ray microbeam was created using the 100 μm capillary. The radiation used was Cu-Kα. Fluorescence was negligible due to the high intensity of the single grain reflections.

The measurements on the non-deformed stent were performed in the middle of the stent nodes (see Figure 17.7). For the dilated and crimped stent, the measurements were performed at two different zones of one of the stent nodes dubbed 'M' (middle zone, see Figure 17.8) and 'E' (edge zone).

The strain-free lattice parameter a_0 was determined with the tensor data from three studied grains of the non-deformed stent number 1. Under the assumption that the tensor component $\sigma_{33} = 0$ MPa (given in the sample system), a grain-specific strain-free lattice parameter for each of the grains studied was calculated. The average of these values was $\bar{a}_0 = (3.5978 \pm 0.0009)$ Å. Using this a_0 the following stress tensors of three grains of the non-deformed stent are (sample system):

$$\text{Grain no. 1:} \quad \Sigma_1^M = \left(\begin{pmatrix} -30 & 0 & -20 \\ & 20 & 10 \\ & & -160 \end{pmatrix} \pm \begin{pmatrix} 15 & 10 & 10 \\ & 15 & 10 \\ & & 10 \end{pmatrix} \right) \text{MPa}$$

$$\text{Grain no. 2:} \quad \Sigma_2^M = \left(\begin{pmatrix} -40 & 10 & 20 \\ & 10 & 30 \\ & & -30 \end{pmatrix} \pm \begin{pmatrix} 40 & 20 & 10 \\ & 40 & 10 \\ & & 30 \end{pmatrix} \right) \text{MPa}$$

$$\text{Grain no. 3:} \quad \Sigma_3^M = \left(\begin{pmatrix} 170 & 20 & 0 \\ & 120 & 20 \\ & & 170 \end{pmatrix} \pm \begin{pmatrix} 30 & 20 & 20 \\ & 20 & 20 \\ & & 20 \end{pmatrix} \right) \text{MPa}$$

The analysis shows that grains number 1 and 2 have only small residual stresses in the range of error. Only the σ_{33} component of grain number 1 shows a significant compressive stress of -160 MPa. Grain number 3 shows large hydrostatic tensile stresses up to 170 MPa. Thus, it becomes evident that single grains may have significant stresses, though other grains show almost no stresses.

Residual stress analysis was also performed on five grains of stent number 3. Using the strain-free lattice parameter a_0 for the determination of the stress tensors, these are (sample system, the index denotes the studied zone):

$$\text{Grain no. 1:} \quad \Sigma_1^M = \left(\begin{pmatrix} 180 & 0 & -10 \\ & 190 & 10 \\ & & 120 \end{pmatrix} \pm \begin{pmatrix} 30 & 30 & 20 \\ & 30 & 20 \\ & & 30 \end{pmatrix} \right) \text{MPa}$$

$$\text{Grain no. 2:} \quad \Sigma_2^M = \left(\begin{pmatrix} 90 & 20 & -30 \\ & 60 & -20 \\ & & 160 \end{pmatrix} \pm \begin{pmatrix} 30 & 20 & 30 \\ & 20 & 20 \\ & & 20 \end{pmatrix} \right) \text{MPa}$$

$$\text{Grain no. 3:} \quad \Sigma_3^M = \left(\begin{pmatrix} 360 & 30 & 110 \\ & 190 & 40 \\ & & 290 \end{pmatrix} \pm \begin{pmatrix} 50 & 30 & 10 \\ & 40 & 20 \\ & & 40 \end{pmatrix} \right) \text{MPa}$$

It can be observed that in zone 'M' of the studied node are tensile stresses ranging up to 360 MPa. The main components lie significantly higher than those of the non-deformed stent.

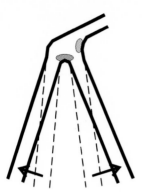

Figure 17.10 Due to the plastic deformation of the stent, zones under tensile and compressive load formed tensile and compressive residual stresses, respectively.

The same behaviour with the opposite sign can be observed in the other zone 'E' at the edge of the node:

$$\text{Grain no. 4:} \quad \Sigma_4^E = \left(\begin{pmatrix} -120 & 30 & 10 \\ & -40 & 10 \\ & & -190 \end{pmatrix} \pm \begin{pmatrix} 40 & 30 & 40 \\ & 40 & 40 \\ & & 40 \end{pmatrix} \right) \text{MPa}$$

$$\text{Grain no. 5:} \quad \Sigma_5^E = \left(\begin{pmatrix} -20 & -80 & 30 \\ & -90 & 40 \\ & & -130 \end{pmatrix} \pm \begin{pmatrix} 20 & 20 & 20 \\ & 20 & 20 \\ & & 20 \end{pmatrix} \right) \text{MPa}$$

The two stress tensors show compressive stresses down to -190 MPa. A simple model which takes the deformation of the stent after its plastic deformation into account is able to explain the results. The deformation is shown in Figure 17.10. Due to this change most of the plastic deformation took place at the nodes. While in zone 'M' the straddling of the two lower stent wires led to a plastic deformation of the node material, under tensile load, the same happened in zone 'E' under compressive load.

17.6   Summary and future prospects

Residual stresses in polycrystalline materials with grains in the range of micrometres and with complex geometries can be analysed using X-ray diffraction in combination with the single crystal measuring and evaluation technique. In the case of coarse grained materials with grain sizes in the millimetre-range, neutron diffraction has already been successfully applied in the past [8]. The stress values obtained by the single crystal measuring and evaluation technique represent the sum of type I and type II stresses of first and second kind. If only the stresses of type I are of interest, measurements on several differently orientated crystallites with a subsequent averaging over the individual stress states is necessary. Here the determination of the different orientation matrices is the time-determining step. An important speed up of this procedure can be expected from the use of two-dimensional position sensitive detectors now available. Angular ranges are covered at appropriate sample to detector distances which allow the simultaneous registration of two reflections, thus enabling a direct determination of the orientation matrix. A wider impact for the development of the single crystal measuring and evaluation technique would be possible by using brilliant synchrotron sources.

References

[1] Dölle, H.; Hauk, V.: Gitterdehnungen in grobkörnigen kubischen Werkstoffen (in German), *Z. Metallkde.*, **71** (1980), 708–713.

[2] Crostack, H.-A.; Reimers, W.: X-ray diffraction analysis of residual stresses in coarse grained materials, in: *Residual Stresses in Science and Technology*, eds.: E. Macherauch, V. Hauk, DGM-Informationsges. Verlag, Oberursel (1987), 289–294.

[3] Busing, W. R.; Levy, H. A.: Angle calculations for 3- and 4-circle X-ray and neutron diffractometers, *Acta Crystallogr.*, **22** (1967), 457–464.

[4] Bilderback, D. H.; Thiel, D. J.: Microbeam generation with capillary optics, *Rev. Sci. Instrum.* **66** (1995), 2059–2063.

[5] Engström, P.; Larson, S.; Rindby, A.; Stocklassa, B.: A 200 μm X-ray microbeam spectrometer, *Nucl. Instrum. Methods* **B26** (1989), 222–226.

[6] Cargill III, G. S.; Hwang, K.; Lam, J. W. *et al.*: Simulations and experiments on capillary optics for X-ray microbeams, *Proc. SPIE Conf.: X-ray Microbeam Techn. and Applications*, San Diego, **2516** (1995), 120–134.

[7] Reimers, W.; Crostack, H.-A.; Wrobel, M.; Eckold, G.: Investigations of large grained samples – examples, in: *Measurement of Residual and Applied Stress using Neutron Diffraction*, eds.: Hutchings, M. T.; Krawitz, A. D., Kluwer Academic Publishers, Dordrecht (1992), 263–276.

[8] Reimers, W.: Analysis of residual stress states in coarse grained and single crystal nickel-base superalloys, in: *Advances in X-Ray Analysis*, eds.: Gilfrich *et al.*, Plenum Press, New York (1997), 211–223

[9] Ruppersberg, H.: Stress fields in the surface region of pearlite, *Mater. Sci. Eng. A* (1997).

[10] Hauk, V.: Die Bestimmung der Spannungskomponente in Dickenrichtung und der Gitterkonstante des spannungsfreien Zustandes (in German), *Härtereitechn. Mitt.*, **46** (1991), 52–59.

[11] Voigt, W: *Lehrbuch der Kristallphysik*, B. G. Teubner, Leipzig, Berlin (1928).

[12] Reuss, A.: Berechnung der Fließgrenze von Mischkristallen auf Grund der plastizitätsbeding-ungen für Einkristalle, *Z. Angew. Math. Mech.* **9** (1929), 49–58.

[13] Eshelby, J. D.: The determination of the elastic field of an ellipsoidal inclusion and related problems, *Proc. Roy. Soc. (London)*, **A241** (1957), 376–396.

[14] Kröner, E.: Berechnung der elastischen Konstanten des Vielkristalls aus den Konstanten des Einkristalls (in German), *Z. Phys.* **151** (1958), 504–518.

[15] Denk, A.; Erbel, R.; Fischer, A. *et al.*: Untersuchungen zum mechanischen Aufweitungsver-halten von marktgängigen koronaren Stentsystemen, in: *Mechanische Eigenschaften von Implantatwerkstoffen, 1. Tagung des DVM-Arbeitskreises Biowerkstoff* (1998), 141–146.

18 Neutron residual stress measurement in welds

S. Spooner

18.1 Introduction

Residual stresses in welds have proven to be an interesting problem for neutron residual stress analysis. Residual stresses are implicated in weld cracking and premature weld failure. Lifetime performance is compromised by the limits of the weld material and quality of the joint. Therefore it is necessary to understand the weld and its manufacture to achieve engineering reliability. A weld joint is full of composition and stress gradients that are the result of a complex thermal history. This means that the strain measurements and their conversion to stress can be complicated. At the same time we would like to make these measurements at appropriate spatial and time resolution.

What does neutron scattering bring to this task? Neutron diffraction is one of several ways to analyse residual stresses in welds. One can use hole drilling with strain gauges, conventional X-ray diffraction, or ultrasonic examination, to name methods in common use. With the exception of ultrasonic examination, these foregoing methods are limited to the evaluation of strain at the surface of welded structures. Neutrons, however, penetrate large depths into a weld non-destructively. The weld structure can be evaluated with spatial resolution less than a millimeter to a depth of many millimeters below the weld surface. With a simple application of neutron residual stress measurement, one can determine the degree and spatial distribution of stress relief following post-weld heat treatment.

In this chapter, we will take a critical look at the method of neutron residual stress measurement. The description of some examples will expose some limitations of the method, especially in relation to weld modeling. Both steady state and pulsed neutron sources are used for residual stress analysis in welds. Steady state neutron sources have been used since the early 1980s and pulsed neutron sources since the 1990s. The examples in this chapter are based on work done at the High Flux Isotope Reactor (HFIR) at Oak Ridge National Laboratory and the NRU reactor at Chalk River. Pulsed source methods are complementary to steady state methods and in fact are essential to overcome some of the limitations of steady state work.

18.2 Neutron residual stress measurements in welds

18.2.1 Composition effects in welds

Fusion welding entails the melting and solidification of a portion of base metal and, in most welds, a filler material introduced in the welding process. The metal in the heat-affected zone (HAZ) experiences a heating transient with each welding pass. While the base metal in the HAZ does not melt, material near the fusion zone approaches the melting point. Although the heating transient is short in duration, there is enough heat to alter the metallurgical condition of the metal. The state of precipitates, local cold work and texture can be changed. In a multiphase material, the heating transient can alter the proportion of the constituent phases. The change in lattice parameter through heating appears as a change in residual strain in weld zone and the HAZ. In ferritic steels this variation is small because the solubility range of carbon in BCC ferrite is limited. In heat treatable aluminium alloys dissolution of precipitates in the HAZ causes a significant change in lattice parameter. The lattice parameter change has to be evaluated separately in zero stress reference material cut from the weld. In these ways composition and mechanical effects may cause d-spacing changes.

18.2.2 Some selected examples of neutron scattering analysis of welding

At the time of planning for residual stress research at Oak Ridge (Figure 18.1 shows the schematic of the Oak Ridge residual stress diffractometer), there appeared the work of Mahin at Sandia National Laboratories with Holden at Chalk River. This was an early effort to validate finite element calculations of residual stress in welds [1, 2] with neutron diffraction. An illustrative test case was that of a partially penetrating stationary gas tungsten arc (GTA) weld in a 304L stainless steel plate. Careful preparation of the weld and measurement of the welding conditions were necessary to assure that the boundary conditions going into

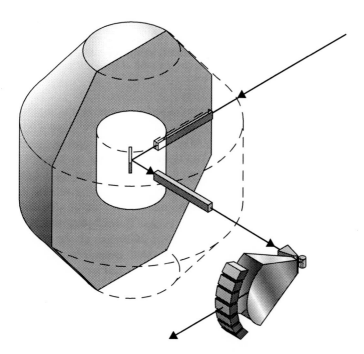

Figure 18.1 This schematic representation of the Oak Ridge residual stress diffractometer shows the reactor beam impinging on the crystal monochromator and the monochromatic beam hitting a test specimen. The beam is diffracted into a fixed set of seven detectors that independently record the Bragg diffracted intensity. Small apertures set close to the test specimen define the gauge volume. The converging structure between the detectors and specimen is a background suppressing collimator.

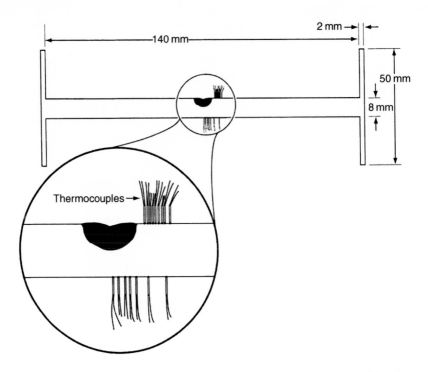

Figure 18.2 The schematic shows a section through the cylindrical weld fixture and the placement of thermocouples. The partial penetration weld is at the centre of the 8 mm plate. The plate is held rigidly within the 50 mm cylindrical boss.

the finite element method (FEM) calculation accurately defined the experimental conditions. The starting material was a plate 50 mm thick. The plate was machined to form a 75 mm radius disk with 8 mm thickness. A 2 mm thick ring 50 mm high was left to act as a restraint and support for the disk as shown in Figure 18.2. The weld was made at the centre of the 8 mm plate with a GTA welder. The arc was struck and held for about 5 s to form a partial penetrating weld pool. The temperature was measured with thermocouples welded to the 8 mm disk. The temperature measurements were used to validate the thermal calculations. The calculated residual stresses were obtained from a coupled thermal–mechanical finite element code (PASTA2D). This two-dimensional code takes advantage of the cylindrical symmetry of the weld geometry to calculate values of the residual stress tensor throughout the volume of the test piece including the fusion zone. Axial, hoop and normal strains were calculated from the FEM code.

The strains along the principal axes were measured by neutron diffraction using the 111 and 200 *d*-spacings. The *d*-spacing shifts were referenced to measurements made in a specimen containing no weld. In stainless steel, the elastic constant is softest for [200] and the hardest for the [111] direction. The two strain measurements would therefore be expected to bracket the computed strains where bulk values for Young's modulus and Poisson's ratio are used.

The gauge volume was defined by an incident beam 1.5 mm wide and 2 mm high and a 1.5 mm wide receiving aperture giving a nominal gauge volume of 4.5 mm^3. The strains were measured on a plane normal to the plate passing through the centre of the fusion zone. A comparison between computation and experiment is shown in Figure 18.3. The

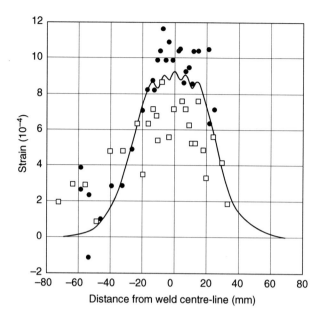

Figure 18.3 The experimental radial strains determined from the 200 Bragg diffraction peak positions are shown in filled circles and from the 111 Bragg peak in open squares. The solid line shows the longitudinal strain calculated by the finite element method. The scatter in the neutron experimental data is typical of data from large grained samples. The 200 strains are larger than the 111 strains because of the difference between the crystallographic elastic constants; the 111 direction being stiffer than the 200 direction. The average of the two strains corresponds closely with the calculated strains where bulk elastic constants are used.

computation is shown as a solid line and, indeed, it is nicely bracketed by the measured strains. The scatter in the neutron strains is as much as 2×10^{-4}. This is about twice the statistical precision estimated from neutron counting statistics. The grain size (150–$200\,\mu$m) appears to be responsible for this variation.

The point in this illustration is that very good validation of the calculation was possible despite the neutron data scatter. The use of neutron scattering validation of calculations was given a early demonstration. Detailed experiment design and strategy is critical for successful validation work.

18.3 Vibratory stress relief

Vibratory treatment is marketed as one of several means of reducing residual stresses in welds. Anecdotal evidence is offered in support of its effectiveness and until the early 1990s had not been evaluated directly by neutron residual stress measurement. A test of vibratory stress relief effects on multiple-pass welds was attempted in one-inch 304 L austenitic stainless steel plates [3]. The test welds were made with a semiautomatic hot-wire gas tungsten arc welding unit. The test welds were made with 14 passes in a 60° V-groove using a type 308 stainless steel filler rod. A 3 mm thick backing plate was tack welded to the bottom of the weld to provide a 6 mm spacing between the plates at the bottom of the grove. The metal plates were clamped then relaxed between weld passes. After the final pass the two plates were essentially coplanar. The distortion generated in the weld of this thickness, if not relieved after each pass, would tear the weld apart before the weld could be completed. The neutron measurements were done at Chalk River, Canada. The diffraction intensities from these thick plates were reduced significantly by beam attenuation and the gauge volumes used were necessarily rather large. The longitudinal strain component was measured with a 5×5 mm^2 incident beam and a 5 mm receiving aperture. The normal and transverse strain components were measured with a 3 mm wide by 10 mm high incident beam and the receiving slit was 3 mm wide. The (111) and (200) Bragg peaks were measured at each test point with a 2.43 Å wavelength from the (311) planes of a Ge monochromator. The stress-free reference sample was a 12 mm cylinder machined from the base metal taken from a remote corner of the plate. The (111) and (200) strains were averaged and put into the continuum elastic equation to convert strain to stress using the bulk Young's modulus and Poisson's ratio.

The location of test points is shown schematically in Figure 18.4(a). Stresses in the fusion zone are not given because zero stress references in the fusion zone had not been determined. The first plate was welded without vibratory treatment. Figure 18.4(b) shows the longitudinal residual stresses in this plate. The second plate was given vibratory treatment during the welding; these stresses are shown in Figure 18.4(c). Differences between the two stress plots are within the estimated errors of the stress determination. Nevertheless, the differences in residual stresses between the two welds were probably comparable to differences between identically prepared welds. A third plate had been welded and then treated after the weld was completed. The few strain measurements made in this plate did not significantly differ from the corresponding strains in the other plates.

18.4   Strain mapping in ferritic and austenitic welds

The general pattern of residual stresses in welded plates has been known for many years. The stresses and strains parallel to the weld or in the longitudinal direction are strongly tensile near the fusion zone and become compressive away from the HAZ. The transverse strains in the plane of the plate and perpendicular to the weld are smaller in amplitude and the normal strains perpendicular to the plate are small and reversed, reflecting a Poisson response to the large in-plane stresses. Detailed mapping of residual stress within a plate was relatively novel in the work done at Oak Ridge. A comparison between strains found in plates of two types of steel is discussed in the following.

Residual stress measurement at Oak Ridge was undertaken first on a multiple-pass weld joining two 0.500 inch thick plates of ferritic steel [4] and then a measurement on a stainless steel weld of the same geometry [5]. Both plates were joined using semiautomatic gas tungsten arc welding. The ferritic alloy was a bainitic 2.25Cr–1Mo steel. A closely matched alloy wire was used for the filler metal. The austenitic alloy was 304L stainless steel with a 308 stainless steel wire for the filler metal.

The plates were 152 mm (6 inch) by 305 mm (12 inch) by 12.7 mm (0.5 inch) thick. The plates were chamfered to form a single 90° V-groove. The plates were held together during welding in a "tent-like" configuration so that the ensuing shape distortion would give a flat welded plate. The ferritic weld was completed in six passes and the austenitic weld required 11 passes.

The ferritic steel measurements were set up on the HB-3 triple-axis spectrometer and the Be (102) reflection was used to give 1.65 Å at a monochromating angle of 77°. Strains were measured with the (211) Bragg reflection alone. The diffraction angle was very close to 90°

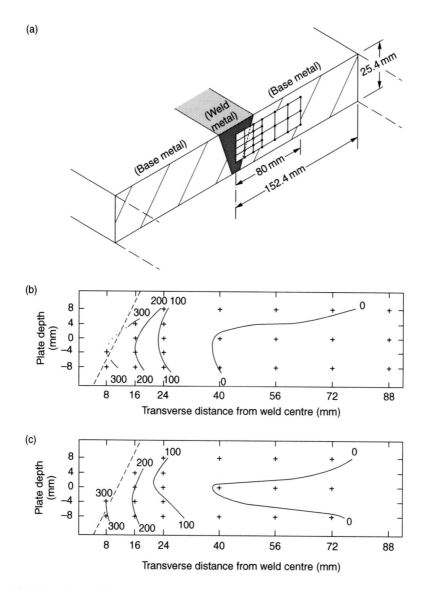

Figure 18.4 The schematic in panel (a) shows the pattern of test points in the weld. Panels (b) and (c) show the residual stress distribution in the weld with no vibratory stress relief treatment and with vibratory stress relief. The magnitudes and distribution of the residual stresses are very similar and differences fall within the experimental uncertainties of the measurement.

scattering angle. This diffraction elastic constant is close to the bulk Young's modulus. The microstructure was bainitic consisting of ferrite and very fine carbide phases. There was a minor texture within the plate and the grain size was less than 10 μm.

HB-3 was also used for the austenitic plate measurements but with the Be (102) monochromating plane to reflect 1.65 Å at a monochromating angle of 77°. The strains were measured with the (311) Bragg reflection at about 86°.

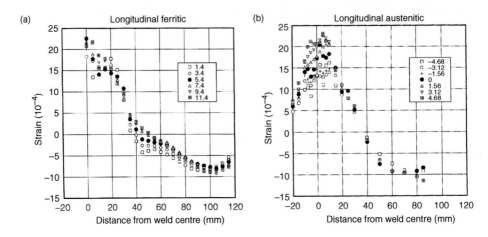

Figure 18.5 The longitudinal strains measured in the 0.5 inch ferritic steel (left) and the austenitic steel (right) plates have similar magnitudes. The filled data points strains are measured at the centre of the plates. The open points are strains measured above and below the plate centre. Measurements in the ferritic plate start at the centre of the weld zone. The measurements in the austenitic plate include both sides of the weld zone centre. A shoulder in the strain at 20 mm from the weld centre in the ferritic plate is not seen in the austenitic weld. The shoulder is attributed to the effects of the bainitic phase transformation in the ferritic steel.

Both the ferritic and austenitic plates were mounted with the weld axis vertical for the measurements of the normal and transverse strains (Figures 18.5–18.7). A long vertical gauge volume used in the ferritic plate measurements was 20 mm high with a 2 mm width in the incident and scattering apertures. The plate was remounted with the weld axis horizontal and the height of the incident beam was reduced to 4 mm to gain resolution transverse to the weld line. The gauge volume used in the austenitic plate was defined by an incident beam 10 mm high and 2 mm wide. The receiving slit was 2 mm wide. The height of the gauge volume used in the longitudinal strain measurements was reduced to 2 mm.

A difference between the two materials appears in a subsidiary peak in the longitudinal strain in the ferritic weld. This shoulder in the longitudinal strain is attributed to the effects of the bainite phase transformation. The base metal in the HAZ transforms to austenite and then retransforms to bainite upon cooling with a significant transformation strain. The strain peak is strongest in the top where the last weld pass is made. The longitudinal strain in the HAZ remains at the nominal yield (0.2%) which suggests that the material has experienced plastic deformation.

18.4.1   Gleeble machine simulation

The study of the ferritic weld was supplemented by an examination of strains in simulated welds made by the Gleeble process [6]. The Gleeble machine heats test bars with very high currents that are clamped at each end by water-cooled copper jaws. The maximum temperature is reached midway between the jaws and thus simulates the HAZ of a weld. The thermal history along the bar is controlled by the balance between the heating of the electrical current and the cooling at the ends of the bar. The bars were 108 mm long with a square cross-section 12.7 mm on a side. The 2.25 Cr–1 Mo alloy test bars were machined from the same stock

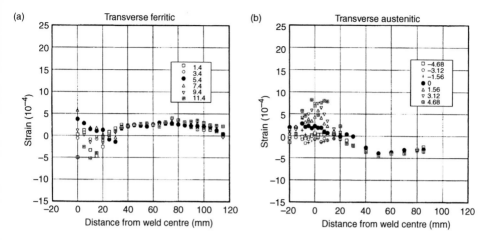

Figure 18.6 The transverse strains are small in both welds with the strains being slightly tensile away from the weld centre in the ferritic plate and slightly compressive in the austenitic plate. The strains at the centre in the ferritic weld zone (closed circles) are tensile while the strains above and below the centre are relatively compressive. In the austenitic plate, the strains above the centre are tensile and below the centre are slightly compressive. Thus there is a peaking of transverse tensile strain in the ferritic weld while there is a monotonic gradient in the transverse strain in the austenitic weld.

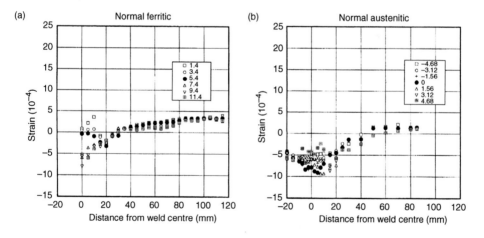

Figure 18.7 The normal strains are relatively small in both welds but compressive strains are present near the weld zone in the austenitic plate.

as the base metal in the above study of ferritic steel. Three bars were prepared. The first bar was heated to 950°C, held 5 s and rapidly air cooled; the second heated to 905°C, held for 60 s and cooled slowly at 100°C/min; and the third bar was heated to 950°C, held for 60 s and rapidly air cooled. The rapid cooling was estimated to be 2500°C/s.

The strains along the length of the bar (longitudinal) and the transverse strains were measured along the centre of the bar. An incident beam 2 mm wide and 2 mm high defined the gauge volume. The receiving slit was 2 mm wide.

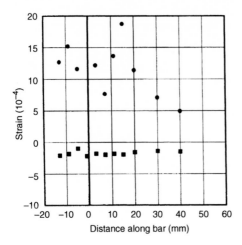

Figure 18.8 The strains along the axis of the Gleeble bar are shown as filled circles and the strain transverse to the axis are shown as filled squares. The heated zone was held at 960°C for 5 s and then air-cooled at a rate exceeding 1000°C/s. The transverse strains are mildly compressive while the longitudinal strains peak at 19×10^{-4} at 15 mm from the centre of the heat zone.

The largest longitudinal strain (exceeds 0.2% strain) occurs in the bar that was held at 960°C for 5 s and rapidly cooled, (as shown in Figure 18.8). The same stresses are reduced markedly (less than 6×10^{-4}) for the bars that were held at 960°C for 60 s in both slow (in Figure 18.9) and fast cooling (in Figure 18.10). In these last two bars transverse strains were compressive, with the fast cooled bar exhibiting a relatively sharp increase in compression at the centre of the weld. What is demonstrated in these tests is that both dwell time at 960°C and cooling rate affect the resulting residual strains. Because the bainitic phase transformation is the dominant source of differences between these Gleeble bars, the possibility of chemical composition variation was checked in pieces cut from the test bars. The maximum to minimum lattice parameter change was equivalent to about 2×10^{-4} and thus comparable to the estimated error of determination of the strains.

The microstructures generated in the Gleeble bars follow the variation found in the weld microstructure along a path transverse to the length of the weld. However, the orientation of the principal axes of the strain tensor found in the Gleeble bar differs significantly from the principal axes found in the weld. The tensile strains are parallel to the direction of heat flow along the length of the Gleeble bar. This direction is the transverse rather than the longitudinal direction in a weld. This outcome is the result of very different mechanical constraints and thermal boundary conditions in the weld and in the Gleeble bar. These effects are important in bainitic steel where there is significant interplay between transformation strains and residual stress.

18.4.2   Reduction of residual stress with post-weld heat treatment

The effects of post-weld heat treatment (PWHT) on the 304L stainless steel plate described earlier were measured following an annealing at 1150°F for 1 h and air-cooling [5]. The incident beam 10 mm high and 2 mm wide defined the gauge volume for the measurement of

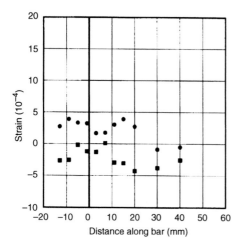

Figure 18.9 The Gleeble sample was held for 1 min at 960°C and given a controlled cooling at 100°C/s. There is more structure in the transverse strains with higher compressive strains while the longitudinal strains are reduced in amplitude compared to Figure 18.8.

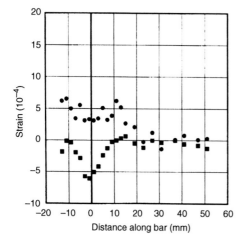

Figure 18.10 The Gleeble sample was held for 1 min at 960°C and air-cooled. The transverse strains show a relatively sharp compressive peak and somewhat higher tensile longitudinal strains. The longer hold time at 960°C permits a higher degree of austenite formation in the hot zone. The subsequent bainite formation thus provides a larger influence on the resulting strain in the Gleeble bar.

the normal and transverse strain components. The longitudinal strain was measured with the 2 mm high gauge volume. The test points were closely spaced in the fusion zone and the HAZ. In the base metal away from the weld, only three depths below the surface were measured and the spacing between the test points was large. Because there was little through-depth variation, the residual strains for each principal axis were averaged and are shown in Figures 18.11 and 18.12. The as-welded specimen shows the longitudinal and transverse stresses in the

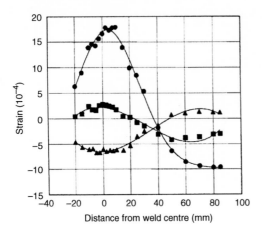

Figure 18.11 Strains are shown for the post-weld heat-treated weld. The fusion zone stresses are tensile and then become compressive beyond about 38 mm from the weld centre. The longitudinal stress is largest followed by transverse and normal stresses.

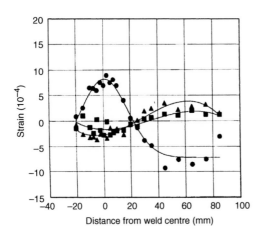

Figure 18.12 Strains are shown for the post-weld heat-treated weld. The normal and transverse stresses have gone to zero while the magnitude of the longitudinal stresses is reduced by a factor of 3 after heat treatment.

fusion zone are tensile and become compressive beyond about 38 mm from the weld centre. The normal strains are compressive at the fusion zone and go tensile. The effect of PWHT, shown in Figure 18.12, is to reduce the longitudinal strain, and the normal and transverse strains become virtually identical, being compressive in the fusion zone. The normal and transverse stresses have almost gone to zero, while the magnitude of the longitudinal stresses is reduced by a factor of 3 after heat treatment.

Contour maps (not shown) of the stresses for the as-welded plate show that the highest stresses occur at the last welding pass. After PWHT the peak stresses "retreat" into the interior of the weld zone. A surface measurement of residual stress therefore probably overestimates the actual stress relief achieved by PWHT.

18.5 Metal–ceramic brazing

Brazing used in forming a ceramic–metal joint develops residual stresses owing to the thermal expansion mismatch between the metal and ceramic parts. If this mismatch is not relieved, joint failure by fracture is a near certainty. Understanding of the residual stresses within these joints is of critical importance. Wang and coworkers at Oak Ridge National Laboratory and Rabin and Williamson at Idaho Engineering National Laboratory [7] carried out detailed strain scanning and finite element calculations for a brazed zirconia–iron joint. A similar study preceded this work [8] and was done at Saclay, France. In both studies the objective was to validate calculations of residual stress and strain. The strain field is cylindrically symmetric and so two-dimensional computations suffice. The anticipated strain fields could be calculated from the elastic response to thermal expansion were it not for the effects of plastic deformation in the metal side of the joint.

The sample was made of a nodular cast-iron cylinder 23 mm in diameter and 10 mm long joined to a 4 mm disk of partially stabilized zirconia of the same diameter. The brazing alloy, Ti–4.5Cu–26.7Ag, was 0.08 mm thick. The zirconia as received consisted of 41% cubic phase, 55% tetragonal and 4% monoclinic phase. The brazing was done in vacuum and the heating to 850°C was done at 20°C/min, held at temperature for 20 min and cooled at 10°C/min. Figure 18.13 shows the sample and the location of measured test points.

A relatively small gauge volume was defined by an incident beam 1.5 mm wide and 2 mm high. The receiving slit was 1.5 mm wide. Strains in iron were measured with the (103) Be monochromator at 90°($\lambda = 1.44$ Å) with a single position-sensitive detector and using the (211) reflection for iron. For zirconia, the (004) pyrolytic graphite monochromator at 97°($\lambda = 2.51$ Å) was used so that a closely grouped set of reflections from zirconia appearing near 90° could be used for strain measurement. Making strain measurements in zirconia was difficult because the precision of determination of the Bragg peak shifts was low. The strains

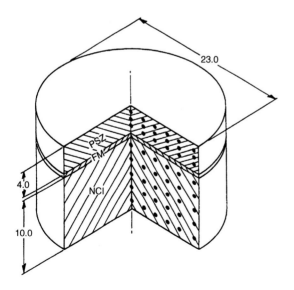

Figure 18.13 The brazing sample is made up from a 4 mm disk of partially stabilized zirconia, a thin foil of silver brazing alloy (4.5Ti–26.7Cu–bal.Ag), and a 10 mm cylinder of nodular cast iron. The location of measurements is indicated on the sectioning plane.

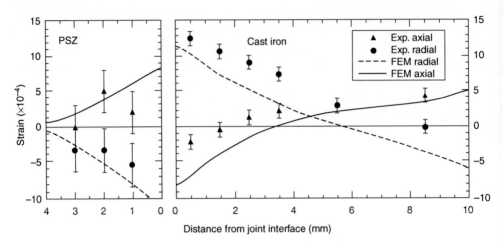

Figure 18.14 The measured strain distributions and the results of a finite element analysis along the centre-line of the specimen. The experimental data are indicated by the solid symbols. The lines are the FEM calculations.

in zirconia and iron are shown in Figure 18.14 along with the computed results. There is adequate agreement in zirconia and, apart from a small uniform shift between the neutron strains and calculated strains, a good agreement in iron. In zirconia, the radial strain is compressive at the interface while the axial strain is tensile. This is balanced at the interface by the strains in iron where the axial strain is tensile and the radial is compressive. The strains in iron deviate a little from linearity as might be expected from only a small plastic relaxation. It would appear that most of the plastic deformation occurred in the brazing layer. Detailed validation of the calculation was limited by the spatial resolution of the neutron measurements, although large-scale agreement is apparent.

18.6   Aluminium welds

The study of Weldalite, a lithium-containing aluminium alloy, was our first experience with neutron stress analysis applied to an aluminium weld. NASA, seeking to increase mission payload by reducing the weight of fuel tanks, had to develop welding technology for joining Weldalite, which has a high strength-to-weight ratio [9]. Achieving reliable and repairable welds in a new alloy is a daunting engineering challenge. Early welding trials indicated that weld cracking needed to be brought under better control. One of the factors under consideration was residual stress in the weld. In the end, X-ray residual stress analysis was deemed adequate for evaluating the process modifications needed to solve the welding problems. In the mean time, aluminium weld test panels were simultaneously examined with neutron diffraction. A Weldalite test panel made from 8 mm thick cold-rolled sheet was welded with a variable polarity plasma arc in two passes; first an autogenous weld followed with a cover pass with an appropriate filler metal. Since test panels were to undergo subsequent mechanical testing there could be no destructive examination of the material after the neutron experiments.

Longitudinal, transverse and normal strains were measured at the mid-length section of the weld. Transverse and normal strains were measured with the weld mounted vertically and a gauge volume defined by a 20 mm high and 1 mm wide incident beam with a 1 mm wide receiving slit. The longitudinal strain measurement required remounting the test panel with the weld line horizontal. The gauge volume for these strain measurements was defined by a 5 mm high and 2 mm wide incident beam with a 2 mm wide receiving slit. The (311) Bragg reflection was used with 1.61 Å beam from (103) Be monochromator crystal planes. A single position-sensitive detector was used.

It was discovered that the diffracted intensity was strongly dependent on the location in the panel. A sharp change in diffracted intensity and diffraction peak width occurs upon entering the HAZ. This indicates clearly that annealing in the HAZ changes the metallurgical condition of the material. An important consequence was that the measurement of the normal strain component in the HAZ was very difficult. There was, as well, a strong through-thickness intensity variation due to texture in locations outside the HAZ.

In Figure 18.15 the longitudinal stress peaks at the edge of the HAZ and all of the stresses go compressive upon approach to the weld centre. This behaviour was very different from the surface X-ray residual stress results. This behaviour signaled clearly that the neutron data had to be corrected for composition variation. At the time, we did not have the liberty of removing a section of the weld to evaluate zero-stress *d*-spacings for true strain measurements. We eventually proceeded on the assumption that the weld was in a plane strain state normally appropriate for a thin plate. Therefore, we allowed a free adjustment of a *d*-spacing correction on the strains until normal stress calculated from the strains at each point went to zero (Figure 18.16). The *d*-spacing shift is plotted as a "chemical shift" in Figure 18.17. This approach to correcting the residual strains brought the neutron longitudinal stresses into substantial agreement with X-ray results.

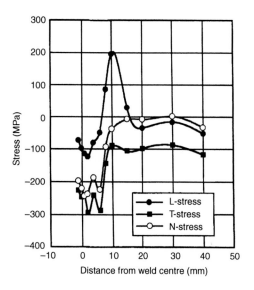

Figure 18.15 The residual stresses calculated from strains referenced to a Bragg angle measured as a point distant from the weld apparently become compressive in the HAZ and weld zone.

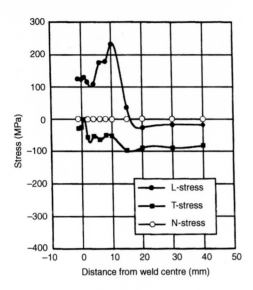

Figure 18.16 Residual stresses are modified by applying an adjustment to the strains so that the resultant normal stress is zero. The modification brings the residual stresses into good agreement with X-ray residual stress measurements and stresses calculated by finite element methods.

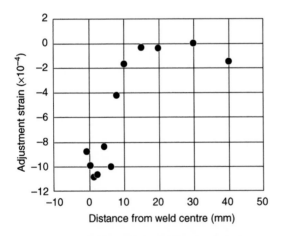

Figure 18.17 The adjustments in strain causing the normal stresses to go to zero indicated a lattice strain that is consistent with a decreased lattice parameter. This, in turn, is consistent with increase in copper content due to precipitates dissolving in the vicinity of the weld zone.

18.7   Experimental validation of calculated residual stresses in a model repair weld

The final example is a neutron validation of a finite element analysis of a weld overlay treated as a repair weld. Iron–aluminium alloys have shown potential as a corrosion-resistant

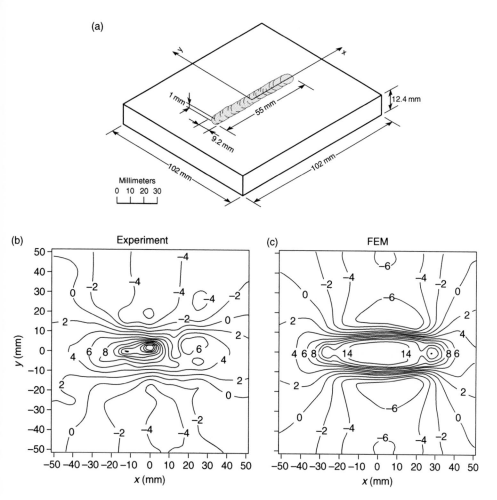

Figure 18.18 The Fe–Al alloy bead melted in a single pass on the cast iron plane is shown in panel (a). The pattern of strains in the direction of the weld pass measured at 4 mm depth is shown in panel (b). The finite element calculation of the longitudinal strains at the same depth shown in panel (c) is in good agreement with the experimental results.

cladding for high temperature applications [10]. Forming a crack-free clad in the form of a weld overlay has proven difficult and so an analysis of residual stresses in the overlay structure was considered relevant. The overlay, Fe–34%Al alloy with Mo, Zr, C and B additions, was applied to a normalized 2.25Cr–1Mo steel plate using the GTA process. The substrate was preheated to 350°C. A single weld pass 55 mm long with a width of 9.2 mm was made on the substrate that was a square 102 mm on a side and 12.4 mm thick. The schematic representation of the sample is shown in Figure 18.18(a). Neutrons of 1.6 Å wavelength from the (102) Be monochromator crystal were used with a single position-sensitive counter. The (211) reflection from ferritic iron was used to make the transverse, longitudinal and normal strain measurements. An incident beam 2 mm high and 2 mm wide defined the gauge volume. The receiving slit was 2 mm wide. Measurements were centred at 4 mm below the surface.

The comparison of strains between neutron measurement and FEM calculation [11] is shown in Figures 18.18(b) and 18.18(c). Figure 18.18(b) gives the measured longitudinal strains and Figure 18.18(c) gives the calculated longitudinal strains. The finite element results and the neutron diffraction data are consistent in terms of the general strain pattern. According to calculation, the maximum and minimum longitudinal strains are about 14×10^{-4} and -8×10^{-4}, respectively, whereas the corresponding measured values are 12×10^{-4} and -6×10^{-4}. The transitions from tension strain zone to compression strain zone are also comparable; the region that exhibits the largest differences is located beneath the weld where the FE model over-predicts the measured value. One possible reason is due to neglecting the effects of weld metal dilution.

A recent study of the strain tensor orientations in this same weld was made in order to compare the measured tensor orientations with the calculated tensor orientations. Strains were measured in the plane of the substrate plate as a function of rotation about the plate normal (Figure 18.19). The principal axis directions of the planar tensor were obtained from a determination of the extrema in the strain amplitude versus $\cos^2(\phi - \phi_0)$. Figure 18.20 shows the comparison by superimposing the measured and calculated principal axes where the magnitude of the vector is proportional to the strain and the direction gives the orientation. Near the weld centre the longitudinal strain is tensile. As the test point moves toward the end

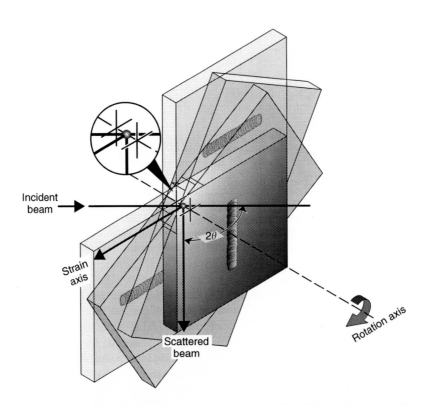

Figure 18.19 The experimental approach to measuring the variation of the in-plane strain with rotation about the plate normal is shown schematically. The test points are on a plane 4 mm below the base plate surface. Seven measurements were made in 30° increments over a range of 180° at each test point.

(a)   (b)

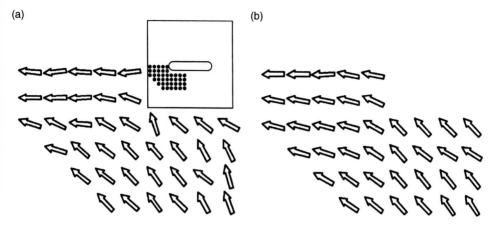

Figure 18.20 The mapping points on the plate are shown in the inset. The directions of the principal axis are shown in (a) by the arrows. The corresponding calculated principal axis directions are shown in (b) and are in good agreement with the measurements.

of the weld, the tensor rotates almost 90°. This turning of the tensor occurs over a wide range and there is only a limited region where the longitudinal and transverse directions are parallel to the principal axes. The significance of these observations is that strains measured without knowledge of the principal axis directions can lead to a gross underestimate of significant strains. Locations having small strains of opposite sign parallel and perpendicular to the weld correspond to positions of large shear strain, for example.

18.8 Discussion

The following discussion of residual stress analysis in welds considers several topics from the point of view of a neutron specialist interested in the most effective use of neutron scattering for the development of welding technology. In the last decade, publications on residual stress in welds have gone from exploratory reports to technically significant work in the welding field. There is now a more complete understanding of the limits and applicability of neutron scattering. This understanding will evolve as pulsed neutron techniques come into more frequent use and as synchrotron radiation applications are found.

The reproducibility of strain measurements has been shown to be adequate to the task of welding residual stress analysis with reproducibility comparable to X-ray and mechanical methods. The key point is to make sure that the strains are an accurate measure of Type I strains. The spatial resolution of residual stress measurement will probably never be completely satisfactory to the materials scientist who is used to atomic scale characterization, but resolution at 1 mm^3 represents an improvement over older residual stress methods and represents a bridge from microscopic to macroscopic technology. One should be alert to the relationship between effective and nominal gauge volumes as an important experimental quality. The precision of strain determination, although largely a matter of diffraction peak intensity, involves important underlying issues such as beam path in the sample, material texture, grain size and background scattering rates which can remove a given case from consideration for neutron analysis. The unique requirement for a zero-stress reference in neutron residual stress arises from the fact that there are no built-in boundary conditions that might substitute for this parameter. When measuring a triaxial residual stress state there has to be a determination of the zero-stress lattice parameter. The necessity for determining the principal axes of the stress and strain tensor has been under-represented in the early studies of relatively thin plates and cylinders. In the many practical cases where the question cannot be simplified, this adds extra experiment time to the job of residual stress analysis. The "ultimate" materials physics question is how to convert measured strain to engineering stress. Progress in this area is likely to result in a series of more accurate approaches having greater complexity. Elastic diffraction coefficients are the experimental answer in practical problems, but the interpretation of elastic diffraction coefficients will require sorting out of elastic and plastic anisotropy effects. Finally, the validation of FEM calculations has been shown. Calculations are going to improve with increasing computer power and more accurate property databases. Neutron residual stress analysis of more complex weld techniques and geometries will be done. It seems obvious that interactive comparison between computed and measured results will become a more common practice in advancing welding technology. A key question is whether experiments will be effectively designed for comparison with calculation.

18.8.1   *The zero-stress reference d-spacing*

The welding process is a severe heat treatment of the material in the region of the joint. The consequence is that the change in *d*-spacings has nothing to do with strain effects. The zero-stress *d*-spacing must be identified for determination of mechanical strain. Destructive examination of the material cannot be avoided in welds where the lattice parameter is strongly dependent on composition. Many ferritic steel welds are relatively insensitive to the effects of heat treatment because substitutional solute concentrations are low and carbon solubility is very small. Reference *d*-spacings can be made on small coupons cut from the HAZ and fusion zone. When the measurement is made with a beam flooding the entire volume of the coupon, the Type I residual stresses are integrated to zero. Type II microstresses can remain, however, since there is no requirement for local strains in the subgroup of diffracting grains to go to zero. The effect of composition on the *d*-spacing is independent of sample orientation, which suggests that directional averaging of the Bragg peaks will give a better determination of the reference *d*-spacing. Annealing heat treatments to produce a stress-free sample cannot be recommended because annealing will further alter material composition. A weld in pure material would be an exception to this recommendation.

18.9   Summary – "Caveat Emptor"

Anyone considering neutron residual stress analysis of welds should be aware of the requirements and preparation needed to achieve a successful application. For many welding problems, be aware of the requirement for destructive preparation of zero-stress reference samples from a companion weld or the weld itself after measurement. Excessive grain size in the fusion zone and the HAZ may render diffraction analysis impractical. Spatial resolution significantly below 1 mm is obtained at a very high cost in neutron experiment time. Dynamic analysis of welding-related processes with a time resolution in minutes is possible, but at present this would be limited to PWHT, for example. Improved data collection rate awaits the availability of higher neutron source intensities and new instrumentation design.

Neutron residual stress analysis of welds explores a new regime of non-destructive subsurface detail. In this new regime, one must consider new questions about the orientation of the principal axes of the strain tensor and the interpretation of elastic and plastic anisotropy. It is possible to make progress in the understanding of residual stresses by resorting to simplifying assumptions as was illustrated in the foregoing examples. The validation of computed residual stress with neutron results is a severe test that challenges both experiment and computation. To keep perspective, however, the accuracy of surface residual stress results by strain relaxation, because of its simplicity, is more accurate and accessible than the residual stress from neutron scattering.

18.10   Opportunities

Possibly the greatest need and opportunity in the field of neutron residual stress analysis is to solidify the technical foundation and limits of this method of weld research. The unique feature of neutron scattering is the ability to characterize, by diffraction, the strain response of welded material in a non-destructive manner. At least two fundamental challenges are met; the requirement to understand the measurement of the triaxial strain state, and to understand how to determine the macroscopic stress (Type I) given the way the diffraction averages the strain response at the macroscopic and microscopic levels.

The weld structures are very messy because of the complex thermal history in which plastic deformation, phase transformation, composition and texture change occur. So far we can only analyse the weld as fabricated. We know very little detail about the strain history leading to the final residual stress state. The analysis of the weld will often require the destruction of the weld to determine, for example, the stress-free reference d-spacings or the constituent phase fractions. The opportunity here is to find new methods of neutron diffraction analysis that are rapid enough to capture some of the strain and phase transformation history of weld formation. In the meantime a more realizable objective is to refine the spatial resolution of diffraction analysis so that the structure of multiple pass weld beads and constituent phase distributions in a weld could be extracted along with the strain state.

18.10.1   *Conventional stress relief codes*

Some insights have been gained from stress mapping of welds subject to residual stress reduction by PWHT. Heat-treatment practices for various engineering materials in various forms and dimensions are contained in the engineering codes for PWHT. Further research into modification of welding processes and PWHT offers the opportunity to define more economical and beneficial practices that reduce the cost of fabricating reliable welds.

Neutron analysis of vibratory and magnetic stress relief of welds has not confirmed the anecdotal evidence for such benefits. The evaluation of a known case of stress relief benefit will require a scrupulously controlled neutron mapping experiment. Small grain size and minimal texture would assure the accuracy of strain measurement and the influence of microstrains (Type II) would have to be fully understood.

18.10.2   *E-beam and laser weld (small HAZ methods)*

The fusion zone and HAZ of an electron beam weld is very small compared to conventional arc welds. Strain field gradients are steeper and the required spatial resolution for residual stress mapping needs to be reduced below a millimeter for useful analysis. The successful use of strain mapping requires the fulfilment of the powder diffraction condition of a sufficient number of grains to form a reasonably continuous Bragg diffraction ring. The condition is very restrictive in large grained material. Beyond the powder limit, one must consider single crystal analysis. In the single crystal regime, one has lost the advantage of easy measurement of average strain from which Type I residual stresses are usually determined. The solution to the large grain size problem is a current subject of research and requires more effort.

18.10.3   Hydrogen effects

The influence of hydrogen on weld cracking is a major problem that ought to be responsive to neutron scattering because of the large scattering of neutrons by hydrogen. The difficulty is that the average hydrogen content is measured in parts per million. Significant scattering effects might be expected at regions of high hydrogen concentration. The usual distributions of such regions are too diffuse to give a measurable scattering signal. An alternative is to consider the subtle effect of hydrogen on the lattice parameter of the hydrogen-containing matrix. However, order of magnitude improvement in d-spacing measurement would appear to be needed.

18.10.4   Calculated residual stresses

Finite element calculations for welds are well established and the most important tool in making effective use of neutron mapping weld analysis. Neutron stress mapping validation of computation has been demonstrated. It is clear that there is room for improvement in the computation and in the analysis of the neutron experiment. On the computation side one could use improvements in the thermo-mechanical properties database and the execution of phase transformation and elasto-plastic models. On the neutron diffraction side, one needs to resolve Type I and Type II strain behaviour. The accuracy of zero-stress reference d-spacings can be improved, especially in cases where the d-spacing is composition sensitive. Quantitative validation of computation results is possibly meaningless without specifying the mechanical and thermal boundary conditions applying to the weld process. Measurement of weld distortion is needed in addition to residual stress. The recording of restraining forces during welding would be an essential part of the experimental effort.

18.10.5   Welding dynamics

Neutron analysis of welding dynamics in real welds is not going to be achieved without a significant improvement in neutron flux and neutron detection technology. The time-scale for the welding process is minutes with solidification processes measured in seconds. Current powder diffraction measurements are made in minutes under favourable conditions. Pulsed neutron sources promise to deliver significant improvements in neutron flux over current neutron sources, and the intrinsic advantage of delivering a powder pattern at a fixed angle will prove helpful. The intrinsic advantage of beam penetration remains a prime justification for the effort.

18.10.6   Relation to other methods

Neutron scattering is nearly unique in its capability to measure strains in depth without destruction of the test specimen. But it is certainly not a portable method and access is limited to the small number of neutron sources around the world. Use of the method must be carefully justified by there being a technically significant outcome from the undertaking. Much of the work with neutron scattering in the past decade has been exploratory and these efforts have sharpened the understanding of limits in the application of neutron analysis to welding problems. It is clear that grain size and morphology and multiphase constituency have to be known in order to make effective use of neutron analysis. But before use of neutrons is decided upon, the use of alternative methods should be considered.

The evaluation of residual stress by measurement of strain relaxation in sectioned materials is an old, simple, reliable and accurate method. One uses bulk elastic coefficients and the strain response depends only on Type I residual stresses. The method is strictly destructive and has limited spatial resolution but may serve for engineering purposes. The Sachs boring method is less destructive, has spatial resolution, is on a scale of millimeters, and is sensitive to the near surface stress state. The portability of the method permits field testing, which is a practical advantage.

The precision and spatial resolution of conventional X-ray residual stress methods are comparable to current neutron diffraction methods but with the important difference that only surface measurements can be done. The X-ray method can be extended into depth but require destructive surface layer removal. Given relative costs, there will always be a use for this portable and convenient residual stress method.

The use of synchrotron radiation for residual stress is in its earliest stages of development. Not enough experience has been accumulated to optimize instrumentation for this area of application. Very useful beam penetration can be achieved in aluminium and steel, making it competitive with neutrons in this respect. However, penetration is achieved with high-energy short-wavelength radiation that gives Bragg diffraction at small angles. Under these circumstances, the gauge volume is needle-like. When the beam is narrow at less than $10\,\mu m$, the long axis can be reduced to less than 1 mm. An additional complication arises because of the high degree of collimation in the beam. The number of grains in the gauge volume selected for diffraction is small. The result is that the same grain-sampling limits encountered in large grained samples with neutron techniques will limit synchrotron radiation experiments requiring grain averaging. Residual strain mapping using very high intensity synchrotron radiation is capable of characterizing small-scale strain variation not achievable with neutron methods. For example, one can analyse grain-to-grain strain tensor variation within a small cluster of grains, and measurement of strain variation within single grains is possible. The information gained at micrometer scale will improve understanding of Type II residual stresses and thereby complement and support the macroscopic scale mapping with neutron methods.

Acknowledgements

I wish to express particular appreciation to Stan David of the Oak Ridge National Laboratory Metals and Ceramics Division Welding group for initiating our efforts on residual stresses in welds. I would like to thank my colleague Cam Hubbard for securing the funding and providing his expertise in diffraction that was essential for building the residual stress program at Oak Ridge. Thanks are due to my ORNL colleagues T. A. Dotson, David Schenk, Michael Wright, Xun Li Wang, Christina Hoffmann, David Wang, Ducu Stoica and Tom Ely. Mihai Popovici and Duco Stoica at the University of Missouri developed the bent Silicon neutron monochromator used most recently in our experiments, and for that we are very grateful. Finally, particular thanks go to my colleagues at Chalk River, Tom Holden and John Root, whose generous assistance started us out in the field.

318   *S. Spooner*

References

[1] Mahin K. W., Winters Jr.,W. S., Krafcik J. K., Holden T. M., Hosbons R. R. and MacEwen S. R., *Proc. Intl. Conf. on Trends in Welding Research*, Gatlinburg, TN, May 1989, ed. S. A. David and J. M. Vitek, pp. 83–89, ASM International (1990).

[2] Mahin K. W., Winters Jr., W. S., Holden T. M., Hosbons R. R. and MacEwen S. R., *Prediction and Measurement of Residual Elastic Strain Distributions in Gas Tungsten Arc Welds*, Sandia Report 90-8647, (1991).

[3] Spooner S., David S. A., Wang X. L., Hubbard C. R. and Holden T. M., *Intl. Conf. Proc. on Modeling and Control of Joining Processes*, Orlando, FL, ed. T. Zacharia, pp. 409–413, Am. Welding Soc. (1994).

[4] Root J. H., Holden T. M., Schroeder J., Hubbard C. R., Spooner S., Dotson T. A. and David S. A., *J. Mater. Sci. Eng.*, **9**, 754–759 (1993).

[5] Spooner S., Wang X. L., Hubbard C. R. and David S. A. *Proc. 4th Intl. Conf. on Residual Stresses*, Baltimore, MD, 1994, pp. 964–969, Soc. Exper. Mech. Inc. (1994).

[6] Spooner S., David S. A. and Hubbard C. R., *Trends in Welding Research, Proc. 4th Intl. Conf.*, Gatlinburg TN, 1995, eds. H. B. Smartt, J. A. Johnson and S. A. David, pp 99–102, ASM International (1996).

[7] Wang X. L., Hubbard C. R., Spooner S., David S. A., Rabin B. H. and Williamson R. L., *Mater. Sci. Eng.*, **A211**, 45–53 (1996).

[8] Pintschovius L., Pyka N., Kubmaul R., Munz D., Eigenmann B. and Scholtes B., *Mater. Sci. Eng.*, **A177**, 55–61 (1994).

[9] Spooner S. and Pardue E. B. S., *Adv. X-ray Anal.*, **39**, 297–303 (1997).

[10] Wang X. L., Spooner S., Hubbard C. R., Maxias P. J., Goodwin G. M., Feng Z. and Zacharia T., *Mater. Res. Soc. Symp. Proc.*, **364**, 109–114 (1995).

[11] Feng Z., Wang X. L., Spooner S., Goodwin G. M., Maziasz P. J., Hubbard C. R. and Zacharia T., *1996 ASME PVP Conference*, Montreal, Canada, **327**, 119–126 (1996).

19 Materials for nuclear fusion applications

R. Coppola and C. Nardi

19.1 Introduction

It is well known that the construction of future fusion reactors will depend on a timely development of materials resistant to the simultaneous effects of high-energy neutron irradiation and intense thermomechanical loads [1, 2]. A schematic picture of the International Thermonuclear Experimental Reactor (ITER) machine is shown in Figure 19.1, where the main components are indicated, while Table 19.1 reports the main radiation effects on some of these components. The damage produced by the fusion reaction

$$D + T \rightarrow n + 14\,MeV$$

will be quantitatively and qualitatively different from the damage experienced in conventional nuclear plants, since high H and He doses will be generated into the structural materials and the displacement rate will be one order of magnitude larger. Furthermore, the pulsed operation mode of these machines will submit the whole structure to low cycle fatigue at service temperatures as high as 600–700°C. Keeping in mind that a huge number of welds and joints will be necessary to fit the complex geometry of these machines, the relevance of stress measurements in this domain is easily appreciated. A review of the various structural materials options currently developed in fusion research and technology can be found in Refs. [2, 3].

This chapter will first review some stress studies carried out, using neutron diffraction, on prototype brazings of interest for plasma facing components such as the 'divertor' (see below). They are developed to protect Cu or Mo cooling pipes with high temperature resistant materials such as composite graphite or tungsten. One of the main concerns in the lifetime prediction of such mock-ups for plasma facing components is the evaluation of the stress field existing in the component after manufacture. This stress field is mainly due to the fabrication operations and to the welding and brazing procedures.

As a second example, some neutron diffraction studies of welded ferritic/martensitic steels are reported. Owing to their resistance to swelling and He effects, these steels appear in fact as one of the most realistic options to construct the 'first-wall' and blanket structures in medium- or long-term projects such as the demonstrative reactor DEMO [4]. A further advantage of these steels arises from the limited quantity of Ni in their composition, limiting the activation products and lowering the radioactive waste disposal cost. These steels are being developed starting from the elemental compositions originally designed for use in fast breeder reactors and by replacing those constituents with high-activation cross sections, such as Ni or Nb, with more favourable ones such as Mn, W or V. Mechanical property

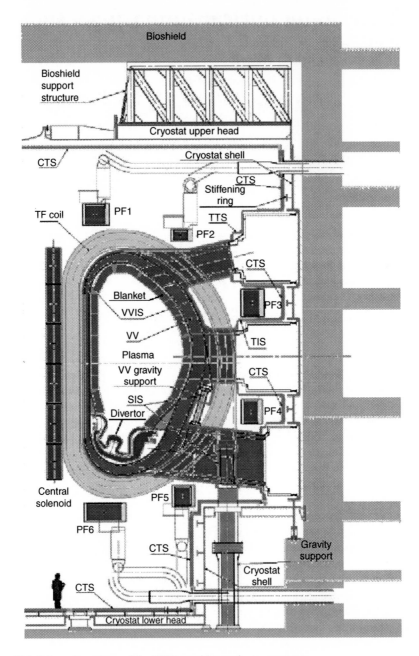

Figure 19.1 Schematic picture of the ITER machine main components.

tests and microstructural examinations are carried out to qualify such materials in view of their technical utilization in fusion relevant conditions. It is more specifically essential to characterize the complex microstructural changes produced by welding and the related stress field.

Table 19.1 Interactions between the plasma and the structure of the reactor, from Ref. [1]

Plasma	First wall	Blanket	Shield	Magnets
Neutrons	Hardening Embrittlement Creep Swelling (n,α)(n,p) Heating Activation	Hardening Embrittlement Creep Swelling (n,T)(n,2n) (n,α)(n,p) Swelling Activation Heating	Activation Heating	Changes in: critical current; resistivity of stabilizer; mechanical properties of insulators; heating
Electromagnetic radiation	Surface Heating ΔT Stresses	Heating	Heating	Heating Changes in mechanical properties of insulators
D, T, α	Sputtering T permeation Hydrides			
Disruptions	Heating Melting Evaporation			

19.2 Brazed divertor components

Near-term fusion reactor prototypes, such as ITER, will require in the vacuum chamber a component, the divertor, specifically designed to contain plasma instabilities and their deleterious effects originating from huge and instantaneous heat changes (see Figure 19.1). The main scope of the divertor system in the plant is in fact first to have an area where the ions of He and highest Z materials are extracted from the plasma, and second to perform the power exhaust from the radiative plasma. This implies high ion bombardment and great energy release on the divertor, thus making it the component where highest local thermal fluxes are present. The divertor is basically an assembly of cooling pipes, in Mo, Cu or Cu alloys, coated by a sacrificial layer of high-temperature resistant material, such as graphite or tungsten. Typical divertor plate sizes are $1 \times 1\,m^2$. It is clear that owing to the extreme service conditions of such components, the choice of the materials and of the method to join them is crucial, and preliminary studies are necessary to investigate them on small samples or 'mock-ups'.

One of the first monoblock divertor concepts, developed for the ITER project, consisted of Mo cooling pipes surrounded by a carbon fibre composite (CFC) graphite armour [5]. Figure 19.2 shows real-scale modules constructed following this concept. The armour and the pipe are joined together by a brazing treatment, usually at about 800–850°C and using alloys such as TiCuAg as fillers. That gives rise to a stress field extremely localized near the interface region and sensitive to thermal treatments also as a consequence of interdiffusion processes. It is therefore essential to investigate the stress field evolution with temperature conditions in these structures, and neutron diffraction provides an appropriate experimental tool for this task.

An initial study [6] was carried out, with the aim of approaching the problem by a simplified brazed joint, on the polycrystalline graphite/Mo brazed platelet sketched in Figure 19.3. The neutron diffraction measurements were carried out, at room temperature, using the

Figure 19.2 Composite graphite coated Mo tubes for 'divertor' components; the coating is joined to the tube by a brazing procedure.

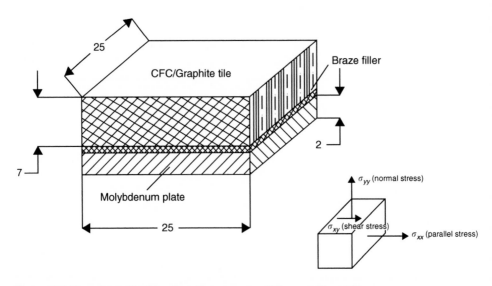

Figure 19.3 Brazed graphite/Mo plate; figures are in millimeters (from [6]).

G5.2 diffractometer at the Laboratoire Léon Brillouin, Saclay (as were most of the works reported in this chapter). Figure 19.4 reports the main stresses measured in graphite, with a diffracting volume of approximately $8 \, mm^3$ (with a space resolution of 2 mm in the brazed layer direction), as a function of the distance from the brazing interface. In Mo, only one point could be tested giving an average stress value of some 86 MPa. These low stress values are in the same order of magnitude of those predicted by finite element method (FEM) calculations

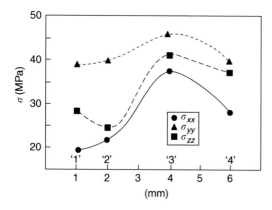

Figure 19.4 Principal stresses in graphite as a function of distance from the brazing for the sample of Figure 19.3 (from [6]).

using the ABAQUS code. In a subsequent study [7] similar room temperature measurements were repeated on a sample obtained by brazing CFC graphite to a Mo small cylinder (similar sizes as the sample of Figure 19.3). In this case, analyzing the data from graphite was complicated by the composite structure of the material, composed of a fibre and a matrix phase, with consequent texture effects. However, by assuming that the fibre contribution to the total intensity distribution of the Bragg peak can be represented by linear background noise (determined by fitting the experimetal data), a more regular trend in the ε versus $\sin^2 \psi$ plots allowed a determination of the principal stresses at two significant points. The values of such stresses, tensile also in this case, are of the same order of magnitude as those determined in polycrystalline graphite.

High-temperature strain measurements were carried out on a castellated mock-up obtained by brazing CFC graphite with an interlayer of Cu, brazed in turn to dispersion strengthened (DS) Cu [8]. A sketch of the sample, designed for optimizing heat dissipation, is shown in Figure 19.5, together with its dimensions; the main axes defined for the neutron diffraction measurements are also shown. In these measurements the linear size of the selected neutron probe was 2 mm, corresponding to a diffracting volume of 10 mm^3. The experimental conditions for graphite were the following: the neutron wavelength was 0.47 nm, with a 2θ angle of approximately 88.7° for the (002) reflection. For DS Cu, the neutron wavelength was 0.286 nm with the angle 2θ at 86.6° using the (111) reflection. The instrumental resolution is about 0.5° in the case of CFC and 0.4° in the case of DS Cu. As discussed above, the diffraction peak from CFC contains a contribution from fibres and a contribution from the matrix, which cannot be distinguished with the available instrumental resolution. Therefore the measured lattice spacing refers to an average over the selected diffracting volume, which is large enough to integrate over these two phases, fibre and matrix, both in the reference powder and in the sample.

The measurements were carried out using a thermocouple furnace, in vacuum, varying the angle ψ for fixed ϕ values. Therefore, only radial and tangential strains could be measured (see Figure 19.5). As the lattice parameters of each of the two investigated materials, CFC graphite and DS Cu, vary with temperature in order to determine the strain values, at high temperature milled powders of both CFC graphite and DS Cu were used to measure the

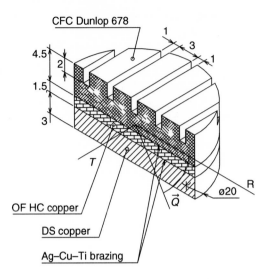

Figure 19.5 Castelled divertor mock-up (from [8]).

respective diffraction peaks at different temperatures. In this way, a calibrated, temperature-dependent value of the unstrained lattice parameter was available to determine the strain

$$\varepsilon(T) = \frac{d(T) - d_0(T)}{d_0(T)}$$

at high temperature in each of the two materials. Strains in graphite were subsequently measured at 30°C, 300°C and 600°C (the maximum available). Then the strains in DS Cu were measured, always keeping the sample at 600°C. The lower temperatures were not investigated in DS Cu since the time required for each measurement was 15–20 h and the whole component underwent a complex thermal treatment with consequent modifications of the initial strain field. It is also noted that the brazing temperature being 850°C, such measurements cannot in any case be meaningfully carried out at temperatures much higher than 600°C.

The results of these measurements are shown in Figure 19.6. The deformation trends exhibit deviations from linearity, likely ascribed to texture effects as seen in previous studies. At room temperature a negligible radial strain is found, while in the tangential direction a tensile strain of approximately 0.0006 is found. By increasing the temperature, the radial strain becomes negative (−0.0006 approximately) while the tangential one becomes negligible. Taking into account the experimental error bars and the dispersion of the points, no significant difference can be detected between 300°C and 600°C data. Concerning the strain evolution in DS Cu (measured at the centre of the component and as close as possible to the brazing surface) the value measured at 600°C coincides with the reference powder value, so that the deformation is zero. This is entirely consistent with what is known on the temperature resistance of DS Cu.

These results may be affected by the combination of the different thermal treatments performed on the whole mock-up, which may have non-negligible consequences especially on the microstructural evolution (precipitates) of DS Cu. The comparison with FEM results is in this case very difficult, the geometry of the sample is a very complex one and the calculations

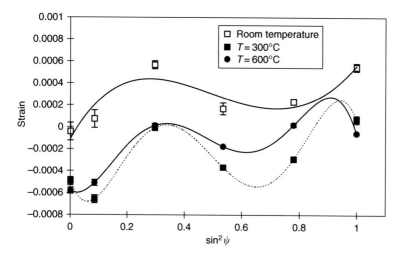

Figure 19.6 Temperature-dependent strain measurements in graphite for the sample of Figure 19.5.

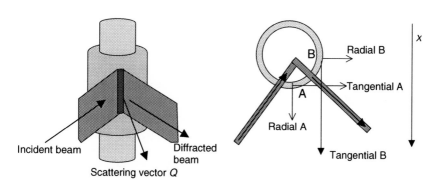

Figure 19.7 Experimental layout for zero strain determination experiment (from [9]).

are extremely sensitive to the hypothesis put on the heat exchange from the castellated volume. Indeed, if the exchange area is evaluated only as the one of the upper surface, a very slow temperature reduction is obtained, while if the full surface of the castellation is evaluated (i.e. with the lateral walls), the time of the temperature reduction drops dramatically.

As a last example an experimental determination of the zero strain temperature by means of neutron diffraction is mentioned [9]. This temperature, which is of utmost importance in engineering calculations to predict mechanical behaviour and stability of such structures, is generally evaluated by numerical methods, however an experimental determination is necessary to validate these theoretical predictions. The measurements were carried out on a sample manufactured by Metallwerke Plansee, made from a pipe in CuCrZr alloy armoured with a CFC (Figure 19.7). The joint between the CFC and CuCrZr alloy was obtained by an intermediate layer of copper, of thickness of about 1 mm.

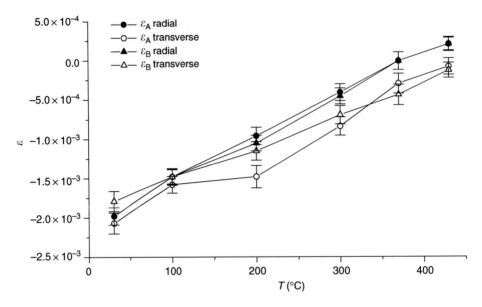

Figure 19.8 Temperature-dependent strains in Cu in the two points of Figure 19.7.

Also, in this case a thermocouple furnace was used for the measurements, which were carried out in He. Two different points (A and B) located at 90° to each other were measured in two directions: radial and tangential. The size of the neutron beam was 20 mm high and 0.5 mm wide. The measurements were performed at different temperatures: RT, 100°C, 200°C, 300°C, 370°C and 430°C. For each temperature, the measuring sequence was the following: point B in the tangential direction, point A in the radial direction, point A in the tangential direction and then point B in the radial direction. The reference lattice parameter was measured from a powder ground from a Cu sample, measured at the same sample measuring temperatures.

The strains determined in the sample using the powder as a reference are shown in Figure 19.8. There is very little difference between the two points in the two directions (radial and tangential) and the strains vanish at about 370°C in the radial direction and at about 430°C in the tangential direction. This equilibrium temperature is determined by fitting the data with a straight line: the correlation coefficient (R^2) is 0.995 and the error on the determination of the equilibrium temperature is at worst 30°C. The results show, moreover, a high reproducibility, also indicated by the quality of the linear fit. The stress-free temperature in the CuCrZr alloy pipe of the monoblock is in the range 370–430°C, which is consistent with what is expected from the thermomechanical behaviour of the component. However, the results show compressive strains in the tube with similar values in radial and tangential directions. This is not in agreement with the common understanding of the problem, since tensile strains are expected in the sample after fabrication. However, an analysis of the results, taking into account the mechanical constants of the different materials and the sample geometry, showed the consistency of the neutron diffraction results and suggested that the experimentally determined strain values can be affected by strains remaining after the fabrication process [9].

19.3 Ferritic/martensitic steels

In demonstrative fusion reactors such as DEMO, the first wall, defining the inner surface of the torus, will be the component more directly exposed to irradiation. It will be basically composed by an assembly of large modules (typical sizes $1 \times 1 \times 10\,m^3$, see Figure 19.9) fitting the torus geometry, each one including a complex system of cooling pipes. The first wall will also have an essential structural function in supporting the inner tube bundle of the module. Currently ferritic/martensitic steels appear as one of the most realistic options to manufacture it: in fact these steels, which have largely been tested in nuclear plants, combine good mechanical properties and resistance to irradiation. However, since some of their conventional consituents such as Nb have high-activation cross sections under 14 MeV neutron irradiation, their chemical composition must be optimized in such a way to obtain low-activation materials (LAM). That implies a series of metallurgical characterization on such developmental steels, including of course study of the stress field under different welding methods.

One of the main low-activation martensitic steels, F82H-mod., developed by the Japanese JAERI-Programme and manufactured by NKK, has been proposed as a reference material for the first-wall and blanket structures for the European Long Term Fusion Material Programme [10, 11]. The main alloying constituents of this steel are the following: Cr 7.7, C 0.09, Mn 0.16, W 2.0, V 0.16 wt% (Nb content restricted to 1 ppm). As a very large number of welds will be necessary to construct those components, the weldablity of this steel, the stress fields associated with different welding methods and their evolution under fusion-relevant thermomechanical loads are important aspects to investigate.

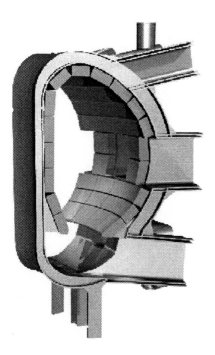

Figure 19.9 Scheme of a first-wall section.

Within this frame a stress study on an electron-beam welded sample of this steel has been carried out [12]. The sample had been obtained from a plate hot rolled and then submitted to standard normalizing and tempering treatment (1040°C for 30 min followed by rapid quench, then 750°C for 1 h) which gives a fully tempered martensite free of δ-ferrite. The sizes of the sample, $6.5 \times 5\,cm^2$ wide and $1.5\,cm$ thick, were defined in order to be large enough to avoid as much as possible boundary effects on the stress field, as well as to be compatible with the maximum allowed neutron path length inside the material. The sample was treated by electron-beam, under a voltage of $60\,kV$ and a current of $150\,mA$, with a speed of $1\,m/min$. The total resulting width of the melted zone and heat-affected zone (HAZ) is of approximately $2.2\,mm$ ($1.2\,mm$ for the former) as was evident from the results of microhardness measurements at different distances from the centre of the weld. Both the melted zone and the HAZ exhibit a microhardness value corresponding to what is expected from a martensitic quench (440–400 Vickers hardness), while the hardness value typical of the tempered material (240) is rapidly recovered for distances larger than $1.5\,mm$ approximately.

X-ray diffraction measurements were carried out using a SetX diffractometer equipped with a ψ goniometer and a PSD detector at ENSAM, Paris. The wavelength was $2.29\,\mathring{A}$ of Cr K-α radiation, the selected Bragg reflection (211) corresponding to a diffraction angle of 156.3°. A circular area $2\,mm$ in diameter was investigated. The stresses were evaluated by using 11 ψ angles. After mechanical polishing, the sample was electrochemically polished using an acetoperchloric solution. The investigated points are shown in Figure 19.10 and the corresponding results are shown in Figure 19.11 (together with those of neutron diffraction).

For the neutron diffraction measurements, the selected reflection was (110), with a neutron wavelength of $2.844\,\mathring{A}$ and a 2θ angle of about 89°. The instrumental linewidth was approximately 0.3°. The Cd diaphragm at the exit of the neutron guide had a diameter of $1.1\,mm$, with a corresponding diffracting volume of approximately $1\,mm^3$ inside the material. The sample was mounted on a Euler cradle and the strain was measured at different values of the φ angle (0°, 45° and 90°) and of the ψ angle, in order to determine the stress tensor. As a reference, unstrained d_0 value the measurement at point 4, far from the HAZ, was used.

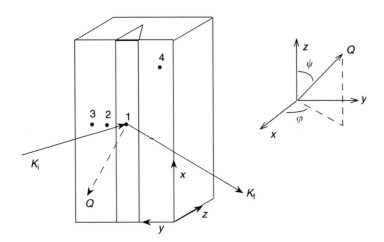

Figure 19.10 Schematic picture of electron-beam welded F82H steel sample with experimental layout and investigated points (from [11]).

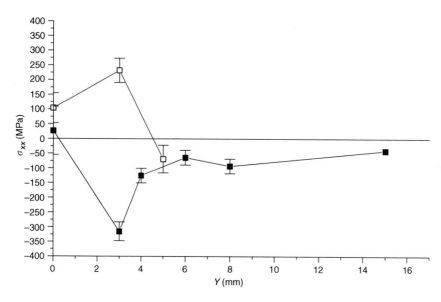

Figure 19.11 xx direction stresses measured in the sample of Figure 19.10 with X-rays (full squares) and neutrons (empty squares) as a function of distance from weld centre.

Comparing the X-ray and neutron diffraction results of Figure 19.11 it is evident that, at the beam centre and 4.5 mm distance, the data are if not in agreement at least compatible, taking account of the differences between the two methods and of the uncertainties associated in the unstrained lattice parameter value assumed for neutron diffraction; furthermore, the microstructure of the melt zone is expected to be quite heterogeneous, especially in the x direction. Far from the HAZ, both neutron and X-ray measurements indicate low stress levels (< -50–100 MPa).

In the intermediate region where experimental studies on other martensitic steels show strong stress gradients [13, 14], there is complete disagreement between X-ray and neutron results. Stress gradients depending on depth inside the material are known to be <10%/mm in martensitic steels welded with electric arc procedures [14]; therefore an explanation of the differences betweeen X-ray and neutron data based on the fact that the two methods probe a depth-dependent stress field is not sufficient to account for the results obtained. On the other hand, it is well known that bainitic or martensitic structure random formation is one of the main effects arising in the HAZ under electron-beam treatments in martensitic steels, with consequences for the stress distribution. Extremely localized bainitic regions can be produced by oscillations of the electron-beam arising from fluctuations in the magnetic permeability of the base material. This can also be inferred from metallographic observations. It is therefore tentatively suggested that the differences between X-ray and neutron data in the HAZ arise from the fact that the neutron beam probes a region where an uncontrolled percentage of bainite, with a different lattice parameter and a different specific volume from the base metal, is present. Furthermore, additional modification of the deformation field in the bulk of the material is produced by the stresses at the boundary of the two phases.

The presence of secondary stresses as a consequence of microstructural phenomena such as Cr segregation is another important field of investigation concerning martensitic steels. In fact the Cr atom distribution can have important consequences on fracture behaviour and ductile-to-brittle transition. Furthermore, the related changes in dislocation density are a key factor in determining the resistance to irradiation effects. X-ray and neutron diffraction line broadening measurements have been carried out [15] to follow microstructural evolution under tempering at 700°C in a martensitic steel originally developed for the Next European Torus project (MANET). Its chemical composition is the following: C 0.10, Cr 10.37, Mo 0.58, Ni 0.65, Mn 0.76, Nb 0.16, V 0.21, Si 0.18 Fe to balance (wt %). All specimens were austenized at 1075°C for 30 min, then cooled with two cooling rates (150°C and 3600°C/min, respectively) and finally tempered at 700°C for different times up to 20 h. Work carried out using scanning electron microscopy, internal friction and small-angle neutron scattering [15–18] showed that during tempering inhomogeneities in Cr distribution arise with consequent changes in microstrain field and precipitates or aggregates occurrence. As an example, Figure 19.12 shows SEM pictures of MANET steel before and after tempering 2 h at 700°C.

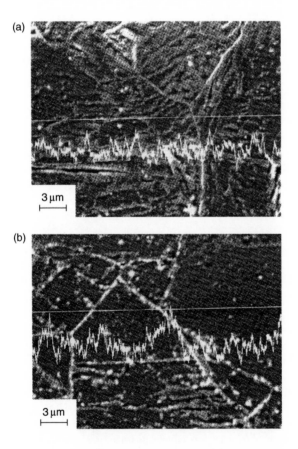

Figure 19.12 SEM pictures of MANET steel as quenched (a) and tempered 2 h at 700°C (b). The Cr microanalytical profiles refer to the points along the indicated white lines (from [15]).

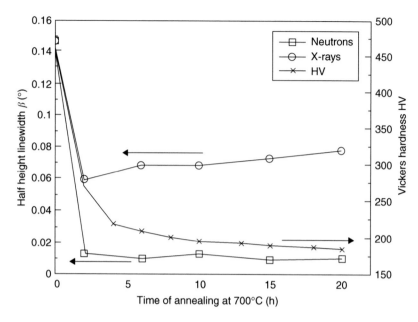

Figure 19.13 Neutron and X-ray diffraction linewidths, together with microhardness, for MANET steel
as a function of tempering time at 700°C (from [15]).

Figure 19.13 reports the comparison of X-ray and neutron diffraction linewidths, as a
function of tempering time, for the same set of samples. Microhardness data are reported as
well. As discussed in Ref. [15], the difference between X-ray and neutron data is attributed
to the fact that, while in the case of neutrons the linewidth is averaged over the whole sample
thickness (1 mm) and therefore the total contribution of local strains like those arising from
the Cr fluctuations shown in Figure 19.12 (with a typical length scale of few micrometers)
is zero, the thickness sampled by X-rays is 20 μm approximately and is therefore affected
by those fluctuations. On the other hand, since immediately after quenching and prior to the
first annealing step the Cr is homogeneously distributed, the initial linewidth values are the
same for neutrons and X-rays.

Neutron diffraction experiments were carried out at the NPI Rez at the high-resolution
three-axis diffractometer TKSN-400 operating at a wavelength of 1.62 Å. The instrument is
equipped with a bent Si(111) monochromator and Si(220) analyser. The high instrumental
resolution of $\Delta d/d = 2.6 \times 10^{-3}$ is achieved by setting the optimum focusing conditions
with respect to the studied Fe(110) reflection. The curve from a well annealed α-Fe sample
was used as an instrumental profile. The effects of the peak broadening were treated in terms
of the integral breadth method of profile analysis: this method enables a separation of the
strain and size contributions to the peak broadening [19].

Figure 19.14(a) documents the recovery of the grain structure with annealing, whereas
Figure 19.14(b) shows a decrease of the relative mean microstrain with annealing time. The
determined values of microstrain result from two main contributions, strain fields around dislocations and Cr aggregates, respectively. Both contributions are competitive when assuming
their evolution with tempering. A decreasing function of mean microstrain indicates that
the effect of dislocations is dominant. The dislocation density (Figure 19.14c) was modeled

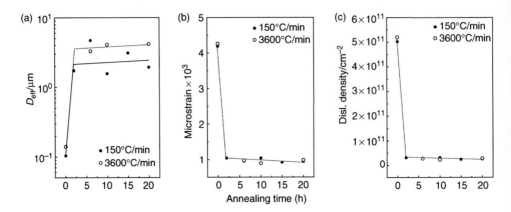

Figure 19.14 Mean size of coherently diffracting blocks (a), mean microstrain (b) and dislocation density (c) in MANET steel as a function of tempering at 700°C [18].

assuming a system of dislocations with Burgers vector $\langle 111 \rangle$ without taking into account the effect of C–Cr aggregates or small precipitates on the total microstrain field. After the first annealing step, the microstrain and dislocation density are relieved and stabilized at least up to 20 h annealing time. SANS measurements [17] revealed that important microstructural changes occurred after the first annealing step and that the average size and relative volume fraction of the C–Cr aggregates do not significantly change in the explored tempering time interval. One could therefore assume that, immediately after quenching, Cr atoms in octahedral positions around interstitial C atoms contribute to a relatively high value of microstrain which is decreased also as a consequence of more complex and more extended C–Cr aggregates after the first annealing step.

19.4   Concluding remarks

These few selected examples have at least given an idea of the complexity of fusion materials research, showing the usefulness of neutron diffraction for an experimental determination of the bulk stress field. Concerning in fact brazed divertor structures, these studies contribute to understanding some fundamental microstructural features of brazed joints. Furthermore, they provide an irreplaceble tool for experimentally determining the strains or the stresses at technologically relevant temperatures and for checking the validity of parameters, such as zero strain temperature, which are needed for constructing large scale components and usually estimated only through numerical analysis. In the case of ferritic/martensitic steels, neutron diffraction is also very useful for metallurgical characterization of new alloys through phase identification and investigation of the stress field in welds; also in this case such fundamental properties must be studied on prototype samples in order to provide the experimental base necessary for the design of the component.

A fundamental point to be clarified by future work is, for a given kind of brazing process, how to correlate numerical predictions of the stress field, which generally refer to ideal samples, and neutron diffraction measurements, which are carried out on real samples and are affected by their thermomechanical history, texture effects, etc. In this regard samples to be tested by neutron diffraction should be free of stresses, except those produced by brazing or at least prepared in stress-controlled conditions. An experiment is underway at the high flux reactor of ILL-Grenoble to investigate the temperature evolution in a reference Cu/W brazed sample. High-resolution measurements are necessary to investigate phase inhomogeneities in the HAZ as a consequence of thermal gradients during the cooling phase. Finally, comparison of neutron and X-ray measurements can be used to investigate segregation phenomena through changes in the microstrain field.

References

[1] Schiller P., Nihoul J., *J. Nucl. Mater.* **155–157** (1988) 41.
[2] Klueh R. L., Ehrlich K., Abe F., *J. Nucl. Mater.* **191–194** (1992) 116.
[3] Aymar R., Proc. 20th SOFT, Marseilles Sept. 1998, B. Beaumont *et al.* (eds), *Fusion Technol.* (1998) p. 4.
[4] Hishinuma A., Kohyama A., Klueh R. L., Gelles D., Dietz W., Ehrlich K., *J. Nucl. Mater.* **258–263** (1998) 193.
[5] Cardella A., Di Pietro E., Brossa M., Guerreschi U., Reale M., Reheis N., Vieider G., *Fusion Technol.* (1992) 211.
[6] Ceretti M., Coppola R., Di Pietro E., Lodini A., Perrin M., Piant A., Rustichelli F., *J. Nucl. Mater.* **212–215** (1994) 1617.
[7] Albertini G., Ceretti M., Coppola R., Di Pietro E., Lodini A., *J. Nucl. Mater.* **233–237** (1996) 954.
[8] Ceretti M., Coppola R., Di Pietro E., Nardi C., *J. Nucl. Mater.* **258–263** (1998) 1005.
[9] Coppola R., Nardi C., Riccardi B., *J. Nucl. Mater.* **283–287** (2000) 1243.
[10] Ehrlich K., Kelzenberg S., Roehrig H. D., Schaefer L., Schirra M., *J. Nucl. Mater.* **212–215** (1994) 678.
[11] Schaefer L., Schirra M., Lindau R., *Fusion Technol.* **2** (1996) 1363–6.
[12] Braham C., Ceretti M., Coppola R., Lodini A., Nardi C., *Proc. ICRS5* (1998) p. 169.
[13] Weisman C. (ed.), *Welding Handbook*, 7th edition, American Welding Soc. (1976).
[14] Montevoni P., Rosellini C., Rep. ENEA (1997).
[15] Coppola R., Lukáš P., Montanari. R., Rustichelli F., Vrána M., *Mater. Lett.* **22** (1995) 17.
[16] Albertini G., Ceretti M., Coppola R., Fiori F., Gondi P., Montanari R., *Physica B* **213/214** (1995) 812.
[17] Coppola R., Fiori F., Magnani M., Stefanon M., *J. Appl. Crystallogr.* **30** (1997) 607.
[18] Coppola R., Lukáš P., Mikula P., Vrána M., *Physica B* **241–243** (1998) 1261.
[19] Macek K., Lukáš P., Janovec J., Mikula P., Strunz P., Vrána M., Zaffagnini M., *Mater. Sci. Eng.* **A 208** (1996) 131.

20 Residual stresses in ceramic materials

R. I. Todd

20.1 Introduction

Ceramic materials often contain residual stresses. These can arise through most of the mechanisms responsible for residual stresses in metallic materials, including thermal expansion mismatches (either at a macroscopic level or on the scale of the microstructure), local dilatations and deviatoric strains occurring during phase transformations (martensitic or diffusional), and differential plastic deformation, such as that occurring during surface grinding.

Residual stresses in ceramics are often greater in magnitude than those found in metallic materials. Thermal microstresses in excess of 2 GPa have been measured in some of the materials reported in this chapter. There are three important reasons for this. First, the relaxation of residual stresses by plastic deformation is difficult owing to the high yield stress of most ceramics. Second, the high stiffness typical of ceramics means that a large stress results from a given strain, imposed for instance by a thermal expansion mismatch or phase transformation. Finally, in the case of thermal residual stresses, the refractory nature of many ceramics allows, and in the case of sintering and other methods of densification, demands, very large thermal excursions which lead to correspondingly large thermal expansion mismatches.

Given their high level, it is not surprising that residual stresses in ceramics can have a profound effect on the mechanical properties and other attributes of the material. The effect can be beneficial, as is the case in a number of toughening mechanisms that have been identified, or detrimental, the most obvious example being the possibility of spontaneous fracture of the material. Either way, it is important that the residual stresses and their effects should be understood, so that the benefits can be maximised and the detrimental effects avoided.

The investigation of residual stresses in ceramics and their effect on properties is certainly an area in which the balance between theory and experiment is heavily biased towards the former at present. This is partly because of the difficulty in measuring residual stresses, although the small size of the commercial market in advanced ceramics compared to metallic materials may also be a factor. The development of neutron diffraction techniques for measuring residual stresses is allowing the balance between theory and experiment to be redressed, however, and there are now reports of experimental measurements of many of the important types of residual stress outlined in the first paragraph of this section. The techniques used are for the most part routine, and are described elsewhere in this book. The objective of the present chapter is to review the available literature with an emphasis on the conclusions of the selected studies, and how they relate to important issues and applications in ceramic materials.

20.2   Thermal microstresses in discontinuously reinforced composites

20.2.1   *Development of models for thermal microstresses in discontinuously reinforced composites*

More papers have been published on thermal microstresses in discontinuously reinforced composites than in any other class of ceramic material. The spatial resolution of neutron diffraction ($>\sim 1\,\mathrm{mm^3}$) is not sufficient to resolve the variation in residual stress on the scale of the microstructure of these materials, so the stresses measured directly are the volume averages of the stresses in a particular component of the composite. In order to understand the influence of the microstresses on toughening, fracture and other properties, however, it is important to have a knowledge of the maximum and minimum stresses, and the spatial variation of stress within the material. This information must be provided by models for the residual stresses. The function of the neutron diffraction experiments is to test and validate these models, and to provide data that they require for their predictions. An obvious example of the information required by models is the relaxation-free temperature change, ΔT, responsible for the residual stresses.

20.2.1.1   *Whisker reinforced ceramics*

Majumdar *et al.* [1] were the first to report the measurement of thermal residual stresses in ceramic composites by neutron diffraction. Thermal strains in a hot pressed alumina/18vol% SiC whisker composite were measured as a function of temperature, and compared with the predictions of two self-consistent elastic models for the stress in a single inclusion embedded in a matrix with the average thermoelastic properties of the composite. One model, based on a long cylindrical whisker, assumed the elastic properties of the whisker to be transversely isotropic, whilst the other, based on the Eshelby ellipsoidal inclusion method, assumed complete elastic isotropy. Both models accounted for the anisotropic thermal expansion of the SiC whiskers, but assumed the alumina matrix to have isotropic thermoelastic properties. The two models gave almost identical predictions, indicating that the effect of elastic anisotropy on the strains in the whisker was small.

Relating the predictions of the models to the experimental results was complicated by the fact that all planes of the same crystallographic type are not in the same orientation relative to the geometrical shape of the whiskers. The whiskers were predominantly cubic (β) SiC, with a ⟨111⟩ direction along the fibre axis. Taking the fibre axis to be normal to the (111) plane, there are three other {111} planes whose normals are in directions inclined to the fibre axis by 70.5°. The strain calculated from the shift of the {111} diffraction peak is therefore a weighted average of the strain in the axial direction and the strain at 70.5° to the axis. For comparison with experiment, the average strains measured from different diffraction peaks were predicted from the models assuming the whiskers to be randomly orientated, although it was noted that this was not in fact the case, the whisker axes tending to lie in the plane of the hot pressed plates. The fact that all the experiments were performed with the diffraction vector at 45° to the hot pressing direction may have circumvented the worst effects of directionality simply by measuring strains in a direction midway between the two extremes of whisker alignment.

Comparisons of the predicted strains as a function of temperature with the experimental results for various SiC diffraction peaks showed reasonable agreement given the approximations made. Noting that both models used indicated a uniform stress state in the whiskers, Majumdar *et al.* were able to estimate the stress components in the whiskers at room temperature to be −2400 MPa (axial) and −1000 MPa (radial and hoop). The matrix tensile hoop stress at the interface with the whisker was estimated to be +1000 MPa. Extrapolation of the results above the maximum temperature used (1000°C) indicated a stress-free temperature of 1350–1400°C.

An extension of this work using similar methods [2] showed excellent agreement between Eshelby-based models and experimental results for the average hydrostatic strain in both the matrix and the reinforcement as a function of temperature and whisker volume fraction. Majumdar *et al.* [1] suggested that those discrepancies that were apparent in more detailed comparisons were largely the result of both the alignment of the whiskers and uncertainties in their crystallography. SiC almost always consists of a mixture of heavily faulted polytypes, differing in the stacking sequence of layers with hexagonal symmetry. The whiskers may, therefore, have contained attributes of both cubic SiC (in which the stacking sequence is in an ABCABC format analogous to that in face-centred cubic metals) and polytypes with hexagonal symmetry. This is particularly important for diffraction strain measurements because it dramatically affects the multiplicity of the different planes which give rise to diffraction, and therefore the weighting of the average strain measured from a particular diffraction peak.

The influence of alignment and crystal structure, as well as thermoelastic anisotropy, was explored by Tomé, Bertinetti and co-workers [3, 4] using self-consistent models based on the Eshelby equivalent inclusion approach coupled with the mean field approximation for composites containing finite volume fractions of inclusions. Various combinations of anisotropic properties, degree of alignment and crystal structure were explored, and an improved agreement with the results of Ref. [1] was demonstrated, particularly when the whiskers were assumed to be mainly hexagonal in crystal structure. Although some discrepancies remained, a more detailed model was not possible without more information about the degree of alignment and crystal structure of the whiskers.

20.2.1.2 *Particulate composites*

The research on whisker-reinforced alumina illustrates some of the difficulties in testing models rigorously using simple peak shift measurements on a single material. The relaxation-free temperature reduction, ΔT, cannot be known accurately *a priori*, so the absolute values of stress do not provide a very useful test of the model. Furthermore, differences in strain between different crystallographic and spatial directions are difficult to model accurately without information concerning the material which may not be easily obtainable. The orientation distribution of non-equiaxed phases, crystallographic texture and, in the case of SiC whiskers, the distribution of polytypes, all come into this category. One approach to providing a more exacting test of models for thermal stresses is to test a range of materials with systematic variations in microstructure, and to correlate the variations in stress measured with those predicted by the models. Todd and Derby [5] studied composites consisting of alumina reinforced with various volume fractions of 3 μm SiC particulate. The particles were approximately equiaxed, and the composites had only a small amount of crystallographic texture. The average thermal stresses in both the matrix and the reinforcement were therefore hydrostatic to a good approximation, and were calculated directly from the measured strains using isotropic values for the thermoelastic constants of the two phases.

The average stresses in both the matrix and the particles varied significantly with SiC volume fraction (Figure 20.1), and the force balance between the two phases was satisfied for all compositions, verifying the validity of the measurements. These experimental results were compared with the predictions of an elastic model for the residual stresses comprising

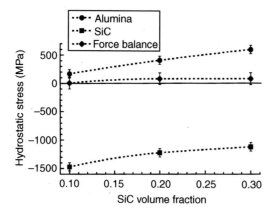

Figure 20.1 Hydrostatic stress in the matrix and reinforcement of alumina/SiC particle composites as a function of SiC content. The 'force balance' is the sum of these stresses weighted by the volume fraction, and should be zero in the absence of an externally applied stress, as is demonstrated by the experimental results.

a spherical SiC particle surrounded by a concentric spherical shell of alumina, the radius of the shell being chosen to give the correct volume fraction of matrix. The exact isotropic elastic solution for this geometry can be derived by adapting Lamé's solution for an internally pressurised spherical shell, and has been used by a number of authors [5, 6]. The SiC particles are predicted to be in uniform hydrostatic compression. The stress state in the matrix is more complicated, but close to the particle the radial stress is predicted to be compressive, and the hoop stress tensile for the range of SiC volume fractions used in the experiments. The model has solutions for the average strain, $\varepsilon_{m/p}$, in the matrix/particle of the form:

$$\varepsilon_{m/p} = j_{m/p}(f)\,\Delta T$$

where the $j_{m/p}$ are functions of the volume fraction of particles and the thermoelastic constants of both phases. Figure 20.2 shows that the experimental results fit this form to within experimental error, and the gradient of the straight line fitting to the results gives a value for ΔT of 1040°C. The idea that thermal residual stresses are relaxed at temperatures above ~ 1100°C is reasonable, since creep, sintering and other high-temperature processes also start to be active in alumina at approximately this temperature.

Given that the spherical particle/shell model gives a reasonable representation of the microstructure of the alumina/SiC$_p$ composites, and gives a sensible value for the relaxation-free temperature reduction, ΔT, we can have some confidence in using it to predict the average thermal stresses in these composites. Todd and Derby [5], however, showed that plots such as Figure 20.2 alone are unable to discriminate between models giving widely differing predictions for the spatial variation of residual stress, which is precisely the information which needs to be supplied by the models for aspects of material behaviour such as fracture to be understood. In contrast, the peak *broadening* which usually accompanies the peak shifts used to measure the average stresses provides a powerful test of the predictions of models in this respect, provided that the broadening results from spatial and directional variations of the residual stresses alone, and not from other possible sources of broadening such as dislocations and particle size effects. Todd [7] has shown that if the distribution of

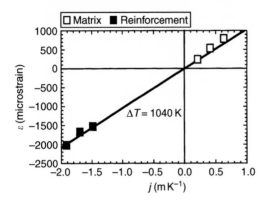

Figure 20.2 Plot of ε against j for the matrix and particles of the same composites as in Figure 20.1.

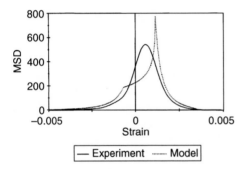

Figure 20.3 MSD for the matrix of an alumina/20% SiC_p composite, and prediction of spherical particle/spherical shell model.

strains in the sample satisfies certain well-defined criteria, the broadened peak (after correction for instrumental effects) can be interpreted directly in terms of the volume weighted distribution function of the strains sampled by the diffracted radiation (or *measured strain distribution*, MSD). The extent to which the strain distribution satisfies the required criteria can be tested experimentally by demonstrating that the same measured strain distribution is obtained from two different orders of the same reflection, or, in the isotropic case, from two peaks corresponding to significantly different interplanar spacings. The same test confirms the absence of particle size effects. The absence of significant dislocation activity must strictly be confirmed by microscopy, but is often the case in ceramic phases that this, and the other criteria necessary for the interpretation in terms of a MSD to be valid, are obeyed.

The SiC particle reinforced alumina composites described above are ideal for this purpose, as both phases have yield stresses sufficiently high to rule out significant dislocation activity, and the grain sizes are much larger than those which would give rise to particle size effects. A MSD for the matrix of an alumina/20 vol% SiC_p composite is shown in Figure 20.3. The definition of the MSD is analogous to a probability distribution function, in that the proportion of the gauge volume with a strain normal to the diffracting planes of between ε and $\varepsilon + d\varepsilon$ is equal to MSD$(\varepsilon)d\varepsilon$. The fact that the centroid of the experimental MSD in Figure 20.3 is at positive strain therefore indicates that the volume averaged strain normal to the diffracting planes is tensile, as was deduced directly from the measurement of diffraction peak shifts described above. The breadth of the distribution is much greater than the average strain, however, so much so that a substantial volume of the matrix sampled by the neutrons contained compressive strains, and large elastic strains $\sim \pm 0.3\%$ are indicated at the extremes of the distribution. These observations have a simple explanation in terms of the spherical particle/spherical shell model, in that some of the neutrons will be diffracted from material which is close to the particle where the stress is much greater than average. Of these neutrons, some will be diffracted from crystal planes whose normals are tangential to the particles, so that the tensile hoop strain will be sampled; others will be diffracted from planes whose normals are in the radial direction, in which case the negative radial strain will be sampled.

It should be noted that although the MSD is a function of the spatial distribution of stress on a microstructural scale, the spatial distribution cannot be inferred directly from the MSD. This is because there is no escaping the fact that the gauge volume is much larger than the scale of the microstructure, so the diffraction peak cannot contain any information about the relative positions of incoherently diffracting domains. The spatial distribution of the stresses must still be provided by models, which can be tested by comparing the experimental MSD with the model's prediction. Figure 20.3 includes the prediction of the spherical particle/spherical shell model for the alumina/20% SiC_p composite for comparison with the experimental MSD. The agreement is poor in the centre of the distribution, but in the tails the model predicts the MSD very well, even correctly predicting the asymmetry in which the compressive strain wing decays to zero more gradually than the tensile wing. This latter observation is important, because the extremes of the distribution represent the highly strained parts of the matrix close to the particles, which are often of greatest interest. In these cases, we conclude that the model provides an excellent description of the stresses of interest. The central region of the MSD represents the matrix material far from the particle where the stress is relatively small. It is entirely to be expected that the spherical particle/spherical shell model gives a poor prediction in these regions since the isolated spherical units do not satisfy the requirement of continuity of the strain distribution in the real composite (which is the reason for the sharp discontinuities in the predicted MSD), and cannot even be stacked to fill space. The MSD therefore indicates the way to develop improved models. Todd and Derby [5], for instance, were able to show that an approximate solution for a spherical particle in a cubic shell, which more nearly filled space, gave a better prediction for the MSD in the central region. Similarly, the MSD could be used in conjunction with more sophisticated models to study experimentally the effect of local volume fraction variations, differing particle shapes and the like.

The same approach can be used for the SiC reinforcement of the composites. The MSDs for the SiC particles are shown in Figure 20.4. The spherical particle/spherical shell model predicts that the strain is uniform and hydrostatic in the SiC particles, so the predicted MSD is a sharp spike at the value of the corresponding hydrostatic elastic strain. As with the matrix MSDs, the experimental results are in agreement with the model's predictions in some respects, but there are discrepancies which point the way to improved models. The experimental MSDs are indeed significantly narrower in breadth than the matrix MSDs, despite the fact that the average strains are much greater. However, the breadth is still sufficient to imply that a small fraction ($\sim$ a few per cent) of the SiC particles have a tensile strain. Part of the broadening may be attributed to local volume fraction variations, but it is difficult to see how this could be responsible for tensile strains in the particles. The most likely explanation for the small regions of tensile strain is in terms of Poissonian expansion rather than tensile stresses. The composites contained a small amount of porosity, almost all of which was adjacent to SiC particles. Since the stress normal to the surface of the SiC adjacent to a pore

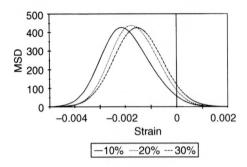

Figure 20.4 MSD for the SiC particles in alumina/SiC$_p$ composites containing 10, 20 and
30 vol% SiC.

must be zero, there is likely to be a Poissonian expansion in this direction owing to the
influence of the presumably compressive stresses in the plane of the surface. The sharp
corners on the particles may also contribute to the broadening. Rao *et al.* [8] have shown that
one of the normal stress components in the particles becomes very large at sharp corners,
whilst the other components remain relatively small. This effect is sufficiently extreme to
cause Poissonian expansions in some directions, and also provides an explanation for the
larger-than-average compressive strains sampled at the other extreme of the MSD. Another
reason for tensile strains has been suggested by finite element modelling of realistic cermet
microstructures by Weisbrook *et al.* [9]. The models showed some regions of tensile strain
in the ceramic particles, which were mostly in compression as with the present system, and
this was attributed to bending moments caused by irregularities in the microstructure.

It is interesting to note that the three MSDs shown in Figure 20.4 are the same shape for all
three SiC contents used. This is in contrast to both the measured and predicted matrix results
in which the breadth of the MSD increases significantly with SiC volume fraction [5]. This
is direct experimental evidence supporting the 'mean field approximation' [10] often used to
adapt the Eshelby ellipsoidal inclusion method, which is valid only for a single particle in an
infinite matrix, for use with finite volume fractions of particles. The approximation assumes
that the effect of a finite volume fraction of particles is to superpose on the single particle
stress field an image stress in both phases, such as would arise if the stress were applied at
the outer boundary of the material. In this system, the elastic mismatch between the matrix
and the particles is small, so the image stress should be uniform throughout the composite.
This explains the similarity in shape of the particle MSDs, since the shape may be regarded
as being determined by local effects such as the particle morphology and the presence of
porosity, whilst the position is determined by the uniform hydrostatic image stress which
simply moves the whole MSD parallel to the strain axis by an amount determined by the
average volume fraction of SiC in the material as a whole.

20.2.2   Applications of thermal microstress measurement in
discontinuously reinforced composites

20.2.2.1   Spontaneous microcracking in alumina/SiC$_p$ composites

The knowledge gained by studies such as those described in the previous section is ultimately
intended to lead to an improved understanding of the influence of thermal microstresses

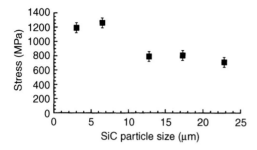

Figure 20.5 Particle stress versus particle size for alumina/20 vol% SiC$_p$ composites.

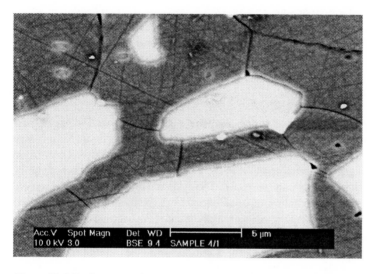

Figure 20.6 Backscattered SEM image of unetched alumina–13 μm SiC composite showing radial crack network in the matrix. The diffuse regions around the SiC particles (lighter phase) are thought to be an electron optical artefact rather than a reaction layer.

on the properties of discontinuously reinforced ceramic composites. Todd *et al.* [11] have reported one such application, in which neutron diffraction was used to study spontaneous microcracking in alumina/SiC$_p$ composites. Figure 20.5 shows the hydrostatic thermal stress in the particles of alumina/20 vol% SiC composites as a function of the average particle size. For composites containing small particles ($\leq 6\,\mu m$) the stress is ~ 1200 MPa, but for larger particle sizes there is an abrupt reduction in stress by about a third. Examination of the composites in the SEM showed that the stress reduction was caused by the occurrence of spontaneous microcracking of the alumina matrix under the action of the thermal residual stresses (Figure 20.6). The microcracks propagated from the particles in the radial direction relative to the particles, as is expected from the tensile hoop stress, and usually extended fully from one particle to the next. It is well known that spontaneous microcracking is expected to occur above some critical particle size, as the elastic strain energy available to cause fracture increases in proportion to the volume of the particles, whereas the surface energy required increases as the area [12]. A fracture mechanics model for microcracking, which made use of both the residual stresses measured using neutron diffraction and the spherical particle/spherical shell model described in Section 20.1.1, was successful in quantitatively predicting the critical particle size observed, and in explaining the abrupt nature of the transition from negligible microcracking to general microcracking throughout the matrix. In the latter case the model shows that, for relatively high SiC contents such as that used (20%), the stress intensity at the tip of a radial crack around a SiC particle increases monotonically as the crack grows, enabling cracking to propagate from one particle to another, as observed (Figure 20.6). This means that cracking need only be nucleated at a few particles for cracking to spread throughout the composite, provided the average particle size is above the critical size.

The influence of the residual stresses and microcracking on fracture of the composites was also studied. The long crack toughness increased monotonically with SiC particle size, reaching $11 \, \text{MPa} \, \text{m}^{1/2}$ for the largest particle size used ($23 \, \mu\text{m}$). Examination of indentation cracks suggested that the toughening was the result of extensive crack bridging associated with intense deflection and bifurcation of the main crack as a result of the residual stresses and, for the larger particle sizes, existing spontaneous microcracks.

20.2.2.2   *Alumina/SiC nanocomposites*

A related application of residual stress measurements concerns alumina/SiC composites in which the SiC particle size is much smaller than the microcracking limit. Nanocomposite ceramics have received considerable attention over the last decade. This interest was originally stimulated by a summary of experimental results by Niihara [13] from a number of composite systems involving at least one submicron phase. The system which caused most interest was the alumina/SiC system, in which the simple addition of only 5 vol% of 300 nm SiC particles to alumina produced a threefold increase in strength. It has since become clear that the major part of the strength improvement claimed was due to refinements of the alumina grain structure (e.g. the avoidance of abnormal grain growth) which can be effected more economically by existing methods (e.g. doping with MgO). At the same time, it has been realised that other aspects of the fracture behaviour of alumina/SiC nanocomposites are novel, and can lead to property enhancements as dramatic as the strength improvements originally claimed by Niihara. These are unified by an apparent strengthening of the alumina grain boundaries, which are notoriously weak in unreinforced alumina. This leads to a change in fracture mode from intergranular to transgranular, and a reduction in the propensity for surface grain pullout during grinding and polishing [14]. This latter manifestation has considerable commercial importance, as it results in improved surface finish or a reduction in the number of polishing tests, and also to a reduction in the erosive [15, 16] and abrasive [17] wear rate for a given alumina grain size by a factor of 3 or more.

A number of mechanisms have been proposed to explain the grain boundary strengthening effect of the SiC additions. Although which of these is responsible for the observed behaviour has yet to be agreed on, it is notable that many rely on the thermal residual stresses which are arguably the most obvious consequence of the SiC additions. In his original paper, Niihara [13] proposed that the formation of subgrains owing to dislocation activity in response to the residual stresses was responsible for the observed reduction in critical defect size. He also suggested that the transition in fracture mode was due to the deflection of grain boundary cracks into the grain interior by their attraction towards the tensile hoop stresses around intragranular particles.

In order to investigate these possibilities, Todd *et al.* [18] measured the stresses in a series of alumina/SiC nanocomposites as a function of SiC volume fraction. The stresses in both the matrix and the nanoparticles were measured using neutron diffraction, with dense polycrystalline alumina as the reference specimen for the matrix, and the SiC powder used in the manufacture of the composites for the particles. The volume averaged hydrostatic stresses for all specimens obeyed the force balance required between the two phases to within experimental error. The stress level was higher than for the corresponding volume fraction of 3 μm SiC [5] (Figure 20.1) approaching 2 GPa in the particles for small volume fractions. The correspondingly higher relaxation-free temperature change, $\Delta T = 1300°C$, may be explained by the fact that many of the particles in the nanocomposites were situated within the grains, where stress relaxation by rapid grain boundary diffusion cannot take place, whereas the 3 μm SiC particles in Ref. [5] were of similar size to the alumina grains, and were therefore always in contact with the matrix grain boundary network. Dislocation relaxation mechanisms may also have been inhibited by the small particle size. Isolated microcracking would also be expected to be less frequent in the nanocomposite, but since this was not found to be severe in the 3 μm SiC 'microcomposite' (Section 20.2.1), it is unlikely that this was a major factor.

Detailed analysis of the results as a function of crystallographic direction revealed that the intergrain stresses, which exist even in pure alumina polycrystals owing to the anisotropic thermal expansion coefficient of the grains, were reduced by the presence of the SiC, possibly by an anisotropic relaxation effect. This is presumably a contributory factor to the apparent grain boundary strengthening, in that the transgranular fracture mode in alumina is thought to be at least partly the result of tensile intergrain stresses acting across grain boundaries. The reduction in these stresses was small for the lowest volume fractions of SiC (a reduction of about 10% for 5 vol% SiC), however, and seems unlikely to be the sole cause of the strengthening.

Recently [19], the experimental neutron diffraction stress measurements have been used to model crack propagation along an alumina grain boundary under the influence of both an externally applied mode I stress and the residual stresses around a SiC particle. The ability of the tensile hoop stresses around intragranular particles to deflect cracks was found to be too weak by an order of magnitude to be able to attract grain boundary cracks into the tougher grain interior, as Niihara [13] had suggested. The same model did, however, suggest an alternative mechanism for the grain boundary strengthening effect of the SiC nanoparticles. Figure 20.7 shows the apparent grain boundary toughness as a crack passes a SiC particle situated on an alumina grain boundary. There is a 50% increase in K_c owing to the compressive *radial* component of the residual stress around the particle as it acts on the crack near to its tip. This offers a direct explanation for the occurrence of transgranular fracture in the nanocomposites, which is supported by both the direct observation of local crack deflection at grain boundary particles [20] and the conclusions of a separate microstructural study which suggested that it is the SiC particles situated on the grain boundaries which are responsible for the strengthening effect [17].

A further neutron diffraction study of residual stresses in nanocomposites [21] has provided direct evidence of the importance of the residual stresses to the improved wear resistance of nanocomposites. Alumina matrix nanocomposites containing 10 vol% of nanosized second phase reinforcements of SiC, Si_3N_4, ZrO_2, TiC and TiCN were produced, and the thermal microstresses in the materials measured using neutron diffraction. The different reinforcements contained hydrostatic stresses ranging from -2800 MPa(Si_3N_4) to $+450$ MPa(ZrO_2). The materials which were most resistant to surface grain pullout by grain boundary fracture

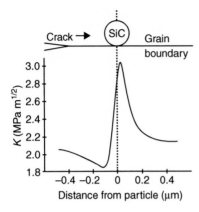

Figure 20.7 Two-dimensional model prediction [19] of stress intensity required to propagate a grain boundary crack in alumina past a SiC particle surrounded by the residual stress state deduced from neutron diffraction measurements.

during a standard grinding treatment were found to be those with large compressive residual stresses in the particles (Si_3N_4 and SiC), as suggested by the model of Figure 20.7. Whilst it is clear that other factors such as interfacial properties, stiffness and chemistry may also be important, these results certainly add support to the work described above in suggesting that the significant property improvements found in ceramic nanocomposites are directly related to the thermal residual stresses within them.

20.2.2.3 Alumina/zirconia composites

Very large microstresses can arise in zirconia-containing materials from the martensitic tetragonal to monoclinic ('t' to 'm') transformation, which is accompanied by a dilatation of $\sim$4%. These stresses are so large that pure zirconia is of little use in structural applications because of the extensive cracking that occurs when the material transforms during cooling. When the spontaneous transformation is suppressed by the addition of stabilising additives and the control of the grain size, however, stress-induced transformation in a limited zone close to the crack tip during propagation is responsible for the potent transformation toughening effect, in which the dilatation in the transformed 'process zone' leads to crack closure. The improvement in mechanical properties arising from transformation toughening has led to a wide range of applications for zirconia ceramics and composites.

The transformation stresses in zirconia have not been studied using neutron diffraction, presumably because of the small volume around the crack tip in which they occur in materials of practical value. Most zirconia-toughened ceramics, however, also contain thermal residual stresses. Modest thermal stresses can be expected in tetragonal zirconia polycrystals (TZP) owing to the anisotropic thermal expansion coefficient of tetragonal zirconia, and in partially stabilised zirconia (PSZ) from the thermal expansion mismatch between the tetragonal precipitates and the cubic matrix. Larger thermal stresses are known to exist in zirconia-toughened alumina (ZTA) and other composites because of the unusually high thermal expansion coefficients of zirconia polymorphs, which exceed that of most ceramic matrices by a considerable margin.

The thermal residual stresses in zirconia are important not only for their direct role in influencing crack propagation and microcracking as in other ceramics, but also for their influence on the all-important 't' to 'm' transformation. Because the transformation leads to a dilatation, it is favoured by the application of a tensile stress. The presence of tensile thermal stresses in t-ZrO_2 therefore raises the martensite start temperature, M_s, and increases the width of the transformation zone, or leads to deleterious spontaneous transformation during cooling. Thermal residual stresses and their effect on the 't' to 'm' transformation have been studied in detail in Al_2O_3–ZrO_2(CeO_2) composites at Oak Ridge National Laboratory, USA [21, 22]. The composites contained between 10 and 40 vol% of ZrO_2 stabilised with 12 mol% CeO_2, and limited variations in grain size were achieved by heat treatment. The ZrO_2 was fully tetragonal at room temperature, and the hydrostatic strain was tensile, as expected from the higher thermal expansion coefficient of ZrO_2 compared to alumina. Considerable anisotropy was observed in the thermal strains in both phases, and this could be understood qualitatively in terms of the anisotropy of the thermal expansion coefficients. The hydrostatic stresses in the ZrO_2 decreased with increasing volume fraction of ZrO_2, but were not a function of grain size. These observations are consistent with simple theoretical considerations, and the stresses lay between the predictions of two extreme cases of the spherical particle spherical shell model described above. At one extreme the ZrO_2 was considered to correspond to the particles and in the other to correspond to the matrix. The latter case was considered because the appearance of the microstructure suggested that the ZrO_2 was starting to become continuous at the higher volume fractions.

The M_s temperatures of the same materials were measured subsequently by monitoring the phase composition during cooling to temperatures as low as 12 K. Both dilatometry and neutron diffraction were used for this purpose, with neutron diffraction being preferred to X-ray diffraction for the same reason as with residual stress measurements, viz. to avoid the influence of misleading surface effects. The results showed that M_s decreased as the volume fraction of ZrO_2 increased, as is to be expected from the accompanying reduction in the tensile residual stress. Increasing the t-ZrO_2 grain size led to an increase in M_s. This was explained by the fact that stress *concentrations* at grain corners, where nucleation of the 't' to 'm' transformation is thought to occur, are expected to increase in severity as the grain size increases [8, 23], despite the fact that the *average* residual stress is not a strong function of grain size. The results agreed well with a model taking account of these factors.

20.3   Thermal microstresses in long fibre composites

Saigal *et al.* [24] measured the thermal residual stresses in composites of reaction bonded silicon nitride (RBSN) reinforced with aligned, SiC monofilaments. Such long fibre composites give the best high temperature toughness of any completely ceramic material, and the thermal residual stresses can have both beneficial and detrimental effects. Compressive interfacial stresses between the fibres and the matrix can increase the frictional force against which work is done during fibre pullout, but the tensile hoop stresses that are expected to accompany this interfacial compression can cause matrix microcracking, and consequent degradation of the mechanical properties.

The residual stresses were measured in composites containing a range of fibre volume fractions. Composites made with coated and uncoated fibres were compared, the coating in the former case being a 3 μm carbon-rich layer. The influence of matrix density was also examined by HIPing some of the standard material, which contained 30–40% porosity, to

full density. Measurements were made using a spallation source and a time-of-flight diffractometer set up to record the axial and transverse strains simultaneously. Loose fibres were
used as the stress-free standard for the SiC, and the stress-free interplanar spacings for the
matrix were estimated by extrapolating the composite results to zero fibre volume fraction.
The experimental results were compared with the predictions of a finite element model.

The results allowed a good understanding of the stresses in the composites to be developed.
The axial stresses in the fibres were tensile with a maximum value of about 650 MPa, but
the radial (interfacial) stresses were compressive with a typical value of −50 MPa. This was
caused by the anisotropic thermal expansion coefficient of the fibres, which was 80% greater
in the axial direction than in the transverse direction, with the thermal expansion coefficient
of the matrix lying between the two extreme values. The magnitude and overall trends in
the measured strains were predicted reasonably well by the finite element model, although
there were some minor discrepancies at the limits of the experimental error. The experimental
scatter in the strain measurements was greater than the predicted effect of varying the matrix
elastic modulus between 50 and 100% of the value at full density, so detailed conclusions
regarding the effect of matrix density on the residual stresses could not be drawn. Comparison
of the specimens made with coated and uncoated fibres showed that the coating had no
significant effect on the stresses. This was confirmed using the finite element model, which
predicted very minor variations, much smaller than the limits of experimental error. The
fibre pushout force measurements were consistent with the idea that the force increased with
the compressive interfacial stress, but again the experimental uncertainties were too great to
confirm this unequivocally, and other effects such as an improvement in interfacial bonding
with increased density may also have played a part.

20.4   Neutron diffraction measurements of 'tempering' stresses in ceramics

The investigations described above have been concerned solely with residual stresses on
a microstructural scale. Longer range stresses also occur in ceramics. Long range stresses
are used to advantage in the process known as 'tempering', in which ceramic components
are treated to produce a compressive surface stress. Ceramics often fail from surface flaws
owing to contact damage, the tendency for bending to cause the maximum stress to be at
the surface, and from the higher stress intensity at surface flaws compared with those in the
specimen interior. A compressive surface stress therefore leads to an increase in strength
because it reduces the tensile stress acting on the critical flaws. Tempering is common in the
glass industry, a familiar example being so-called 'toughened' glass, in which a compressive
surface stress is introduced by cooling the surface of hot glass sheet to produce a solid skin
while the interior can still flow. The contraction associated with the subsequent cooling and
solidification of the interior puts the surface into compression, increasing the strength of the
glass substantially.

There is considerable interest in using the same principle to raise the strength of crystalline
ceramics. An elegant method of tempering alumina has been devised by Marple and Green
[25] in which the surface of a partially sintered component is infiltrated with hydrolysed
ethyl silicate. When subsequently sintered to full density the silicate reacts with the alumina,
forming a Mullite/alumina composite surface layer. Mullite has a lower thermal expansion
coefficient than alumina, so during cooling the pure alumina core of the component contracts
more than the alumina/Mullite composite, putting the surface layer into compression. Root
et al. [26] have measured the stresses at various positions in cylinders of alumina tempered

using a range of infiltration times. The diffracted peak intensities were also used to estimate the proportion of Mullite at a particular sampling position. The surface layers contained microstresses owing to the thermal expansion mismatch between the Mullite and alumina grains, as well as the long range surface stress responsible for the tempering effect. Since a suitable reference for the Mullite phase could not be produced, the micro- and macrostresses could not be separated, so an accurate measurement of the tempering stress was not possible. Nevertheless, the variation in the components of total stress (micro + macro) with distance from the surface was consistent with the expectation of a compressive surface stress combined with an increasing tensile component of microstress in the alumina as the Mullite content increased. The results can be used to provide an estimate of the long range compressive surface stress if it assumed that the component of macrostress normal to the surface is zero at the surface, and that the microstresses are isotropic. This leads to typical values of the order of 100 MPa for the tempering stress, and in the simplest case, in which the critical flaw size, stiffness and toughness are unaffected by the tempering process, this is the magnitude of the strength increase that can be expected.

An alternative method of producing compressive stresses in load-bearing components has been investigated by Üstündag *et al.* [27]. Model specimens consisting of an alumina disc surrounded by a hoop of the spinel compound $NiAl_2O_4$ were subjected to reducing treatments in $CO/CO_2/N_2$ atmospheres at temperatures in the range 1100–1300°C. The equilibrium reduction reaction partially reduces the $NiAl_2O_4$ to α-alumina and Ni metal:

$$NiAl_2O_4 \rightarrow Ni + \alpha\text{-}Al_2O_3 + \tfrac{1}{2}O_2.$$

The reduction is accompanied by a significant shrinkage (17.6% by volume), which would be expected to cause large compressive stresses in the alumina disc in the absence of other effects. A smaller, additional compressive stress would result during cooling from the reduction temperatures owing to the larger thermal expansion coefficient of the Ni/alumina composite compared to the alumina disc. The total in-plane stress for this case was estimated using a finite element model to be -1200 MPa (-1100 MPa from the reduction reaction, -100 MPa from the coefficient of thermal expansion mismatch). Neutron diffraction measurements of the in-plane stresses showed a maximum compressive stress of only -103 ± 40 MPa, however, and this was in accord with X-ray diffraction measurements made on the surfaces of the discs. The results clearly indicate that the majority of the potential residual stress was relaxed. Density measurements and a detailed microstructural examination showed that the actual shrinkage was much smaller than the theoretical value used in the calculations owing to an increase in porosity during reaction, and the presence of metastable phases occupying larger volumes than expected. The metastable phases also had the effect of diminishing, or even reversing the thermal expansion mismatch during cooling. Creep deformation is also likely to have contributed to the relaxation of the stresses at the higher reduction temperatures.

20.5   Conclusions

Whilst the number of neutron diffraction studies of residual stresses in ceramics is limited, the papers reviewed suggest that this is set to be a growth area. The work on model development in Section 20.1 demonstrates the feasibility and success of neutron diffraction in making accurate bulk measurements of thermal stresses in ceramic materials. Despite the relatively small number of studies to date, the wide range of influences residual stresses can exert on the mechanical properties of ceramics is already evident, with examples including the dependence of pullout stress on residual clamping force in long fibre composites, spontaneous microcracking with large reinforcements, grain boundary strengthening with nanoscale reinforcements, nucleation of transformation toughening by tensile thermal stresses, and strength increases through the tempering of bulk ceramic components. It is not difficult to think of other situations in which residual stresses are considered important but have yet to be explored experimentally. The research described in this chapter is therefore likely to represent the beginning of a sustained effort to characterise and understand residual stresses in ceramics, so that their harmful effects can be avoided and their potential benefits fully exploited.

References

[1] Majumdar S., Kupperman D. and Singh J., *J. Am. Ceram. Soc.* **71**, 858 (1988).
[2] Majumdar S. and Kupperman D., *J. Am. Ceram. Soc.* **72**, 312 (1989).
[3] Tomé C. N., Bertinetti M. A. and Macewen S. R., *J. Am. Ceram. Soc.* **73**, 3428 (1990).
[4] Bertinetti M. A., Turner P. A., Bolmaro R. E. and Tome C. N., *J. Am. Ceram. Soc.* **79**, 1466 (1996).
[5] Todd R. I. and Derby B., 1993, in *Residual Stresses in Composites*, TMS-AIME, Warrendale, PA, edited by E. V. Barrera and I. Dutta, pp. 147–160; Todd R. I. and Derby B., 1993, in *Euro-Ceramics*, vol. II, Deutsche Keramische Gesellschaft e. V., Cologne, edited by G. Ziegler and H. Hausner, pp. 1425–1429.
[6] Hsueh C.-H., *J. Mater. Sci.* **21**, 2067 (1986).
[7] Todd R. I., paper in preparation.
[8] Rao S. S., Tsakalakos T. and Cannon W. R., *J. Am. Ceram. Soc.* **75**, 1807 (1992).
[9] Weisbrook C. M., Gopamaratnam V. S. and Krawitz A. D., *Mater. Sci. Eng.* **A201**, 134 (1995).
[10] Mori T. and Tanaka K., *Acta Metall.* **23**, 571 (1973).
[11] Todd R. I., Morsi K. and Derby B., paper in preparation.
[12] Davidge R. W. and Green T. J., *J. Mater. Sci.* **3**, 629 (1968).
[13] Niihara K., *J. Ceram. Soc. Jpn* **99**, 974 (1991).
[14] Zhao J., Stearns L. C., Harmer M. P., Chan H. M., Miller G. A. and Cook R. F., *J. Am. Ceram. Soc.* **76**, 503 (1993).
[15] Walker C. N., Borsa C. E., Todd R. I., Davidge R. W. and Brook R. J., *Br. Ceram. Proc.* **53**, 249 (1994).
[16] Sternitzke M., Dupas E., Twigg P. and Derby B., *Acta Mater.* **10**, 3963 (1997).
[17] Winn A. J. and Todd D., *Br. Ceram. Trans.* **98**, 219 (1999).
[18] Todd R. I., Bourke M. A. M., Borsa C. E. and Brook R. J., *Acta Mater.* **45**, 1791 (1997).
[19] Philpot B., 1998, Final year project report, University of Manchester.
[20] Jiao S., Jenkins M. L. and Davidge R. W., *J. Microscopy* **185**, 259 (1997).
[21] Wang X.-L., Hubbard C. R., Alexander K. B., Becher P. F., Fernandez-Baca J. A. and Spooner S., *J. Am. Ceram. Soc.* **77**, 1569 (1994).
[22] Alexander K. B., Becher P. F., Wang X.-L. and Hsueh C.-H., *J. Am. Ceram. Soc.* **78**, 291 (1995).
[23] Chiu Y. P., *J. Appl. Mech.* **44**, 587 (1977).
[24] Saigal A., Kupperman D. S., Singh J. P., Singh D., Richardson J. and Bhatt R. T., *Composites Eng.* **3**, 1075 (1993).
[25] Marple B. R. and Green D. J., *J. Am. Ceram. Soc.* **71**, C471 (1988).
[26] Root J. H., Sullivan J. D. and Marple B. R., *J. Am. Ceram. Soc.* **74**, 579 (1991).
[27] Üstündag E., Zhang Z., Stocker M. L., Rangaswamy P., Bourke M. A. M., Subramanian S., Sickafus K. E., Roberts J. A. and Sass S. L., *Mater. Sci. Eng.* **A238**, 50 (1997).

Index

Lightning Source UK Ltd.
Milton Keynes UK

175289UK00001B/24/A